METHODS IN MOLECULAR BIOLOGY™

Series Editor
John M. Walker
School of Life Sciences
University of Hertfordshire
Hatfield, Hertfordshire, AL10 9AB, UK

For other titles published in this series, go to
www.springer.com/series/7651

Quorum Sensing

Methods and Protocols

Edited by

Kendra P. Rumbaugh

Department of Surgery, Texas Tech University Health Sciences Centre, Lubbock, TX, USA

Humana Press

Editor
Kendra P. Rumbaugh
Department of Surgery
Texas Tech University Health Sciences Center
Lubbock, TX
USA
kendra.rumbaugh@ttuhsc.edu

ISSN 1064-3745 e-ISSN 1940-6029
ISBN 978-1-4939-6162-7 ISBN 978-1-60761-971-0 (eBook)
DOI 10.1007/978-1-60761-971-0
Springer New York Dordrecht Heidelberg London

Printed on acid-free paper

Humana Press is part of Springer Science+Business Media (www.springer.com)

Foreword I

Great! Wow, a whole book on methods for the studying of quorum sensing. I have witnessed the birth of this field and the explosion of research in the area. I believe we are just touching the surface in terms of learning about quorum sensing as it pertains to behaviors fundamental to biology, learning about microbial strategies for successful competition, and harnessing our new knowledge to somehow advance the human condition. Books like this will help the field continue to grow.

I first learned about cell density-dependent expression of luminescence in *Vibrio harveyi* and *Vibrio fischeri* during the summer of 1973 at an MBL summer course in Woods Hole. Ken Nealson and Anatol Eberhard were instructors in that course, and they, together with Woody Hastings, had uncovered the interesting phenomenon of cell density-dependent expression of luminescence. I believed that this represented a social activity in bacteria and decided that this was what I would focus on as a postdoc. When I finished my PhD in 1977, I moved to Hastings' laboratory. By then Nealson had established his own laboratory in La Jolla. So there were all of two laboratories working on what we then called autoinduction, Nealson on *V. fischeri* and Hastings and I on *V. harveyi*. With Hastings, I found that *V. harveyi* not only responded to a signal it made itself, but also to signals produced by a number of other marine bacteria. We called this alloinduction, induction by others. I struggled to identify the signal and ultimately decided that *V. harveyi* might respond to lots of different molecules.

The problem seemed too hard. At the same time, Nealson together with Eberhard identified the *V. fischeri* acyl-HSL signal and, with a graduate student JoAnne Engebrecht and with Mike Silverman, cloned the genes responsible for signal production and signal reception. I switched to *V. fischeri*! Not too much later, Bonnie Bassler arrived in Mike Silverman's laboratory, and they switched to *V. harveyi*! Bassler did what I could not. She identified the *V. harveyi* alloinducer, which she calls Autoinducer-2.

In a relatively short period of scientific time since, there have been thousands of publications on these types of gene regulation, which, for better or for worse, we termed quorum sensing in 1994. The level of the science in our field is high, and I am delighted to see that there is now a book on methods to study quorum sensing of the *V. fischeri* type, *V. harveyi* type, and other types of quorum sensing. My hat is off to Kendra Rumbaugh for taking the challenge and to all of the authors who willingly participated in this effort.

E. Peter Greenberg, Ph.D.

Foreword II

Over the last two decades, our understanding of microbial behavior has undergone a paradigm shift such that we now appreciate that simple unicellular organisms possess an extraordinary capacity for cooperative social behavior. Consequently, communication and teamwork are just as important as competition in the race to colonize new niches, exploit available food resources, and combat host defenses. Such a talent for primitive multicellular behavior depends on the deployment and recognition of diffusible signal molecules through "quorum sensing," a signal transduction mechanism for coordinating gene expression at the population level.

Quorum sensing directs interactions not only within a signal producing microbial population, but also between different species and between pathogenic and beneficial microbes and higher organisms in the context of symbiosis, growth promotion, and pathogenicity. It has major impacts on agriculture, ecology, medicine, and industry where such systems control the adaptive behavior of microbes in the context of plant–microbe interactions with respect to pathogenicity, plant growth promotion, biocontrol, food spoilage, marine and industrial plant biofouling, and in the colonization and infection of animal hosts.

Quorum sensing signal molecules per se may also exert diverse biological effects on the tissues and organs of higher organisms. Intriguingly, the converse is also true – microbes, plants, and animals have all evolved mechanisms for sensing, mimicking, or destroying quorum sensing signals. Quorum sensing has also emerged as a valid target for the development of novel agents such as small molecules, antibodies, and enzymes capable of controlling bacterial behavior through the blockade of bacterial cell-to-cell communication. Indeed, quorum sensing has been described as "surely the most consequential molecular microbiology story of the last decade" (1).

Quorum Sensing: Methods and Protocols provides a unique and timely series of articles and methodologies written by leading experts in the field. This book offers an ideal opportunity for practical engagement in the science of quorum sensing through chapters which present biological and physico-chemical approaches to signal molecule detection, synthesis, analysis, and functional evaluation alongside strategies for investigating the impact of signals on higher organisms and for determining the efficacy of novel disruption strategies. Dr. Rumbaugh is to be commended for her foresight and enthusiasm in driving forward the publication of a book which should become essential reading for anyone interested in the practical science of bacterial cell-to-cell communication.

Paul Williams, Ph.D.

Reference

1. Busby, S., and de Lorenzo, V. (2001) Cell regulation: putting together pieces of the big puzzle, *Curr Opin Microbiol* 4, 117–118.

Preface

"I think that a multiple of bacteria are stronger than a few and thus by union are able to overcome obstacles too great for the few." This perceptive comment was made by the plant pathologist E. F. Smith in 1905 (1), 89 years before the term quorum sensing (QS) was coined (2). Now we know that Smith was correct and bacteria *are* capable of acting in "union" by synchronizing the transcription of subsets of genes, which then modulates their behavior. QS has changed our view of bacteria from simple, solitary egocentrics to social, networking altruists that use cell-to-cell signaling to build sophisticated biofilms, cause infections, or resist killing by antibiotics.

When I started graduate school in 1996, Quorum Sensing was a new buzz word. I quickly became infatuated with the idea that bacteria could "talk" to each other and focused my graduate work on examining the role of quorum sensing in *Pseudomonas aeruginosa* infections. Like any burgeoning field, the discoveries were exciting and fast-paced, and it seemed like new developments from the Greenberg, Williams, and Iglewski groups were published every month.

Since 1996, the QS field has grown from a few dozen laboratories, investigating the pathways, proteins, and chemicals that facilitate signaling in bacteria, to hundreds of groups that have integrated evolutionary biology, computer science, mathematics, engineering, and metagenomics to create an ever-expanding and dynamic field. Although my own fascination with "talking bacteria" continues today, my research now focuses on understanding how bacteria use quorum sensing signals to affect host cells, or, in other words, on determining what bacteria have to "say" to us. It seems we have only scratched the surface of these complex communication systems and that bacteria still have many new and fascinating secrets yet to tell us.

The aim of *Quorum Sensing: Methods and Protocols* is to provide an in-depth set of diverse protocols that span this broad field. The protocols presented here are broken up into three general categories: (1) detecting, isolating, and characterizing the QS signals made by both Gram− and Gram+ bacteria, (2) determining the function of QS signals in vivo, and (3) developing QS disruption strategies.

I would like to thank John M. Walker, the series editor, and Humana for the opportunity to make this contribution to the field. Most of all, thanks to all the authors who contributed their precious time and imparted their expert knowledge and their laboratories' valuable techniques to this collection of protocols. In preparation for this book, I asked a few of the seminal contributors to the QS field to reflect on how they have seen the field grow and change. Thanks for your insightful comments, Pete and Paul!

Lubbock, TX *Kendra P. Rumbaugh, Ph.D.*

References

1. Smith, E. F. (1905) Bacteria in relation to plant disease, Carnegie Institution, Washington, DC.
2. Fuqua, W. C., Winans, S. C., and Greenberg, E. P. (1994) Quorum sensing in bacteria: the LuxR-LuxI family of cell density-responsive transcriptional regulators, *J Bacteriol* 176, 269–275.

Contents

Contributors

ALI ALAGELY • *Soil and Water Sciences Department, Genetics Institute, Institute of Food and Agricultural Sciences (IFAS), University of Florida, Gainesville, FL, USA*
NAOMI BALABAN • *Department of Biomedical Sciences, Cummings School of Veterinary Medicine, Tufts University, North Grafton, MA, USA*
MARY B. BARNES • *Tulane National Primate Research Center, Covington, LA, USA*
THOMAS BJARNSHOLT • *Department of International Health, Immunology, and Microbiology, Faculty of Health Sciences, University of Copenhagen, Copenhagen, Denmark;*
Department of Clinical Microbiology, Rigshospitalet, Copenhagen, Denmark
MIGUEL CÁMARA • *Centre for Biomolecular Sciences, School of Molecular Medical Sciences, University of Nottingham, Nottingham, UK*
LOUISE DAHL CHRISTENSEN • *Department of International Health, Immunology, and Microbiology, Faculty of Health Sciences, University of Copenhagen, Copenhagen, Denmark*
WEIHUA CHU • *Department of Biology, Texas State University-San Marcos, San Marcos, TX, USA; Department of Microbiology, School of Life Science and Technology, China Pharmaceutical University, Nanjing, P. R. China*
MAIR E.A. CHURCHILL • *Department of Pharmacology, University of Colorado Denver School of Medicine, Aurora, CO, USA*
AURÉLIE DEVEAU • *Department of Microbiology and Immunology, Dartmouth Medical School, Hanover, NH, USA*
ERIC DÉZIEL • *INRS-Institut Armand-Frappier, LavalQC, Canada*
STEPHEN P. DIGGLE • *School of Molecular Medical Sciences, Centre for Biomolecular Sciences, University of Nottingham, Nottingham, UK*
DRAGOMIR I. DRAGANOV • *Division of Epidemiology, Department of Internal Medicine, The University of Texas Southwestern Medical Center, Dallas, TX, USA*
PETER EDMONSON • *Zen Sensing LLC., Decatur, GA, USA*
WALTER FAST • *Division of Medicinal Chemistry, College of Pharmacy, University of Texas, Austin, TX, USA*
MATTHEW P. FLETCHER • *School of Molecular Medical Sciences, Centre for Biomolecular Sciences, University of Nottingham, Nottingham, UK*

MARTIN FUSSENEGGER • *Department of Biosystems Science and Engineering, ETH Zurich, Basel, Switzerland*
MARIA VAN GENNIP • *Department of International Health, Immunology, and Microbiology, Faculty of Health Sciences, University of Copenhagen, Copenhagen, Denmark*
MICHAEL GIVSKOV • *Department of International Health, Immunology, and Microbiology, Faculty of Health Sciences, University of Copenhagen, Copenhagen, Denmark*
BRIAN K. HAMMER • *School of Biology, Georgia Institute of Technology, Atlanta, GA, USA*
DEBORAH A. HOGAN • *Department of Microbiology and Immunology, Dartmouth Medical School, Hanover, NH, USA*
WILLIAM HUNT • *Zen Sensing LLC, Decatur GA, USA*
TIM HOLM JAKOBSEN • *Department of International Health, Immunology, and Microbiology, Faculty of Health Sciences, University of Copenhagen, Copenhagen, Denmark*
KIM D. JANDA • *Worm Institute for Research and Medicine (WIRM), The Skaggs Institute for Chemical Biology, La Jolla, CA, USA; Departments of Chemistry and Immunology & Microbial Science, The Scripps Research Institute, La Jolla, CA, USA*
PALMY R. JESUDHASAN • *Food Safety & Environmental Microbiology Program, Texas A&M University, College Station, TX, USA*
GUNNAR F. KAUFMANN • *Departments of Chemistry and Immunology & Microbial Science, The Scripps Research Institute, La Jolla, CA, USA*
VLADIMIR V. KRAVCHENKO • *Department of Immunology & Microbial Science, The Scripps Research Institute, La Jolla, CA, USA*
DIANE M. KUBITZ • *Antibody Production Core Facility, The Scripps Research Institute, La Jolla, CA, USA*
FRANÇOIS LÉPINE • *INRS-Institut Armand-Frappier, Laval, QC, Canada*
COLIN A. LOWERY • *The Skaggs Institute for Chemical Biology, La Jolla, CA, USA; Departments of Chemistry and Immunology & Microbial Science, The Scripps Research Institute, La Jolla, CA, USA*
VATSALA MAITIN • *Department of Family and Consumer Science, Molecular and Cellular Nutrition Program, Texas State University-San Marcos, San Marcos, TX, USA*
LAUREN MASHBURN-WARREN • *Section of Molecular Genetics and Microbiology, The University of Texas at Austin, Austin, TX, USA*
ALEXANDER V. MAYOROV • *The Skaggs Institute for Chemical Biology and Departments of Chemistry and Immunology & Microbial Science, The Scripps Research Institute, La Jolla, CA, USA*

ROBERT J.C. MCLEAN • *Department of Biology, Texas State University-San Marcos, San Marcos, TX, USA*
GREGORY C. PALMER • *Section of Molecular Genetics and Microbiology, The University of Texas at Austin, Austin, TX, USA*
JUNGUK PARK • *Departments of Chemistry and Immunology & Microbial Science, The Scripps Research Institute, La Jolla, CA, USA*
SURESH D. PILLAI • *Food Safety & Environmental Microbiology Program, Texas A&M University, College Station, TX, USA*
SATHISH RAJAMANI • *Department of Microbiology and Immunology, Dartmouth Medical School, Hanover, NH, USA; Life Sciences Institute, Ann Arbor, MI, USA*
RICHARD SAYRE • *Donald Danforth Plant Science Center, St. Louis, MO, USA*
JEFFREY W. SCHERTZER • *Section of Molecular Genetics and Microbiology, The University of Texas at Austin, Austin, TX, USA*
MARTIN SCHUSTER • *Department of Microbiology, Oregon State University, Corvallis, OR, USA*
HIRUY M. SIBHATU • *Department of Pharmacology, University of Colorado Denver School of Medicine, Aurora, CO, USA*
KIRTHIRAM K. SIVAKUMAR • *Food Safety & Environmental Microbiology Program, Texas A&M University, College Station, TX, USA*
DESMOND STUBBS • *Zen Sensing LLC., Decatur, GA, USA*
HIROAKI SUGA • *Chemical Biology and Biotechnology Lab, Research Center for Advanced Science and Technology (RCAST), The University of Tokyo, Tokyo, Japan*
SINE LO SVENNINGSEN • *Biomolecular Sciences Section, Institute of Biology, University of Copenhagen, Copenhagen, Denmark*
JOHN F. TEIBER • *Division of Epidemiology, Department of Internal Medicine, The University of Texas Southwestern Medical Center, Dallas, TX, USA*
MAX TEPLITSKI • *Soil and Water Sciences Department, Genetics Institute, Institute of Food and Agricultural Sciences (IFAS), University of Florida, Gainesville, FL, USA*
PEI W. THOMAS • *Division of Medicinal Chemistry, College of Pharmacy, University of Texas, Austin, TX, USA*
CHARIS L. UHLSON • *Department of Pharmacology, University of Colorado Denver School of Medicine, Aurora, CO, USA*
RICHARD J. ULEVITCH • *Department of Immunology & Microbial Science, The Scripps Research Institute, La Jolla, CA, USA*
DHIRAJ A. VATTEM • *Molecular and Cellular Nutrition Program, Department of Family and Consumer Science, Texas State University-San Marcos, San Marcos, TX, USA*

Wilfried Weber • *Department of Biosystems Science and Engineering, ETH Zurich, Basel, Switzerland*
Marvin Whiteley • *Section of Molecular Genetics and Microbiology, The University of Texas at Austin, Austin, TX, USA*
Paul Williams • *Centre for Biomolecular Sciences, School of Molecular Medical Sciences, University of Nottingham, Nottingham, UK*

Part I

Detection, Isolation, and Characterization of Quorum Sensing Compounds

Chapter 1

Bioassays of Quorum Sensing Compounds Using *Agrobacterium tumefaciens* and *Chromobacterium violaceum*

Weihua Chu, Dhiraj A. Vattem, Vatsala Maitin, Mary B. Barnes, and Robert J.C. McLean

Abstract

In most bacteria, a global level of regulation exists involving intercellular communication via the production and response to cell density-dependent signal molecules. This cell density-dependent regulation has been termed quorum sensing (QS). QS is a global regulator, which has been associated with a number of important features in bacteria including virulence regulation and biofilm formation. Consequently, there is considerable interest in understanding, detecting, and inhibiting QS. Acyl homoserine lactones (acyl HSLs) are used as extracellular QS signals by a variety of Gram-negative bacteria. *Chromobacterium violaceum*, a Gram-negative bacterium commonly found in soil and water, produces the characteristic purple pigment violacein, the production of which is regulated by acyl HSL-mediated QS. Based on this readily observed pigmentation phenotype, *C. violaceum* strains can be used to detect various aspects of acyl HSL-mediated QS activity. In another commonly used bioassay organism, *Agrobacterium tumefaciens,* QS can be detected by the use of a reporter gene such as *lacZ.* Here, we describe several commonly used approaches incorporating *C. violaceum* and *A. tumefaciens* that can be used to detect acyl HSLs and QS inhibition.

Key words: Quorum sensing, *N*-acyl homoserine lactone, Violacein, *Chromobacterium violaceum*, *Agrobacerium tumefaciens*, Violacein

1. Introduction

1.1. Quorum Sensing Detection in Gram-Negative Bacteria

Although first considered to be a curiosity associated with light production in *Vibrio fischeri* (1), quorum signaling (QS) is now recognized as a major component of gene regulation and intercellular communication in bacteria (2, 3) and is now considered to be involved in other aspects of microorganisms including competition and structure (4). QS is an environmental sensing

Kendra P. Rumbaugh (ed.), *Quorum Sensing: Methods and Protocols*, Methods in Molecular Biology, vol. 692, DOI 10.1007/978-1-60761-971-0_1,

system that allows bacteria to monitor their own population density and to couple cell population density with gene expression. As part of their normal metabolic activities, bacteria produce several small diffusible signaling molecules. These signal molecules often positively regulate their own synthesis and as a consequence were originally described as autoinducers (AI). The genes associated with the production of these signals include an acyl HSL synthase and an acyl HSL-binding transcriptional regulator. In *V. fischeri*, these genes are *luxI* and *luxR*, respectively. Similar genes in other acyl HSL-regulated organisms are considered to be *luxI* and *luxR* homologues, even though the QS-mediated gene functions in these organisms do not involve bacterial luminescence (5). As AI-mediated signaling is typically associated with population density-dependent activity, the term QS has been used (6). Of the various QS signals, perhaps the best known are the acyl HSLs, of which the *N*-3-oxo-hexanoyl homoserine lactone (3-oxo-C_6-HSL) represents the signal associated with *V. fischeri* (7). Other representative acyl HSLs include the *N*-butyryl HSL (C_4-HSL) and *N*-3-oxo-dodecanoyl HSL (3-oxo-C_{12}-HSL) both associated with the opportunistic pathogen, *Pseudomonas aeruginosa* (8), *N*-3-oxo-octanoyl HSL (3-oxo-C_8-HSL) associated with *Agrobacterium tumefaciens* (9), and *N*-hexanoyl HSL (C_6-HSL) associated with *Chromobacterium violaceum* (10). Acyl HSL-based QS is reviewed in refs. (5, 11).

Acyl HSLs are associated with a number of important microbial activities including virulence gene regulation (12, 13), antibiotic resistance (14), and aspects of biofilm formation in *P. aeruginosa* (15, 16). QS interference by brominated furanones and other compounds (17–19) as well as acyl HSL-degrading enzymes has shown promise as novel antibacterial treatment strategies due to their effectiveness against highly resistant biofilm populations (20, 21). As a result, there is considerable interest in the identification of QS and QS inhibitors (QSIs). There are a number of chemical approaches including chromatography and mass spectrometry that can be used to characterize acyl HSLs (cf (22–25)). However, the equipment and expertise needed for these approaches can be quite significant. In contrast, QS bioassays (reviewed in ref. (26)) are relatively inexpensive and thus allow screening for QSI and acyl HSLs in regions of the world including places with high biodiversity, such as the tropics, and limited financial resources.

Bioassay organisms for QS will have a transcriptional response regulator (*luxR* homologue in the case of acyl HSLs) coupled to a reporter gene allowing a readily observable phenotype. In the case of the widely used *A. tumefaciens* acyl HSL bioassays (27, 28), the *traR* gene (*luxR* homologue) is coupled to *lacZ*. In this fashion, acyl HSLs can be detected on the basis of β-galactosidase activity often using X-gal. The normal phenotype associated with QS in *A. tumefaciens* is gene conjugation (27), which is not as readily visualized. In contrast to *A. tumefaciens*, several organisms including *Serratia marscesens* (29), *Pseudomonas aureofaciens* (30), and *C. violaceum* (10, 31) naturally produce pigmented compounds in response to QS. We shall focus on the use of *A. tumefaciens* and *C. violaceum* in this chapter.

1.2. Agrobacterium tumefaciens as a Biosensor

A. tumefaciens is a Gram-negative opportunistic plant pathogen that causes crown gall formation on plants through the transmission of DNA fragments into the nuclei of infected plants (27). Conjugation-based gene transfer is under the regulation of 3-oxo-C_8-HSL-dependent QS. The genes responsible for synthesis and response regulation of this gene are *traI* and *traR*, respectively, which are homologues of *luxI* and *luxR* (32). During their investigations of QS-related functions of *A. tumefaciens*, Winans, Fuqua, Zhu, and colleagues developed several reporter strains, notably A136 (pCF218)(pCF372) (referred in text as A136; also referred in literature as WCF47(pCF218)(pCF372)) (27, 33) and KYC55 (pJZ372)(pJZ384)(pJZ410) (referred in text as KYC55) (28) (listed in Table 1), which have become widely used

Table 1
Commonly used bioassay strains

Strain	Use	References
Agrobacterium tumefaciens		
A136 (pCF218)(pCF372) (also referred to as WCF47(pCF218)(pCF372))	Detection of broad range of acyl HSLs (3-oxo-C_4 to 3-oxo-C_{12}-HSLs, C_5–C_{10}-HSLs)[a,b]	(27, 33)
KYC55 (pJZ372)(pJZ384)(pJZ410)	Detection of broad range of acyl HSLs (3-oxo-C_4 to 3-oxo-C_{18}-HSLs, C_4-C_{18}-HSLs)[a,c]	(28)
KYC6	3-oxo-C_8-HSL overproducer, positive control for bioassay	(41)
Chromobacterium violaceum		
ATCC 12472	Wt, used in quorum signal inhibition screens, indirect acyl HSL detection	(10, 31)
ATCC 31532	C_6-HSL overproducer, used as positive control for CVO26 bioassay	(10)
CVO26	Detection of C_4- and C_6-HSLs	(10)

[a]For details, see refs. (28, 33). All strains grow at 30°C
[b]For plasmid maintenance, grow on LB + spectinomycin (50 μg/ml) and tetracycline (4.5 μg/ml)
[c]For plasmid maintenance, grow on LB + spectinomycin (50 μg/ml) and tetracycline (4.5 μg/ml) and gentamycin (15 μg/ml)

in the detection of acyl HSLs. These strains lack the *traI* gene (acyl HSL synthesis), overexpress the *traR* gene, and employ *lacZ* as a reporter gene. They require supplementation with a β-galactosidase substrate such as 5-bromo-4-chloro-3-indolyl β-d-galactopyranoside (X-gal) in order to visualize acyl HSL recognition. These two strains recognize a wide range of acyl HSLs (28, 33) with the KYC55 strain being able to detect a wider range of acyl HSLs with increased sensitivity (28). It has been our experience as well that both strains will detect their cognate acyl HSL (3-oxo-C_8-HSL) at sub-pmol concentrations, with other acyl HSLs being detected in the micromolar to nanomolar range. *A. tumefaciens* bioassays have been incorporated into acyl HSL detection by reverse-phase thin layer chromatography (TLC) (34) and high performance liquid chromatography (22). Alternatively, assays for β-galactosidase (LacZ gene product) can be employed to gain quantitative data (28).

1.3. Chromobacterium violaceum as a Biosensor

C. violaceum is a Gram-negative bacterium that produces a purple pigment, violacein, under regulation of C_6-HSL-dependent QS. The C_6-HSL synthesis gene and response regulator genes are *cviI* and *cviR*, respectively, and are homologues of *luxI* and *luxR* (35). Several strains are commonly used for bioassays (Table 1). These strains allow the direct detection of short-chain acyl HSLs through the induction of pigmentation in a strain CVO26 (10), which is unable to produce acyl HSLs but is fully capable of producing violacein in response to its cognate signal molecule (C_6-HSL) or the short-chain C4-HSL. As a positive control for the bioassay, a C_6-HSL-producing, nonpigmented strain (ATCC 31532), is used in association with the CVO26 bioassay strain (31). Using a plate streaking protocol (Subheading 3.2.1), the pigmentation is readily visible after overnight culture (Fig. 1a) and is absent if C_6 HSL is not present (Fig. 1b). We have observed that some pigmentation will occur in older (>48 h) CVO26 cultures. Acyl HSLs can be extracted from cultures (process described below), synthesized in the lab (24) or alternatively purchased from commercial sources such as Sigma-Aldrich. Several investigators have modified the *C. violaceum* assay by extracting the violacein with a solvent (typically acetone or ethanol) and then measuring absorption using a spectrophotometer (35). This approach enables one to gain quantitative data from the bioassay.

Acyl HSL-based QS is very specific in that the response regulator (*cviR* in the case of *C. violaceum*) will only respond to the cognate acyl HSL (C_6-HSL) or a closely related acyl HSL such as C_4-HSL. Other acyl HSLs will bind ineffectively to the CviR receptor and competitively interfere with its ability to activate genes associated with violacein production. Overall this is seen as a loss in pigmentation. An example using an overlay assay (Subheading 3.2.2) is shown in Fig. 2.

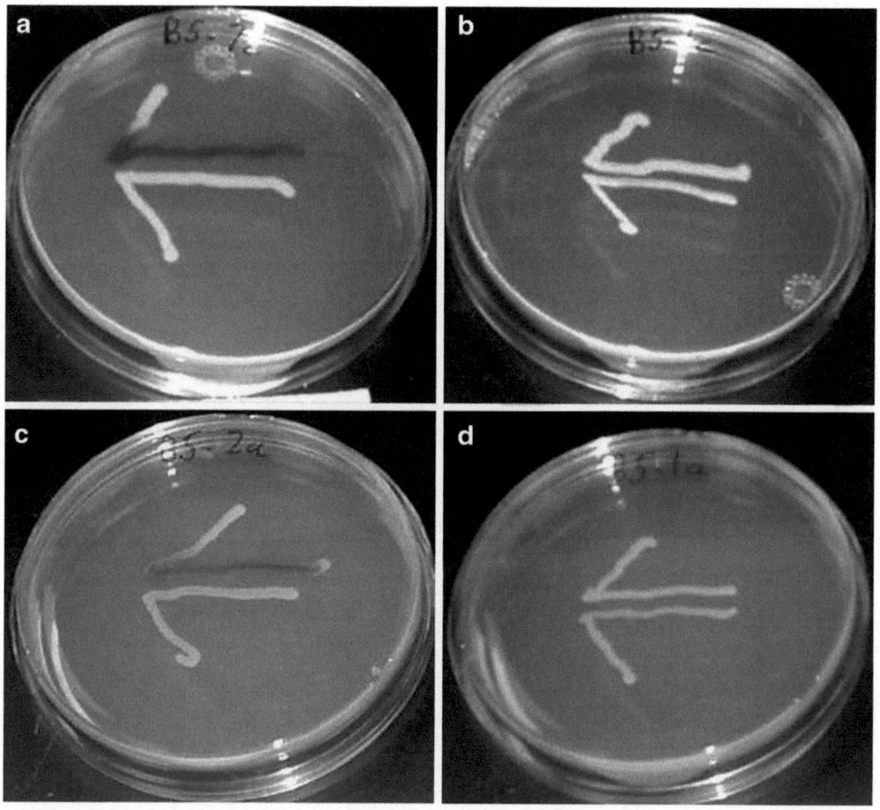

Fig. 1. Cross-feeding plate assay showing indicator strain *Chromobacterium violaceum* CV026 (*top culture* in both panels) producing violacein in response to *C. violaceum* 31532 (C_6-HSL over producer, positive control) (**a**) (10, 31). No violacein is produced when strain CV026 (*cviI* mutant) is streaked against itself (**b**). The *Agrobacterium tumefaciens* plate assay is shown in (**c**) and (**d**). (**c**) *A. tumefaciens* A136 (bioassay strain, *top culture*) expressing *lacZ* in response to *A. tumefaciens* KYC6 (3-oxo-C_8-HSL overproducer). (**d**) Negative control when strain A136 (*traI* mutant) is streaked against itself.

Here we describe several approaches in which *A. tumefaciens* and *C. violaceum* can be used for QS detection. The first section describes extraction and concentration of acyl HSLs using an ethyl acetate protocol (33, 34). Chemical extraction is not required but can increase the sensitivity of bioassays and also provides material for downstream applications including the detection of individual acyl HSLs from environmental samples. The second section describes the *A. tumefaciens* bioassays, which are generally used for detecting C_6–C_{12}-HSLs, although C_{14}–C_{18}-HSLs have been detected at micromolar concentrations. The third section describes the *C. violaceum* acyl HSL bioassay for detecting C_4- and C_6-HSLs on the basis of pigment induction in the reporter strain, CVO26. The fourth section describes a pigmentation inhibition assay, using the type strain 12472, which can be used as an indirect detection approach for other acyl HSLs (other than C_4- or C_6-HSLs). Alternatively this fourth approach can be used as an initial screen for QS inhibiting compounds or

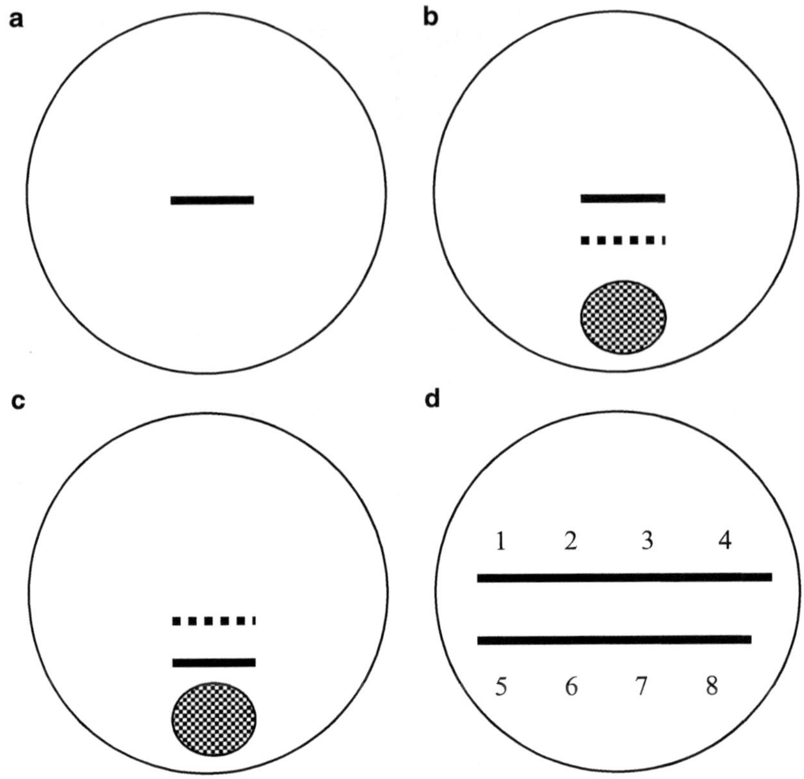

Fig. 2. Inoculation strategies for bioassays on Petri plates. (**a**) Shows inoculation of a test organism prior to an overlay assay (Subheading 3.2.2). (**b**) and (**c**) Show the inoculation pattern of a test organism or extract (*patterned small circle*) and varied location of acyl HSL-producing (*dashed line*) and acyl HSL-detecting (*solid line*) strains to differentiate interference with acyl HSL production or acyl HSL response (details in Subheading 3.2.3). (**d**) Shows an inoculation strategy for high throughput screens of test organisms (indicated by *numbers*) near a bioassay organism (*solid line*).

acyl HSL degrading (quorum quenching) enzymes (31). The fifth section describes a TLC protocol that can be used to detect acyl HSLs (34). Here, *A. tumefaciens* A136 (pCF218)(pCF372) is typically used as a biosensor.

2. Materials

2.1. Acyl HSL Extraction (33, 34)

1. Centrifuge capable of 4,000 × g and solvent-resistant tubes.
2. Ethyl acetate (containing 0.1% (v/v) acetic acid) (store in dark in volatile chemical storage).
3. Overnight broth culture (stationary phase) of organism to be investigated (see Note 1).
4. Bath sonicator (for environmental biofilm samples, see Note 2).
5. Small glass vial for storing extracted acyl HSLs (scintillation vials with autoclavable polycarbonate lids are excellent general purpose containers in this regard).

6. Water aspirator or vacuum source (use water trap to avoid contamination of vacuum line).
7. Rotary evaporator.
8. Freezer (preferably –80°C).

2.2. Agrobacterium tumefaciens Assay for Acyl HSLs

1. *A. tumefaciens* A136 (pCF218)(pCF372) (biosensor strain), maintain on LB + spectinomycin (50 µg/ml) and tetracycline (4.5 µg/ml) (27, 33).
2. *A. tumefaciens* KYC55 (pJZ372)(pJZ384)(pJZ410) (alternate, highly sensitive biosensor strain), maintain on LB + spectinomycin (50 µg/ml) + tetracycline (4.5 µg/ml) + gentamycin (15 µg/ml) (28).
3. Luria-Bertani medium (LB medium).
4. LB agar plates (1.5% (w/v) agar).
5. LB soft agar (containing 0.3% (w/v) agar).
6. Glass or metal spreading rod and 250–500 ml beaker filled to a depth of 2–3 cm with 100% ethanol (alcohol sterilize spreading rod before use by dipping in 100% ethanol and flame, see Note 3).
7. 20 mg/ml X-gal (in dimethylformide) (add 50 µl and spread over surface of agar prior to inoculating bacteria for bioassay).
8. Other agar as appropriate for organism to be investigated (see Note 4).
9. 50°C Water bath (needed for overlay assay).
10. Test tubes with caps and rack for preparing soft agar (5 ml aliquots).
11. Vortex mixer.
12. Pipette and sterile tips (adjustable 10–200 µl).

2.3. Chromobacterium violaceum Acyl HSL Detection

1. *C. violaceum* CVO26 (biosensor strain) (10) (see Note 5).
2. *C. violaceum* ATCC 31532 (C_6-HSL-overproducer, used as positive control) (10).
3. LB medium.
4. LB agar plates (1.5% (w/v) agar).
5. LB soft agar (containing 0.3% (w/v) agar).
6. Other agar as appropriate for organism to be investigated (see Note 6).
7. Other materials as shown in Subheading 2.2 (above).

2.4. Indirect C. violaceum Assay for Acyl HSLs and QS Inhibition

1. *C. violaceum* ATCC 12472 type strain used for pigmentation inhibition (31).
2. *P. aeruginosa* PAO1 (used as positive control) (31).
3. Other materials as listed in Subheading 2.2 (above).

2.5. TLC Detection of Acyl HSLs

1. Ethyl acetate (containing 0.1% (v/v) acetic acid) (store in dark in volatile chemical storage).
2. C18 TLC plate (Si-C18F TLC Plate Reversed Phase Octadecyl, JT Baker).
3. C_6-HSL and C_8-HSL standards (these and other acyl HSLs can be purchased from Sigma-Aldrich), dissolve in ethyl acetate (5 pmol/μl).
4. 17.5 cm × 16.0 cm × 6.2 cm TLC glass tank (Aldrich).
5. Paper towels or large Kimwipes (we use these to line tank during run, to prevent an uneven migration of solvent front) ("smile" or "frown" pattern).
6. Methanol/water mixture (60:40) – need at least 200 ml.
7. Laboratory adjustable temperature hot plate.
8. Labeling tape (approximately 2 cm wide).
9. Plastic tub and cover (big enough to incubate TLC plate with indicator bacteria).
10. Incubator (30°C).
11. *A. tumefaciens* A136 bioassay strain.
12. X-gal solution: 20 mg/ml in dimethylformamide.

3. Methods

3.1. Ethyl acetate extraction (33, 34) (see Note 7)

1. Grow 20 ml broth culture to stationary phase. If biofilm from pebbles or other surfaces is desired, sonicate pebbles in sterile H_2O for 15 min, and remove 20 ml of the sonicated liquid to a centrifuge tube.
2. Centrifuge to remove bacterial cells, at 3,000 × *g* for 10 min at 4°C.
3. Transfer supernatant to clean bottle.
4. Extract the cell-free supernatant three times with three volumes of ethyl acetate.
5. Pool the ethyl acetate fractions (top layers) and evaporate to dryness using a rotary evaporator (with water bath set at 40°C).
6. Suspend residue in 1 ml ethyl acetate and transfer to a small glass vial. Again evaporate the ethyl acetate, this time using a Pasteur pipette attached to a vacuum source (aspirator, linked to a water tap works fine for this) that is placed above the liquid (careful: it is easy to suck up liquid here).
7. Resuspend the residue in 100 μl ethyl acetate and store at –80°C until needed.

3.2. Acyl HSL Reporter Agar-Based Plate Bioassays (36, 37)

3.2.1. Plate-Based Assay (See Note 8)

1. If using *A. tumefaciens*-based bioassay, pipette 50 μl X-gal solution onto agar and spread across surface with alcohol-sterilized spreading rod, prior to inoculating with bacteria.
2. Streak test organism and bioassay organism beside each other on LB agar plate using a "T" shape. The cross of the "T" would represent strain CVO26 and the vertical line would represent the test organism. Following 16–48 h incubation at 30°C, you should see a positive blue coloration (due to X-gal hydrolysis) in the case of *A. tumefaciens* A136 or KYC55 bioassay; purple violacein (in the case of *C. violaceum* bioassay) production in the bioassay in the region closest to the intersection of the two strains and a lack of pigmentation away from this intersection. An alternative streaking design is shown in Fig. 1 (see Note 9).
3. As a positive control in the *C. violaceum* bioassay, streak CVO26 (bioassay strain) and *C. violaceum* 31532 (C_6-HSL overproducer) in this same manner. For *A. tumefaciens* bioassay, first spread 50 μl X-gal solution on surface of LB agar, then streak the bioassay strain (A136 or KYC55) and KYC6 (3-oxo-C_8-HSL overproducer) in this same manner. For a negative control, we simply streak the reporter strains (CVO26, A136, or KYC55) against itself (both as reporter and test) as these strains do not produce acyl HSLs.

3.2.2. Soft Agar Overlay Assay (31)

This approach gives more sensitivity than the plate-based assay (Subheading 3.2.1) and also enables the observation of potential competition between the test organism and the bioassay organism (*C. violaceum*), which may indicate production of an antibacterial compound (see Note 10). This test typically takes 3 days.

1. Day 1: Streak the test organism onto suitable agar using the pattern shown (Fig. 2a), then grow in incubator overnight (see Note 11).
2. As a positive control, streak *P. aeruginosa* onto LB agar in the same pattern and grow overnight at 37°C. If desired, a negative control would consist either of strain 12472 on the plate or alternatively strain 31532 (C_6-HSL overproducer).
3. Grow an overnight LB broth culture of *C. violaceum* (30°C with shaking).
4. If some other substance (example would be a plant leaf or other tissue) is to be examined, then place the object onto a plate of LB agar and immediately proceed to steps 5 and 6 (agar overlay).
5. Day 2: Prepare and autoclave test tubes containing 5 ml LB soft agar (0.3% (w/v) agar). Store tubes in 50°C water bath once they come out of the autoclave.

6. It is important that step 6 be done as quickly as possible as *C. violaceum* is susceptible to heat. For each test plate (from step 1 or 2), mix 5 μl overnight *C. violaceum* 12472 culture with 5 ml soft agar, vortex for 5 s, then pour mixture over plate. Incubate at 30°C overnight.
7. Day 3: Examine and photograph plates. A positive result is indicated by a loss of pigmentation in the vicinity. An example is shown in Fig. 3c.

3.2.3. Alternative Plate-Based Assays

The aforementioned assays in Subheadings 3.2.1 and 3.2.2 are useful for screening various microorganisms and other tissues for acyl HSLs and QS inhibiting materials. QS bioassays also lend themselves to the examination of other substances including plant and food extracts (39). Here, one can examine pigment induction in an acyl HSL-responsive strain (CVO26), or pigmentation inhibition (strain 12472) (see Note 12). Here the test material can be placed onto a small sterile disk of filter paper (we typically use a three-hole office punch to generate these and then autoclave before use).

1. Prepare overnight broth cultures of *Chromobacterium* strains as described in Subheading 3.2.2 (use CVO26 for pigment induction and 12472 for pigment inhibition).
2. Take 100 μl overnight culture and spread over surface of LB agar (forming a bacterial lawn).
3. Using alcohol sterilized forceps, place filter paper disk (containing test substance) onto plate. The substance can be added to the filter paper disk with a pipette after the disk has been placed onto the agar.

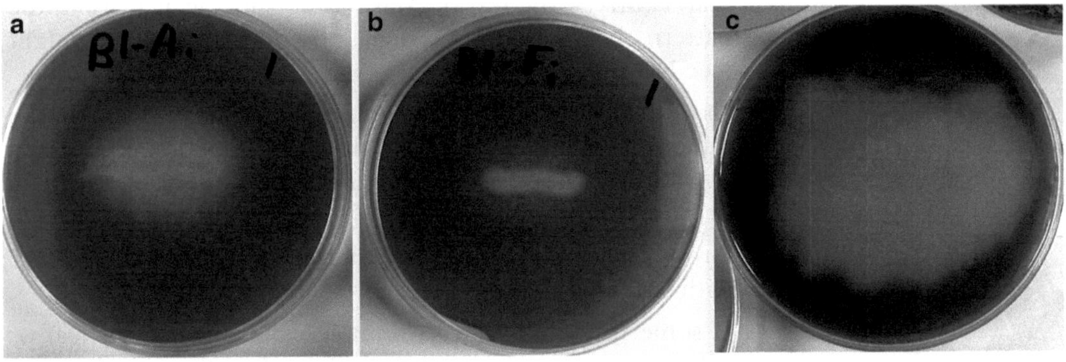

Fig. 3. *Chromobacterium violaceum* pigmentation inhibition assay is used to detect QS disruption in *C. violaceum* 12472 ((10) as modified (31)). Here, acyl HSLs or enzymes that disrupt the binding of the *C. violaceum* cognate acyl HSL (C6 HSL) cause a loss of pigmentation in the vicinity of the test organism in the *center* of the plate. (**a**) (positive control), *Pseudomonas aeruginosa* PAO1 acyl HSLs (C_4-HSL and 3-oxo-C_{12}-HSL) interfere with the *C. violaceum* cviR receptor. Pigmentation inhibition is absent [negative control (**b**)] in the vicinity of C_6-HSL-producing *C. violaceum* strain 31532, although the 31532 strain itself is nonpigmented. (**c**) Shows the pigmentation inhibition, due to an uncharacterized quorum disrupting compound produced by an environmental isolate (42) (RJC McLean and C Fuqua, unpublished). Details are in text.

4. Incubate overnight at 30°C and examine for pigmentation. Again the effect should be most noticeable in the vicinity of the test material.

5. As an alternative, one can use two reporter strains simultaneously (40) and investigate whether QS inhibition might be due to interference with acyl HSL production (*luxI* effect) or a transcription response (*luxR* effect). In the case of the *C. violaceum* bioassay, two reporters are used: 31532 (C_6-HSL overproducer) and CVO26 (acyl HSL biosensor). Overnight cultures of the reporters are prepared as describe above and the samples inoculated as shown in Fig. 2b, c.

   The test material is shown as a darkened circle within the Petri plate, the bioassay strain CVO26 is shown as a solid line, and the C_6-HSL-producing strain 31532 shown as a dashed line. Normally, these two strains will cross-feed each other (positive control for acyl HSL detection) such that strain CVO26 will produce violacein in the presence of 31532. Using the first pattern (Fig. 2b), the test material is closest to strain CVO26 and so pigmentation inhibition results could be interpreted as transcription inhibition (*luxR* effect). In the case of the second pattern (Fig. 2c), the acyl-HSL producing strain, 31532, is closest to the test material and so a pigmentation inhibition effect could be interpreted as interference with C_6-HSL production (*luxI* effect) (40).

6. A schematic of a plate-based protocol for a high throughput bioassay is shown in Fig. 2d. Here, the bioassay strain is shown as a solid line, and the test organisms (indicated by numbers) can be spotted near the bioassay strain. Due to possible interactions between test organisms during the high throughput assays, we recommend that positive results be confirmed using the plate-based assay (Subheading 3.2.1) or overlay assay (Subheading 3.2.2).

3.2.4. TLC Detection of Acyl HSLs

1. The resuspended sample and standards are added to a C18 TLC plate. Apply 1 or 2 μl dot-wise, approximately 2 cm apart, along a line which is 2 cm from the bottom of the plate.

2. For standards, use C6 and C8 standards, 5 pmol/μl.

3. Line chromatography chamber with white paper towel or filter paper. Add methanol/water (60:40, vol/vol), 200 ml. Moistened towels will become saturated with the solvent, and humidify chamber, thereby preventing uneven solvent front migration.

4. Place TLC plate into the chamber and cover with glass lid. Develop chromatogram until solvent front is approximately 15 cm from the starting line.

5. Remove TLC plate and let solvent evaporate in the fume hood. Then place labeling tape (upright orientation) around edge of plate to form a wall and prevent spills from the indicator bacteria–agar mixture.
6. Overlay dried plate with indicator bacterium as follows:
    a. From a fresh 5 ml overnight culture, inoculate 50 ml media and grow to late exponential phase (see Note 13).
    b. Add this 50 ml culture to: 100 ml media plus 1.12 g melted agar plus 10 mg X-gal (500 μl of 20 mg/ml stock in dimethylformamide), tempered at 45°C.
    c. Prewarm TLC plate by putting on heater (low setting) for approximately 5 min (allows bacteria–agar mixture to spread evenly over TLC plate). Immediately pour mixture over prewarmed plate and spread evenly to cover. Layer should be approximately 3 mm thick.
    d. Allow the agar to solidify, then place in a plastic tub (avoids dehydration during incubation). Cover and place in a 30°C incubator for 12–18 h. Check for blue spots. The most hydrophobic (largest) acyl HSLs will be near the origin (bottom of the TLC plate) with the smaller (less hydrophobic) acyl HSLs migrating a longer distance. Sample results are shown in Fig. 4.
    e. Plastic tub can be disinfected after use with bleach or 70% (v/v ethanol).

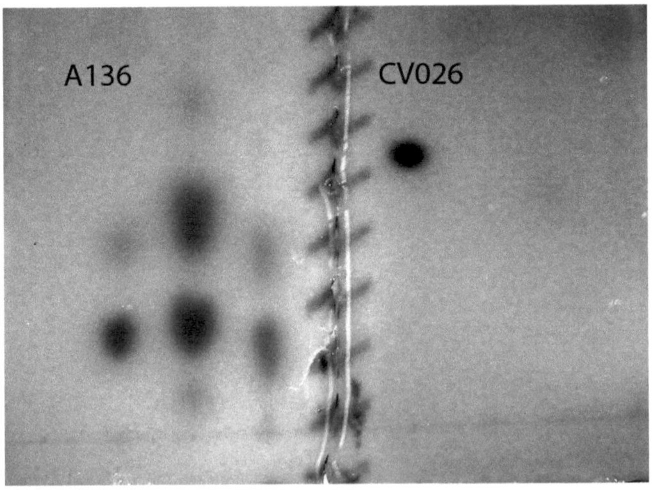

Fig. 4. TLC analysis of acyl HSL extracts from several strains of *Pseudomonas aeruginosa*. Of note, *Agrobacterium tumefaciens* A136 (*left*) detected many more acyl HSLs than did the *Chromobacterium violaceum* CV026 (*right*).

4. Notes

1. Alternatively if naturally occurring biofilms are to be investigated, then prepare small glass containers with water – fill 1/2 full with water, cover, and autoclave. Container size is chosen based on environmental sample size.
2. We have found it important to use glass rather than plastic as the latter substance (plastic) dampens the shear forces generated by the sonicator and reduces biofilm disruption.
3. Caution flammable – if ethanol in beaker catches fire, cover top of beaker with a suitable object, such as a book, in order to quickly smother the flame.
4. As a cautionary note, ensure that this medium does not inhibit *A. tumefaciens.*
5. For long-term storage of *C. violaceum* strains, grow an overnight culture in broth with shaking at 30°C, then mix 1.2 ml of the overnight strain with 0.4 ml sterile 50% (v/v) glycerol, place into a 2 ml sterile cryogenic storage tube, and store at −80°C. To revive the frozen culture, simple scrape some cells from the top of the frozen culture onto LB and incubate overnight at 30°C. (Cautionary note: For short-term (2–3 days) storage, leave at room temperature (20–25°C). *Chromobacterium* strains do not survive very well at 4°C.
6. As a cautionary note, ensure that this medium does not inhibit *C. violaceum.*
7. If using a solvent to extract acyl HSLs (typically acidified ethyl acetate) or potential QS inhibiting compound (usually an organic solvent such as acetone), test the solvent on the bioassay strain to ensure that any pigmentation or viability alteration due to the solvent is measured. As a safety matter, wear gloves (we recommend nitrile gloves) and perform any solvent-based studies in a fume hood.
8. Works well if both bioassay (*C. violaceum* or *A. tumefaciens*) and test organism grow on same medium.
9. Although antibiotics are required to maintain the plasmids in strains A136 or KYC55, these strains are sufficiently stable to be grown without antibiotics for the duration of the bioassay.
10. Pigmentation inhibition (Subheading 3.2.2) can give a useful indication of potential QS inhibition. However, the CviR transcriptional activator (LuxR homologue) in *C. violaceum* can be blocked by other acyl HSLs (31) and so identification of inhibitory substances should not be made on the basis of a *C. violaceum* bioassay alone. We recommend that potential

QS inhibitors be screened for acyl HSLs with a broad range bioassay system (*A. tumefaciens* A136 or KYC55 strains mentioned above).

11. For freshwater aquatic isolates, we typically use R2A agar (38).
12. We have found the *C. violaceum* CVO26 bioassay to respond to low micromolar to high nanomolar concentrations to its cognate acyl HSL (C_6-HSL). This strain is best suited to

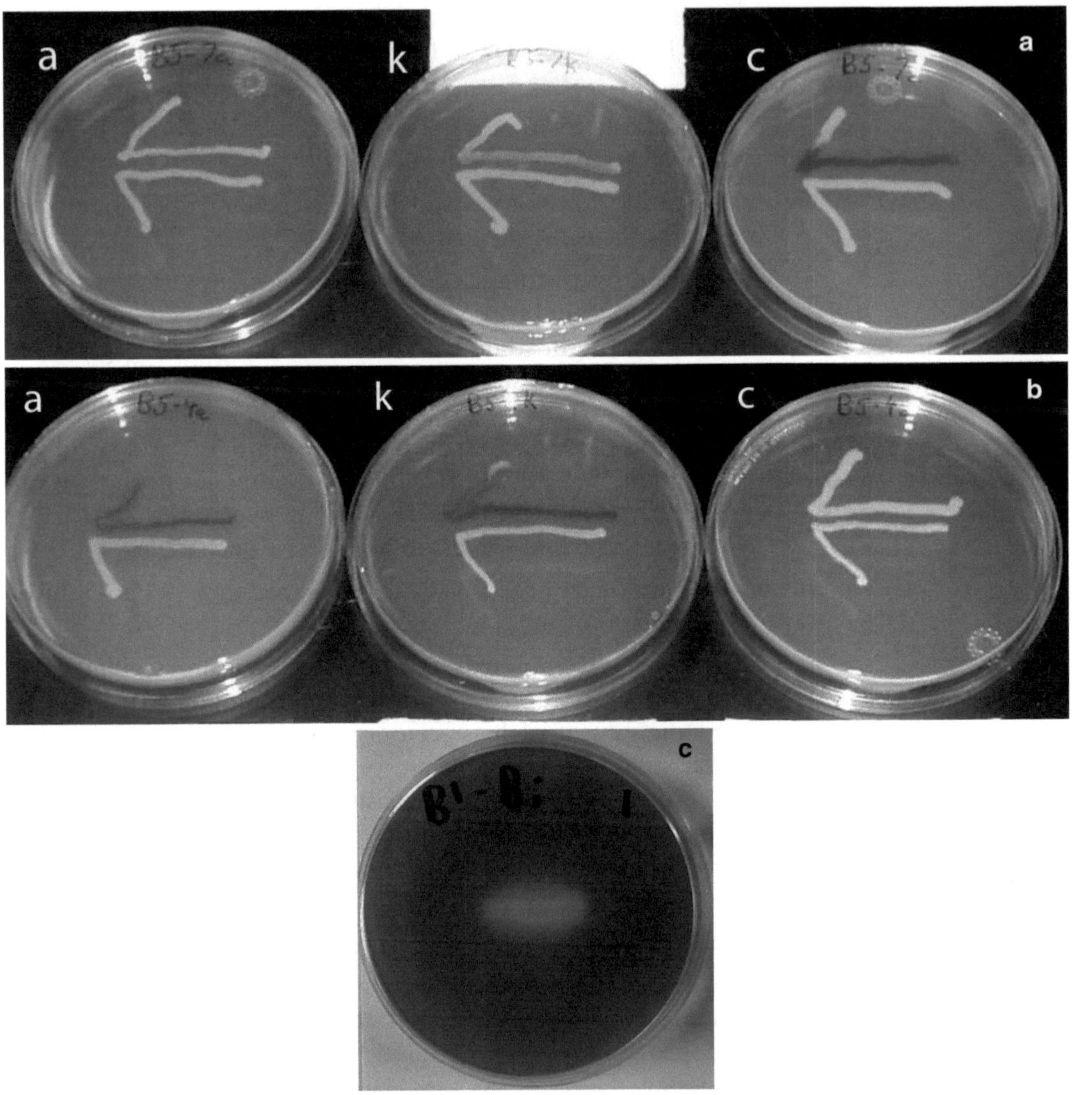

Fig. 5. Comparison of *Chromobacterium violaceum* with *Agrobacterium tumefaciens* biosensor activity. (**a**) Shows reaction of *A. tumefaciens* A136 (pCF 218)(pCF 372) (denoted by a) (27), *A. tumefaciens* KYC55 (pJZ372)(pJZ384)(pJZ410) (denoted by k) (28), and *C. violaceum* CV026 (denoted by c) (10) biosensors (*top streak*) to C_6-HSL production by *C. violaceum* 31532 (*bottom streak*) in a plate bioassay (Subheading 3.2.1). (**b**) Shows response of same three biosensor strains to 3-oxo-C_8-HSL production by *A. tumefaciens* strain KYC6 (41). However, the 3-oxo-C_8-HSL, produced by *A. tumefaciens*, can be detected indirectly by *C. violaceum* 12472 pigmentation inhibition (**c**, Subheading 3.2.2), bioassay.

detecting short-chain acyl HSLs (C_4- and C_6-HSLs). In contrast, another acyl HSL bioassay strain, *A. tumefaciens* A136 (pCF218)(pCF372) developed by Fuqua and Winans (27), is much more sensitive that CVO26, in that it can detect its cognate acyl HSL (3-oxo-C_8-HSL) in the low picomolar range. As well that particular *A. tumefaciens* strain will detect a wider variety of acyl HSLs with acyl groups ranging in size from 6 to 12 carbons (33). Zhu and Winans developed an even more sensitive *A. tumefaciens* strain, designated KYC55 (28) capable of detecting long-chain acyl HSLs (C_6- to C_{14}-HSLs). Both *A. tumefaciens* biosensors employ *lacZ* as a reporter gene and require supplementation of X-gal to the medium. A comparison of the sensitivity of the *C. violaceum* CVO26 assay to the aforementioned *A. tumefaciens* bioassays is shown in Fig. 3. Here, the CVO26 assay readily shows the presence of C_6-HSL, whereas this acyl HSL is weakly detected by strain KYC55 and not detected by strain A136 (Fig. 3a). In contrast, the two *A. tumefaciens* biosensors readily detect their cognate acyl HSL (3-oxo-C_8-HSL), whereas CVO26 does not. See Fig. 5.

13. It is important that steps 6b and 6c be done quickly as *A. tumefaciens* is sensitive to elevated temperatures.

Acknowledgements

We are grateful to Clay Fuqua and Steve Winans for providing these strains and introducing us to quorum signaling. Work in RJCM's laboratory is supported by a grant from the Norman Hackerman Advanced Research Program (003615-0037-2007).

References

1. Nealson, K. H., Platt, T., and Hastings, J. W. (1970) Cellular control of the synthesis and activity of the bacterial luminescent system, *J. Bacteriol.* 104, 313–22.
2. Ng, W. L. and Bassler, B. L. (2009) Bacterial quorum-sensing network architectures, *Annu. Rev. Genet.* 43, 197–222.
3. Whiteley, M., Lee, K. M., and Greenberg, E. P. (1999) Identification of genes controlled by quorum sensing in *Pseudomonas aeruginosa*, *Proc. Natl. Acad. Sci. USA.* 96, 13904–9.
4. Schertzer, J. W., Boulette, M. L., and Whiteley, M. (2009) More than a signal: non-signaling properties of quorum sensing molecules, *Trends Microbiol.* 17, 189–95.
5. Fuqua, C., Winans, S. C., and Greenberg, E. P. (1996) Census and consensus in bacterial ecosystems: the LuxR–LuxI family of quorum-sensing transcriptional regulators, *Annu. Rev. Microbiol.* 50, 727–51.
6. Fuqua, C., Parsek, M. R., and Greenberg, E. P. (2001) Regulation of gene expression by cell-to-cell communication, *Annu. Rev. Genet.* 35, 439–68.
7. Eberhard, A., Burlingame, A. L., Eberhard, C., Kenyon, G. L., Nealson, K. H., and Oppenheimer, N. J. (1981) Structural identification of autoinducer of *Photobacterium fischeri* luciferase, *Biochemistry.* 20, 2444–9.

8. Pearson, J. P., Passador, L., Iglewski, B. H., and Greenberg, E. P. (1995) A second *N*-acylhomoserine lactone signal produced by *Pseudomonas aeruginosa*, *Proc. Natl. Acad. Sci. USA*. 92, 1490–4.
9. Zhang, L., Murphy, P. J., Kerr, A., and Tate, M. E. (1993) *Agrobacterium* conjugation and gene regulation by *N*-acyl-l-homoserine lactones, *Nature*. 362, 446–8.
10. McClean, K. H., Winson, M. K., Fish, L., Taylor, A., Chhabra, S. R., Camara, M., Daykin, M., Lamb, J. H., Swift, S., Bycroft, B. W., Stewart, G. S. A. B., and Williams, P. (1997) Quorum sensing and *Chromobacterium violaceum*: exploitation of violacein production and inhibition for the detection of *N*-acylhomoserine lactones, *Microbiology*. 143, 3703–11.
11. Williams, P. (2007) Quorum sensing, communication and cross-kingdom signalling in the bacterial world, *Microbiology* 153, 3923–38.
12. Rumbaugh, K. P., Griswold, J. A., Iglewski, B. H., and Hamood, A. N. (1999) Contribution of quorum sensing to the virulence of *Pseudomonas aeruginosa* in burn wound infections, *Infect. Immun*. 67, 5854–62.
13. Wu, H., Song, Z., Givskov, M., Döring, G., Worlitzsch, D., Mathee, K., Rygaard, J., and Hoiby, N. (2001) *Pseudomonas aeruginosa* mutations in *lasI* and *rhlI* quorum sensing systems result in milder chronic lung infection, *Microbiology*. 147, 1105–13.
14. Bjarnsholt, T., Jensen, P. O., Burmolle, M., Hentzer, M., Haagensen, J. A. J., Hougen, H. P., Calum, H., Madsen, K. G., Moser, C., Molin, S., Hoiby, N., and Givskov, M. (2005) *Pseudomonas aeruginosa* tolerance to tobramycin, hydrogen peroxide and polymorphonuclear leukocytes is quorum-sensing dependent, *Microbiology*. 151, 373–83.
15. Davies, D. G., Parsek, M. R., Pearson, J. P., Iglewski, B. H., Costerton, J. W., and Greenberg, E. P. (1998) The involvement of cell-to-cell signals in the development of a bacterial biofilm, *Science*. 280, 295–8.
16. Shrout, J. D., Chopp, D. L., Just, C. L., Hentzer, M., Givskov, M., and Parsek, M. R. (2006) The impact of quorum sensing and swarming motility on *Pseudomonas aeruginosa* biofilm formation is nutritionally conditional, *Mol. Microbiol*. 62, 1264–77.
17. de Nys, R., Givskov, M., Kumar, N., Kjelleberg, S., and Steinberg, P. D. (2006) Furanones, *Prog. Mol. Subcell. Biol*. 42, 55–86.
18. Adonizio, A. L., Downum, K., Bennett, B. C., and Mathee, K. (2006) Anti-quorum sensing activity of medicinal plants in southern Florida, *J. Ethnopharmacol*. 105, 427–35.
19. Defoirdt, T., Miyamoto, C. M., Wood, T. K., Meighen, E. A., Sorgeloos, P., Verstraete, W., and Bossier, P. (2007) The natural furanone (5Z)-4-bromo-5-(bromomethylene)-3-butyl-2(5H)-furanone disrupts quorum sensing-regulated gene expression in *Vibrio harveyi* by decreasing the DNA-binding activity of the transcriptional regulator protein luxR, *Environ. Microbiol*. 9, 2486–95.
20. Rasmussen, T. B. and Givskov, M. (2006) Quorum-sensing inhibitors as anti-pathogenic drugs, *Int. J. Med. Microbiol*. 296, 149–61.
21. Givskov, M., de Nys, R., Manefield, M., Gram, L., Maximilien, R., Eberl, L., Molin, S., Steinberg, P. D., and Kjelleberg, S. (1996) Eukaryotic interference with homoserine lactone-mediated prokaryotic signalling, *J. Bacteriol*. 178, 6618–22.
22. Moré, M. I., Finger, L. D., Stryker, J. L., Fuqua, C., Eberhard, A., and Winans, S. C. (1996) Enzymatic synthesis of a quorum-sensing autoinducer through the use of defined substrates, *Science*. 272, 1655–8.
23. Makemson, J., Eberhard, A., and Mathee, K. (2006) Simple electrospray mass spectrometry detection of acylhomoserine lactones, *Luminescence*. 21, 1–6.
24. Eberhard, A. and Schineller, J. B. (2000) Chemical synthesis of bacterial autoinducers and analogs, *Methods Enzymol*. 305, 301–15.
25. Charlton, T. S., de Nys, R., Netting, A., Kumar, N., Hentzer, M., Givskov, M., and Kjelleberg, S. (2000) A novel and sensitive method for the quantification of *N*-3-oxoacyl homoserine lactones using gas chromatography-mass spectrometry: application to a model bacterial biofilm, *Environ. Microbiol*. 2, 530–41.
26. Steindler, L. and Venturi, V. (2007) Detection of quorum-sensing *N*-acyl homoserine lactone signal molecules by bacterial biosensors, *FEMS Microbiol. Lett*. 266, 1–9.
27. Fuqua, C. and Winans, S. C. (1996) Conserved cis-acting promoter elements are required for density-dependent transcription of *Agrobacterium tumefaciens* conjugal transfer genes, *J. Bacteriol*. 178, 435–40.
28. Zhu, J., Chai, Y., Zhong, Z., Li, S., and Winans, S. C. (2003) Agrobacterium bioassay strain for ultrasensitive detection of *N*-acylhomoserine lactone-type quorum-sensing molecules: detection of autoinducers in *Mesorhizobium huakuii*, *Appl. Environ. Microbiol*. 69, 6949–53.
29. Glansdorp, F. G., Thomas, G. L., Lee, J. K., Dutton, J. M., Salmond, G. P. C., Welch, M., and Spring, D. R. (2004) Synthesis and stability of small molecule probes for *Pseudomonas aeruginosa* quorum sensing modulation, *Org. Biomol. Chem*. 2, 3329–36.

30. Wood, D. W. and Pierson, L. S. (1996) The phzI gene of *Pseudomonas aureofaciens* 30–84 is responsible for the production of a diffusible signal required for phenazine antibiotic production, *Gene.* 168, 49–53.
31. McLean, R. J. C., Pierson, L. S., and Fuqua, C. (2004) A simple screening protocol for the identification of quorum signal antagonists, *J. Microbiol. Methods.* 58, 351–60.
32. Fuqua, W. C. and Winans, S. C. (1994) A *lux*R-*lux*I type regulatory system activates *Agrobacterium* Ti plasmid conjugal transfer in the presence of a plant tumor metabolite, *J. Bacteriol.* 176, 2796–806.
33. Zhu, J., Beaber, J. W., Moré, M. I., Fuqua, C., Eberhard, A., and Winans, S. C. (1998) Analogs of the autoinducer 3-oxooctanoyl-homoserine lactone strongly inhibit activity of the TraR protein of *Agrobacterium tumefaciens*, *J. Bacteriol.* 180, 5398–405.
34. Shaw, P. D., Ping, G., Daly, S. L., Cha, C., Cronan, J. E., Jr., Rinehart, K. L., and Farrand, S. K. (1997) Detecting and characterizing *N*-acyl-homoserine lactone signal molecules by thin layer chromatography, *Proc. Natl. Acad. Sci. USA.* 94, 6036–41.
35. Blosser, R. S. and Gray, K. M. (2000) Extraction of violacein from *Chromobacterium violaceum* provides a new quantitative bioassay for *N*-acyl homoserine lactone autoinducers, *J. Microbiol. Methods.* 40, 47–55.
36. McLean, R. J. C., Whiteley, M., Stickler, D. J., and Fuqua, W. C. (1997) Evidence of autoinducer activity in naturally-occurring biofilms, *FEMS Microbiol. Lett.* 154, 259–63.
37. Stickler, D. J., Morris, N. S., McLean, R. J. C., and Fuqua, C. (1998) Biofilms on indwelling urinary catheters produce quorum-sensing signal molecules *in situ* and *in vitro*, *Appl. Environ. Microbiol.* 64, 3486–90.
38. Bates, C. L., Forstner, M. R. J., Barnes, M. B., Whiteley, M., and McLean, R. J. C. (2006) Identification of heterotrophic limestone-adherent biofilm isolates from the Edwards Aquifer, Texas, *Southwest. Nat.* 51, 299–309.
39. Vattem, D. A., Mihalik, K., Crixell, S. H., and McLean, R. J. C. (2007) Dietary phytochemicals as quorum sensing inhibitors, *Fitoterapia.* 78, 302–10.
40. McLean, R. J. C., Bryant, S. A., Vattem, D. A., Givskov, M., Rasmussen, T. B., and Balaban, N. (2008) Detection in vitro of quorum-sensing molecules and their inhibitors, in *The Control of Biofilm Infections by Signal Manipulation* (Balaban, N., Ed.) pp. 39–50, Springer-Verlag, Heidelberg.
41. Fuqua, C., Burbea, M., and Winans, S. C. (1995) Activity of the *Agrobacterium* Ti plasmid conjugal transfer regulator TraR is inhibited by the product of the traM gene, *J. Bacteriol.* 177, 1367–73.
42. McLean, R. J. C., Barnes, M. B., Windham, M. K., Merchant, M. M., Forstner, M. R. J., and Fuqua, C. (2005) Cell–cell influences on bacterial community development in aquatic biofilms, *Appl. Environ. Microbiol.* 71, 8987–90.

Chapter 2

Detection of 2-Alkyl-4-Quinolones Using Biosensors

Stephen P. Diggle, Matthew P. Fletcher, Miguel Cámara, and Paul Williams

Abstract

2-Alkyl-4-quinolones (AQs) such as 2-heptyl-3-hydroxy-4-quinolone (PQS) and 2-heptyl-4-quinolone (HHQ) are quorum sensing signal molecules. Here we describe two methods for AQ detection and quantification that employ thin layer chromatography (TLC) and microtitre plate assays in combination with a *lux*-based *Pseudomonas aeruginosa* AQ biosensor strain. For TLC detection, organic solvent extracts of bacterial cells or spent culture supernatants are chromatographed on TLC plates, which are then dried and overlaid with the AQ biosensor. After detection by the bioreporter, AQs appear as both luminescent and green (pyocyanin) spots. For the microtitre assay, either spent bacterial culture supernatants or extracts are added to a growth medium containing the AQ biosensor. Light output by the bioreporter is proportional to the AQ content of the sample. The assays described are simple to perform, do not require sophisticated instrumentation, and are highly amenable to screening large numbers of bacterial samples.

Key words: *Pseudomonas aeruginosa*, Biosensor, 2-Alkyl-4-quinolones, 2-Heptyl-3-hydroxy-4-quinolone, 2-Heptyl-4-quinolone, *pqsA*

1. Introduction

In *Pseudomonas aeruginosa*, cell–cell communication (quorum sensing, QS) is known to control the production of extracellular virulence factors and promote biofilm maturation. The QS system consists of two *N*-acylhomoserine lactone (AHL) regulatory circuits (*las* and *rhl*) linked to a 2-alkyl-4-quinolone (AQ) system (1, 2). In the *las* system, the *lasI* gene product directs the synthesis of *N*-(3-oxo-dodecanoyl)-L-homoserine lactone (3-oxo-C_{12}-HSL), which interacts with the transcriptional regulator LasR and activates target promoters. In the *rhl* system, *rhlI* directs the synthesis of *N*-(butanoyl)-L-homoserine lactone (C_4-HSL), which

Kendra P. Rumbaugh (ed.), *Quorum Sensing: Methods and Protocols*, Methods in Molecular Biology, vol. 692,
DOI 10.1007/978-1-60761-971-0_2, © Springer Science+Business Media, LLC 2011

interacts with the cognate regulator RhlR and activates target gene promoters. The *las* and *rhl* systems are hierarchically connected and have been found to regulate the timing and production of multiple virulence factors including elastase, alkaline protease, exotoxin A, rhamnolipids, pyocyanin, lectins, superoxide dismutases, and biofilm formation (1).

Pesci et al. demonstrated the addition of a spent culture medium extract from a *P. aeruginosa* wild type (PAO1)-induced expression of the elastase gene *lasB* in a PAO1 AHL-deficient *lasR* mutant (3). These data suggested that a non-AHL signal produced by the bacterium was capable of activating *lasB*. It was also shown that the novel signal required a functional *rhl* system for its bioactivity, as *lasB* could not be activated in an *rhlI/rhlR* double mutant by PAO1 spent culture extracts. The molecule responsible for the non-AHL-mediated QS signalling pathway was purified and chemically identified as 2-heptyl-3-hydroxy-4-quinolone and termed the pseudomonas quinolone signal (PQS) (3) (Fig. 1). PQS belongs to the AQ family of compounds, which were first chemically identified in the 1940s and studied for their antibacterial properties. In addition to PQS, other major molecules produced by this organism that belong to this family include 2-heptyl-4-quinolone (HHQ) (Fig. 1), 2-nonyl-4-quinolone (NHQ), and 2-heptyl-4-quinolone *N*-oxide (HQNO) (4). More recently, it has been shown that AQs are produced by other genera of bacteria including *Burkholderia pseudomallei*, the causative organism of melioidosis (5), suggesting that AQ signalling may be more widespread than previously thought.

For detection of AQs, both NMR (nuclear magnetic resonance) and LC–MS (liquid chromatography–mass spectroscopy) have been used (6, 7). These methods are extremely sensitive but

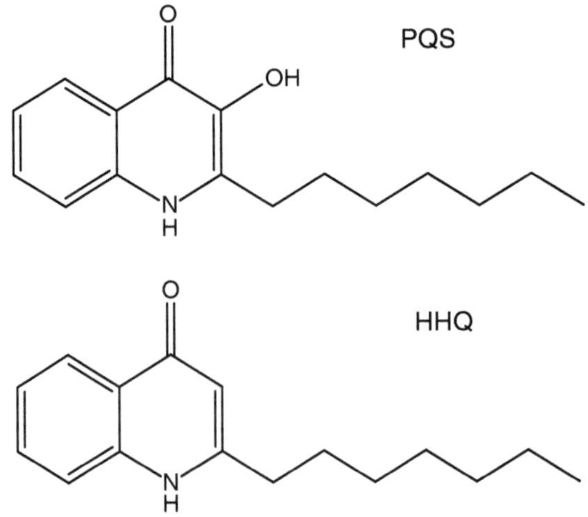

Fig. 1. Structures of the AQs PQS and HHQ.

rely on access to expensive instrumentation and require considerable expertise. An alternative is to use thin layer chromatography (TLC) to detect AQs and previously TLC assays have been used to detect PQS under UV light (8–10). However, these are not discriminatory in complex bacterial supernatants and are not particularly suitable for 2-alkyl-4-quinolones such as HHQ, which fluoresces much more weakly under UV light than PQS. A third option is via the use of a specific AQ biosensor (11–13) and this chapter will describe this method of AQ detection.

2. Materials

2.1. Bacterial Bioreporters, Growth Media, and AQ Compounds

1. The bioreporter used in this assay is PAO1Δ*pqsA* CTX-*lux::pqsA* (*13*). This strain is *P. aeruginosa* PAO1 with a chromosomal deletion in the *pqsA* gene. It is AQ negative but responds to the presence of AQs such as PQS and HHQ. The reporter gene construct in this strain is the *pqsA* promoter (which is sensitive to both PQS and HHQ) fused to a CTX-*luxCDABE* cassette. The resulting construct was conjugated into PAO1Δ*pqsA* resulting in a stable chromosomal single copy reporter. The strain can be maintained in tetracycline 125 μg/ml and stored at −80°C in 25% (vol/vol) glycerol.
2. PQS standard, 10 mM in methanol (PQS MW 259).
3. HHQ standard, 10 mM in methanol (HHQ MW 243).
4. Luria-Bertani (LB) medium: 1% (wt/vol) bacto-tryptone, 0.5% (wt/vol) yeast extract, 1% (wt/vol) sodium chloride in distilled water.
5. LB agar: LB media with addition of 2% (wt/vol) agar technical no. 3.
6. Tetracycline, 50 mg/ml in methanol.

2.2. Extraction of AQs

1. Potassium phosphate monobasic (KH_2PO_4).
2. Methanol-HPLC gradient grade.
3. Ethyl acetate-HPLC gradient grade.
4. Acetone-HPLC gradient grade.
5. Dichloromethane-HPLC gradient grade.
6. Glacial acetic acid-analytical grade.
7. Rotary evaporator (e.g. R-114, Buchi).

2.3. Detection of AQs Using TLC

1. Normal phase 20 × 20 cm silica 60_{F254} TLC plates (Merck).
2. TLC developing tank (Sigma-Aldrich, 27.5 × 27.5 × 7.5 cm).
3. Soft top agar: 0.65% (wt/vol) agar technical no. 3, 1% (wt/vol) tryptone, 0.5% (wt/vol) sodium chloride.

4. UV transilluminator (e.g. Vilber Lourmat TFX-20.M, 312 nm).
5. Luminograph photon video camera (e.g. LB 980, EG & G Berthold).
6. X-ray film (e.g. Pierce CL-Xposure 18 × 24 cm film).

2.4. Detection of AQs Using Microtitre Plates

1. 96-Well plate luminometer (e.g. Anthos Lucy1 or Tecan GENios Pro).
2. 96-Well, black, clear bottom microtitre plates (e.g. Costar-Corning Inc.).

3. Methods

3.1. Preparing Bacterial Cultures for AQ Extraction

1. Under sterile conditions streak out a 10 μl loop of the test bacterium, *P. aeruginosa* PAO1 (positive control), the AQ-negative mutant PAO1Δ*pqsA* (AQ negative control), and the PQS-negative (but HHQ positive) PAO1Δ*pqsH* (PQS negative control) onto fresh LB agar plates. Grow overnight at 37°C.
2. On the following day, inoculate 5 ml of LB medium with single colonies of the relevant strains. Grow the cultures overnight at 37°C with shaking at 200 rpm. It is not essential that LB medium is used for this stage, any nutrient medium can be used, although *P. aeruginosa* strains produced high concentrations of AQs in LB. If an individual strain requires antibiotics for selection, then these can be added at this stage.
3. The following day, the optical densities (OD) of the cultures should be measured at 600 nm (OD_{600}). Using these readings, standardise the cultures to approximately OD 1.0 by diluting with the growth medium used in step 2.
4. Using sterile techniques, transfer 0.25 ml of standardised culture into 250 ml Erlenmeyer flasks containing 25 ml LB broth (or alternative growth medium). A small volume of culture in a large flask allows good aeration of the medium. Incubate at 37°C, with shaking at 200 rpm for 8 h (see Note 1).

3.2. AQ Extraction of Bacterial Cultures

1. Transfer a defined volume of each culture (in this example 10 ml) to a 50 ml centrifuge tube and centrifuge at 10,000 × *g* for 10 min. Extract the cells (see step 2) or the supernatant (see step 3). It is possible to carry out both procedures together.
2. For extraction of AQs from cells, centrifuge the culture at 10,000 × *g* for 10 min, remove the supernatant (save if required for supernatant extraction) and resuspend the cells in 10 ml of PBS buffer. Centrifuge again at 10,000 × *g* for

10 min and discard the supernatant. Repeat the media wash steps twice to remove all traces of the supernatant AQs from the cells. Add 10 ml of methanol to the cell pellet and vortex until fully resuspended. Allow to stand for 10 min to allow the cells to lyse before centrifuging again at 10,000 × *g* for 10 min. Filter the extract through sterile Minisart (Sartorius) 0.2 μM filters into clean centrifuge tubes to remove all cell debris from the extraction mixtures. At this stage, it is possible to store the cell extractions in the freezer at −20°C for several days if required.

3. For supernatant extraction, centrifuge the culture at 10,000 × *g* for 10 min and filter the supernatants through sterile Minisart (Sartorius) 0.2 μM filters into clean centrifuge tubes to remove any cells. Add 10 ml of acidified ethyl acetate (glacial acetic acid 0.01% (vol/vol) in ethyl acetate) to the supernatant and vortex for 30 s so the two phases are well mixed. Transfer the extraction mixtures into a separating funnel that has been previously washed with acetone and allow the extraction mixtures to settle and the two phases to separate. Transfer the top organic layer into a fresh centrifuge tube. Repeat the extraction procedure on the bottom layer twice before discarding. Pool the collected organic layers. If time is limiting, supernatant extraction mixtures can be stored in the freezer at −20°C for several days.
4. For both the cell and supernatant procedures, transfer the extraction mixtures to 50 ml round bottom flasks that have been previously washed with acetone and rotary evaporate the mixtures to dryness.
5. Add 0.5 ml of methanol to the round bottom flasks and agitate for 30 s before transferring the liquid to 2 ml glass sample vials. Repeat this step with two further additions of 0.5 ml methanol and pool each sample in each vial. If time is limiting, both cell and supernatant extraction mixtures can be stored in the freezer at −20°C for several days.
6. Dry down the extraction mixtures in the sample vials under a stream of nitrogen gas. Dry cell and supernatant extraction residues can be stored in the freezer at −20°C for several months.

3.3. Preparation of TLC Plates and Running of Samples

1. Prepare normal phase silica 20 × 20 cm 60_{F254} TLC plates by soaking in a 5% (wt/vol) solution of KH_2PO_4 for 30 min before activating at 100°C for 1 h. A hybridization oven can be used to do this.
2. Draw a faint line in pencil approximately 1 in. from the bottom of the silica TLC plate to use as a guide for spotting sample extracts.

3. Reconstitute sample extracts in 100 μl of methanol and spot 5 μl of each onto the TLC plate. The amount spotted onto the TLC depends on the concentration of AQs in the sample. *P. aeruginosa* produces high amounts of AQs and in this case, 5 μl is a good starting quantity. As positive controls, 2 μl of 10 mM stock solutions of PQS and HHQ (or other AQs) in methanol can be spotted onto the TLC plate. Space each spot at 2 cm intervals along the line. A hairdryer can be used to dry samples during spotting to give a tighter spot.
4. When dry, place the TLC plate into a developing tank and run the TLC using a mixture of dichloromethane:methanol (95:5) as the mobile phase until the solvent front is 1–2 cm from the top of the plate. The TLC plate can be visualised using a UV transilluminator at 312 nm and photographed at this point.
5. Allow the TLC plate to dry and apply autoclave tape to both the underside of the TLC plate and around the edges so that the tape creates a well at least 0.5 cm deep around the TLC plate into which the nutrient agar containing the biosensor will be poured. Make sure the autoclave tape is firmly pressed down and forms a tight seal (Fig. 2).

3.4. Overlay of TLC Plates and Detection of AQs Using a Bioreporter

1. Streak out a 10 μl loop of the AQ biosensor (PAO1Δ*pqsA* CTX-*lux*::*pqsA*) (*13*) onto fresh LB agar plates containing tetracycline 125 μg/ml and grow overnight at 37°C.
2. The following day, inoculate a single colony into 5 ml LB medium containing tetracycline 125 μg/ml and grow overnight at 37°C with shaking at 200 rpm.
3. The following day, gently melt 100 ml of soft top agar in a microwave and allow to cool to ~50°C. Add 1 ml of the overnight culture to the soft top agar and mix gently (see Note 2).
4. Pour the agar mixture slowly into the well made around the TLC plate, being careful to minimise bubble formation in the agar and on the TLC plate (see Note 3).
5. Allow the agar to solidify around a Bunsen flame (to help keep sterile) and then incubate the plate at 37°C for 6 h to view light production, or overnight to view pyocyanin production. Visualise the plates for light production using a luminograph photon video camera (or similar) or develop using X-ray film (e.g. Pierce CL-XPosure film). Alternatively simply view production of the blue/green phenazine pigment pyocyanin by eye (Fig. 3) (see Note 4).

3.5. Testing for AQ Production Using a Microtitre Plate Assay

1. Prepare crude bacterial culture supernatants by growing the test bacterium as described in Subheading 3.1.
2. Remove 5 ml of culture and spin at 10,000×*g* for 5 min before collecting the supernatant and passing it through a

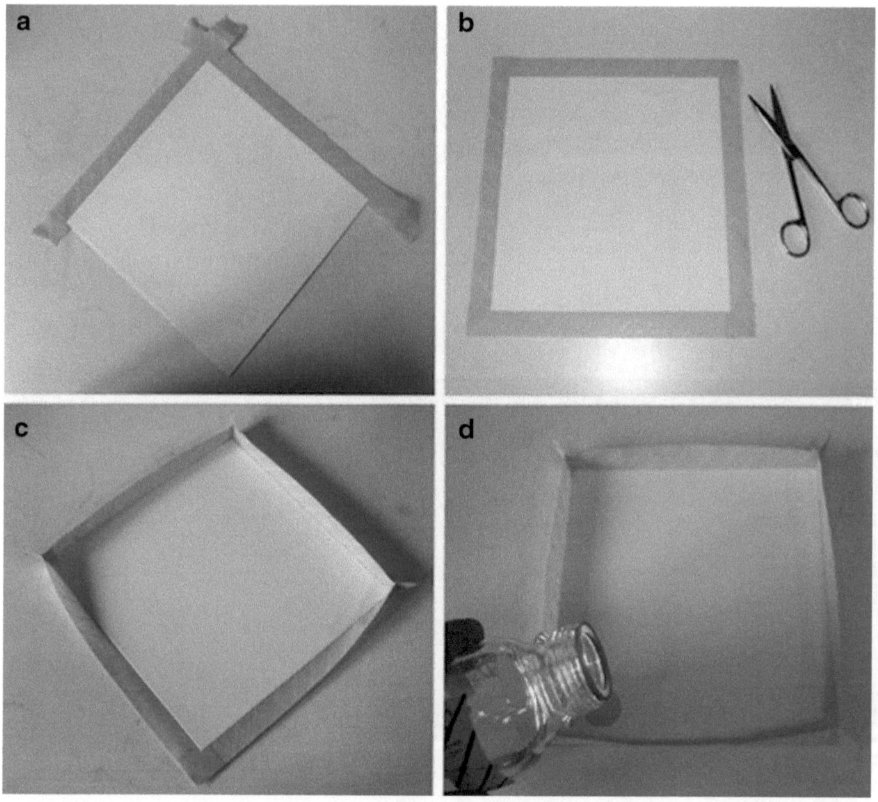

Fig. 2. Preparation of the biosensor bacteria agar TLC plate overlay. (**a**) Attach a strip of autoclave tape to the aluminium backing along each edge of the TLC plate, pressing down firmly to ensure a good seal. (**b**) Neatly trim the excess autoclave tape using scissors. (**c**) Pinch the autoclave tape at each corner of the TLC plate, creating a barrier of autoclave tape along each side of the TLC plate to create a well. (**d**) Place the TLC plate prepared as shown in (**c**) into an appropriate dish and pour into the well the molten soft top agar containing the biosensor bacteria.

sterile Minisart (Sartorius) 0.2 μm filter into clean tubes (see Note 5). If time is limiting, this supernatant extract can be frozen for a few days.

3. Grow the AQ biosensor overnight and dilute with LB medium to OD_{600} 1.0. Further dilute this standardised biosensor culture with LB medium to give (a) 1 in 50 and (b) 1 in 100 dilutions.

4. Sterilise a 96-well plate, e.g. under strong UV light for 15 min (see Note 6). For each well, mix 100 μl of the sterile test bacterial supernatant with 100 μl of the 1 in 50 dilutions of the biosensor to give a final culture dilution of 1 in 100. For each negative control well, add 200 μl of the 1 in 100 dilutions of the AQ biosensor. A positive control well containing a PQS or HHQ standard at a concentration of 25 μM can be added or alternatively a *P. aeruginosa* culture supernatant (100 μl) can also be included in a well during the assay.

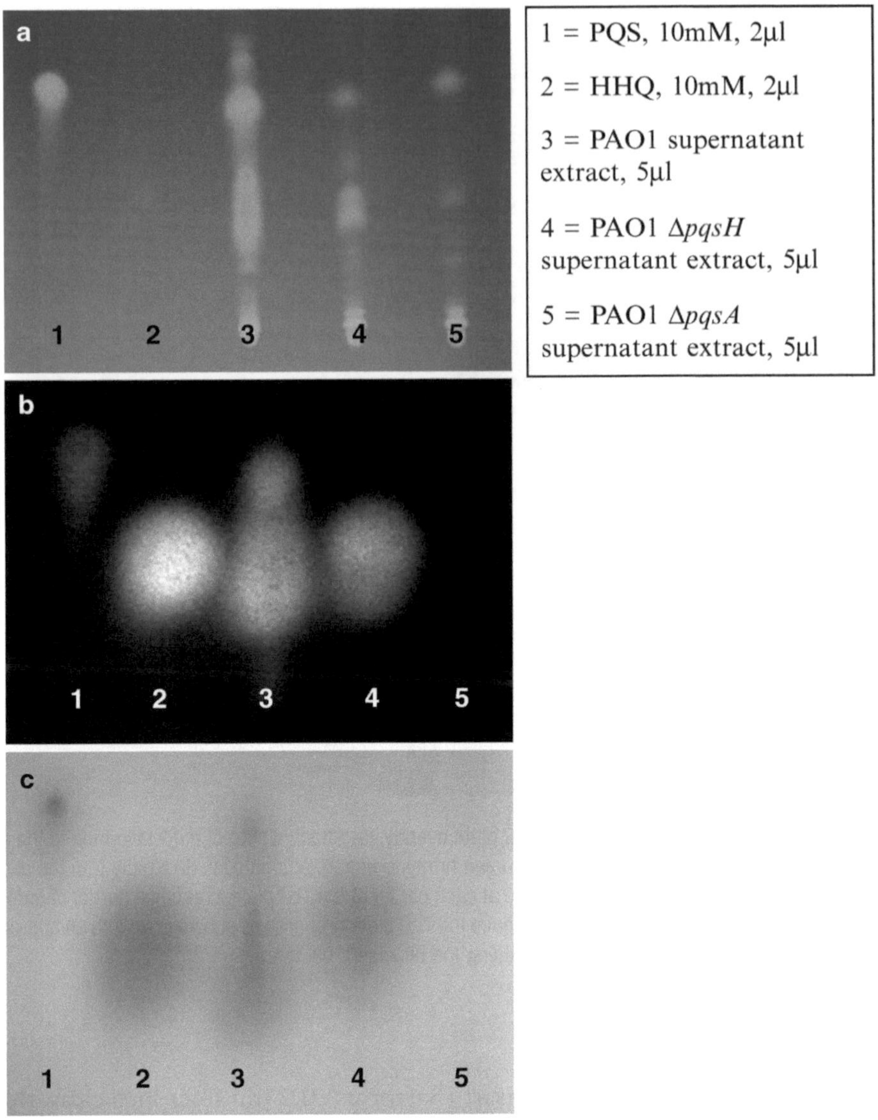

Fig. 3. TLC assay for AQs. (**a**) TLC plate run with standards of PQS and HHQ and supernatant extracts of PAO1, PAO1Δ*pqsH*, and PAO1Δ*pqsA* and visualized under UV at 312 nm. (**b**) Overlay of TLC plate with soft top agar containing biosensor bacteria showing the production of light in response to AQs visualized using a luminograph photon video camera. (**c**) Overlay of TLC plate with soft top agar containing the biosensor bacteria showing production of the blue/green pigment, pyocyanin, in response to AQs. TLC lanes: (1) PQS 10 mM, 2 μl, (2) HHQ 10 mM, 2 μl, (3) PAO1 supernatant extract, 5 μl, (4) PAO1Δ*pqsH* supernatant extract, 5 μl, (5) PAO1Δ*pqsA* supernatant extract, 5 μl. The AQ biosensor emits light over spots identified as PQS and HHQ in PAO1 and HHQ only in PAO1Δ*pqsH*. No light spots are observed for the AQ-negative *pqsA* mutant.

5. Monitor bioluminescence and OD at 37°C using a combined spectrophotometer/luminometer, e.g. the Anthos LUCY1 controlled by the Stingray software (Dazdaq). This measures OD and bioluminescence from all wells every 30 min for 24 h. Luminescence is recorded as relative light units (RLU)

per unit of OD. If an automated combined spectrophotometer/luminometer is unavailable, readings can be taken manually at defined time points by growing the bacterial cultures under specific conditions and measuring the OD and bioluminescence of culture samples using a spectrophotometer and tube luminometer (e.g. EG & G Junior), respectively.

4. Notes

1. Growth of *P. aeruginosa* cultures for 8 h in LB medium results in the bacteria reaching mid-stationary phase, which is sufficient for high quantities of AQs to be produced. If using an alternative growth medium, the time of incubation may have to be altered to accommodate the bacterial cells reaching the stationary phase of growth. We recommend that stationary phase cells be used when attempting to detect AQs produced by bacterial species.
2. Make sure the agar has cooled sufficiently before adding the bacterial reporter strain. Addition at too high a temperature will attenuate bacterial growth.
3. Pour the agar promptly or it will begin to solidify. Bubbles can be removed by gently passing a Bunsen burner flame over the surface of the agar.
4. Both PQS and HHQ will activate light production in the reporter, as both control the expression of the *pqsA* gene. In addition, both PQS and HHQ are required for the production of pyocyanin.
5. In addition to cell-free culture supernatants, the solvent extracted culture extracts described in Subheading 3.2 may also be analysed via this method. Simply dilute 5 μl of the solvent extract in 100 μl of LB and add to 100 μl of 1 in 50 dilutions of the AQ biosensor per well.
6. Specialised 96-well plates need to be used when monitoring bioluminescence. The plates need to have black sides to reduce light scatter between wells and have clear plastic bottoms so that an automated luminometer can detect and measure light output accurately.

Acknowledgments

We gratefully acknowledge the Royal Society (SPD) and BBSRC (MPF) for funding.

References

1. Williams, P., and Cámara, M. (2009) Quorum sensing and environmental adaptation in *Pseudomonas aeruginosa*: a tale of regulatory networks and multifunctional signal molecules. *Curr. Opin. Microbiol.* **12**, 182–191.
2. Popat, R., Crusz, S. A., and Diggle, S. P. (2008) The social behaviours of bacterial pathogens. *Br. Med. Bull.* **87**, 63–75.
3. Pesci, E. C., Milbank, J. B., Pearson, J. P., McKnight, S., Kende, A. S., et al. (1999) Quinolone signaling in the cell-to-cell communication system of *Pseudomonas aeruginosa*. *Proc. Natl. Acad. Sci. USA.* **96**, 11229–11234.
4. Dubern, J. F., and Diggle, S. P. (2008) Quorum sensing by 2-alkyl-4-quinolones in *Pseudomonas aeruginosa* and other bacterial species. *Mol. Biosyst.* **4**, 882–888.
5. Diggle, S. P., Lumjiaktase, P., Dipilato, F., Winzer, K., Kunakorn, M., et al. (2006) Functional genetic analysis reveals a 2-alkyl-4-quinolone signaling system in the human pathogen *Burkholderia pseudomallei* and related bacteria. *Chem. Biol.* **13**, 701–710.
6. Lépine, F., Déziel, E., Milot, S., and Rahme, L. G. (2003) A stable isotope dilution assay for the quantification of the Pseudomonas quinolone signal in *Pseudomonas aeruginosa* cultures. *Biochim. Biophys. Acta.* **1622**, 36–41.
7. Lépine, F., Milot, S., Déziel, E., He, J., and Rahme, L. G. (2004) Electrospray/mass spectrometric identification and analysis of 4-hydroxy-2-alkylquinolines (HAQs) produced by *Pseudomonas aeruginosa*. *J. Am. Soc. Mass Spectrom.* **15**, 862–869.
8. Diggle, S. P., Winzer, K., Chhabra, S. R., Worrall, K. E., Cámara, M., and Williams, P. (2003) The *Pseudomonas aeruginosa* quinolone signal molecule overcomes the cell density-dependency of the quorum sensing hierarchy, regulates *rhl*-dependent genes at the onset of stationary phase and can be produced in the absence of LasR. *Mol. Microbiol.* **50**, 29–43.
9. Calfee, M. W., Coleman, J. P., and Pesci, E. C. (2001) Interference with Pseudomonas quinolone signal synthesis inhibits virulence factor expression by *Pseudomonas aeruginosa*. *Proc. Natl. Acad. Sci. USA.* **98**, 11633–11637.
10. Gallagher, L. A., McKnight, S. L., Kuznetsova, M. S., Pesci, E. C., and Manoil, C. (2002) Functions required for extracellular quinolone signaling by *Pseudomonas aeruginosa*. *J. Bacteriol.* **184**, 6472–6480.
11. Fletcher, M. P., Diggle, S. P., Crusz, S. A., Chhabra, S. R., Cámara, M., and Williams, P. (2007) A dual biosensor for 2-alkyl-4-quinolone quorum-sensing signal molecules. *Environ. Microbiol.* **9**, 2683–2693.
12. Fletcher, M. P., Diggle, S. P., Cámara, M., and Williams, P. (2007) Biosensor-based assays for PQS, HHQ and related 2-alkyl-4-quinolone quorum sensing signal molecules. *Nat. Protoc.* **2**, 1254–1262.
13. Diggle, S. P., Matthijs, S., Wright, V.J., Fletcher, M. P., Chhabra, S. R., et al. (2007) The *Pseudomonas aeruginosa* 4-quinolone signal molecules HHQ and PQS play multifunctional roles in quorum sensing and iron entrapment. *Chem. Biol.* **14**, 87–96.

Chapter 3

FRET-Based Biosensors for the Detection and Quantification of AI-2 Class of Quorum Sensing Compounds

Sathish Rajamani and Richard Sayre

Abstract

Intercellular small molecular weight signaling molecules modulate a variety of biological functions in bacteria. One of the more complex behaviors mediated by intercellular signaling molecules is the suite of activities regulated by quorum sensing molecules. These molecules mediate a variety of population-dependent responses, including the expression of genes that regulate bioluminescence, type III secretion, siderophore production, colony morphology, biofilm formation, and metalloprotease production. Given their central role in regulating these responses, the detection and quantification of QS molecules has important practical implications. Until recently, the detection of QS molecules from Gram-negative bacteria has relied primarily on bacterial reporter systems. These bioassays though immensely useful are subject to interference by compounds that affect bacterial growth and metabolism. In addition, the reporter response is highly dependent on culture age and cell population density. To overcome such limitations, we developed an in vitro protein-based assay system for the rapid detection and quantification of the furanosyl borate diester (BAI-2) subclass of autoinducer-2 (AI-2) QS molecules. The biosensor is based on the interaction of BAI-2 with the *Vibrio harveyi* QS receptor LuxP. Conformation changes associated with BAI-2 binding to the LuxP receptor change the orientation of cyan and yellow variants of GFP (CFP and YFP) fused the N- and C-termini, respectively, of the LuxP receptor. LuxP-BAI2 binding induces changes in fluorescence resonance energy transfer (FRET) between CFP and YFP, whose magnitude of change is ligand concentration dependent. A set of ligand-insensitive LuxP-mutant FRET protein sensor was also developed for use as control biosensors. The FRET-based BAI-2 biosensor responds selectively to both synthetic and biologically derived BAI-2compounds. This report describes the use of the LuxP-FRET biosensor for the detection and quantification of BAI-2.

Key words: Autoinducer, Quorum sensing, LuxP, Ligand, BAI-2, DPD, FRET, Biosensor, GFP, CFP, YFP, Dissociation constant, Quantification, Fluorescence

1. Introduction

Vibrio harveyi bioassays for the autoinducer 2 (AI-2) class of quorum sensing (QS) compounds have been used for over a decade to monitor QS signals in biological samples. *V. harveyi*

Kendra P. Rumbaugh (ed.), *Quorum Sensing: Methods and Protocols*, Methods in Molecular Biology, vol. 692, DOI 10.1007/978-1-60761-971-0_3, © Springer Science+Business Media, LLC 2011

uses a two-component sensor kinase system to detect autoinducer 1 [AI-1, *N*-(3-hydroxybutanoyl)-l-homoserine lactone] and the boron derivative of autoinducer 2 [BAI-2, (2*S*,4*S*)-2-methyl-2,3,3,4-tetrahydroxytetrahydrofuran-borate]. These QS compounds regulate the expression of genes involved in bioluminescence, type III secretion, siderophore production, colony morphology, and metalloprotease production (1–4). Until recently, the detection of the BAI-2 class of QS compounds was based on BAI-2-induced bioluminescence using *V. harveyi* bioassays. The BAI-2 bioassay system, however, was notoriously difficult to standardize, takes several hours to complete, and is subject to substantial environmental and biological perturbations (5, 6). Fluctuations in culture pH, metabolites, and growth inhibitor concentrations can all affect BAI-2 bioassays. These shortcomings necessitated the need for the development of a more rapid, ligand-specific assay for the detection and quantification of BAI-2. With this knowledge, we developed an in vitro LuxP-FRET-based biosensor (CLPY) consisting of a cyan fluorescent protein (CFP) and a yellow fluorescent protein (YFP) fused to the surface-exposed N- and C-termini of the BAI-2 receptor protein LuxP, devoid of its N-terminal periplasmic targeting peptide (23 amino acids).

LuxP belongs to a large family of bacterial periplasmic-binding proteins (bPBP) (7, 8). The LuxP class of bPBPs is highly conserved among many *Vibrio* species including several potential human pathogens (9). The bPBPs are ideally suited for the development of FRET-based biosensors with their N- to C-termini distances between 10 and 100 Å (10, 11). They typically have two globular protein domains tethered by a flexible hinge region that encompasses the ligand-binding site (8). Structure–function analyses of bPBPs have demonstrated that the binding of the ligand induces substantial conformational changes in the receptor protein (12–14) including changes in the protein radius of gyration and the distance between the N- and C-termini of the protein (7, 8, 12–14).

The CLPY biosensor (MW: 98 kDa, Fig. 1) is conveniently expressed in *Escherichia coli* using an inducible T5-promoter/lac operator vector construct and purified using an N-terminal 6x Histidine tag. Herein, we describe a rapid, highly sensitive, BAI-2 biosensor, CLPY, and accompanying control biosensors (M2CLPY and M3CLPY) for the FRET-based detection and quantification of BAI-2 from biological samples.

2. Materials

2.1. Bacterial Strains and Plasmids

1. *E. coli* BL21 (*luxS*$^-$) and *V. harveyi* strains BB120 (wild type), MM30 (*luxS*$^-$), and MM32 (*luxS*$^-$, *luxN*$^-$) (generously provided by Dr. Bonnie L. Bassler – Princeton University).

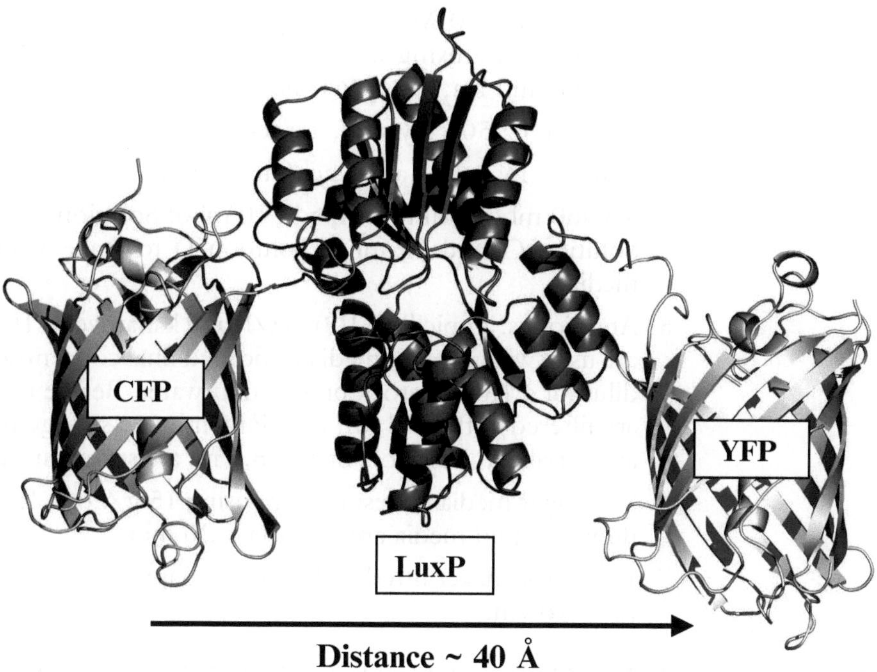

Fig. 1. LuxP-protein biosensor (CLPY). Schematic representation of LuxP-protein biosensor with CFP and YFP proteins attached to the N- and C-termini of LuxP. The calculated distance between the flurophores of ~40 Å is optimal for measuring conformational changes within LuxP. The ensemble measure of LuxP-BAI-2 binding induced distance and conformational changes in LuxP is observed by proportional decrease in the CLPY FRET ratio (YFP/CFP fluorescence ratio) changes.

2. Strain BL21 (*luxS*⁻) carrying plasmid constructs pQE30-CLPY (wild type LuxP biosensor), LuxP-mutant biosensors: pQE30-M2CLPY (Q77A and S79A) and pQE30-M3CLPY (Q77A, S79A, and W82F) (9).

2.2. Bacterial Culture Media

1. Unless otherwise stated, media, stocks, and solutions are made with deionized distilled water prepared using Milli-Q® water purification system (Millipore).
2. Luria-Bertani (LB) medium: 5.0 g/l yeast extract, 10.0 g/l bactotryptone, and 10.0 g/l sodium chloride are dissolved in water and autoclave.
3. Luria-Marine (LM) medium (15): 5.0 g/l yeast extract, 10.0 g/l bactotryptone, and 20.0 g/l sodium chloride are dissolved in water and autoclave.
4. Autoinducer Bioassay (AB) medium (2): Prepare solutions A, B, and C first as detailed below.

   Solution A: 0.3 M NaCl (17.53 g/l), 0.05 M $MgSO_4·7H_2O$ (12.32 g/l), 0.2% Casamino acids (2.0 g/l) in 960 ml water. Adjust pH to 7.5 with 10.0 M KOH, autoclave, and cool to room temperature.

Solution B: 1.0 M NaH_2PO_4–Na_2HPO_4 buffer (pH 7.0) is prepared by mixing autoclaved stocks of 1.0 M NaH_2PO_4 (39 ml) and Na_2HPO_4 (61 ml).

Solution C: 50% glycerol, autoclave.

Solution D: 0.1 M arginine, filter sterilize.

To 960 ml of Solution A, add 10 ml of Solution B, 20 ml of Solution C, and 10 ml of Solution D to make 1.0 l of AB media.

5. Antibiotics: ampicillin (100 mg/l) and kanamycin (100 mg/l) are used with growth media to needed final concentrations by diluting a 1,000× stock prepared in water. Before use, stocks are filtered with 0.22 μm PVDF membrane syringe filter disk and sterilized. Aliquots of 1.0–5.0 ml are stored at −20°C.
6. Solid agar media plates: Prepare using 15 g/l of select agar in desired liquid media and autoclave 20 min.
7. Incubator shaker fitted with desired adaptors for bacterial culture growth.

2.3. Protein Overexpression and Purification

1. 1.0 M isopropyl thiogalactoside (IPTG) is dissolved in distilled water and filter sterilized with 0.22 μm PVDF membrane syringe filter disk.
2. Lysozyme (50 mg) in 1.0 ml of distilled water, stored in 25 μl aliquots at −20°C.
3. 0.1 M phenylmethyl sulfonyl fluoride (PMSF) in isopropanol, stored at −20°C (see Note 6).
4. Following stocks in water are prepared and autoclaved: 1.0 M NaH_2PO_4, 1.0 M Na_2HPO_4, and 3.0 M NaCl.
5. A buffer stock of 1.0 M NaH_2PO_4–Na_2HPO_4 (pH 8.0) is prepared by mixing appropriate volumes of 1.0 M NaH_2PO_4 and 1.0 M Na_2HPO_4. For 100 ml of stock, add 94.7 ml of Na_2HPO_4 and 5.3 ml of NaH_2PO_4 and check the final pH using pH electrode.
6. 2.5 M imidazole in water, adjust pH to 7.5 with HCl and filter sterilize. Store at 4°C.
7. Buffer A (Column equilibration/wash buffer): 25 mM NaH_2PO_4–Na_2HPO_4 (pH 8.0), 35 mM NaCl, and 10 mM imidazole.
8. Buffer B (Lysis buffer): To buffer A solution, add 15 mM 2-mercaptoethanol (2-ME), 1 mM PMSF, and 0.2 mg/ml lysozyme (see Note 6).
9. Buffer C (Elution buffer): 25 mM NaH_2PO_4–Na_2HPO_4 (pH 8.0), 35 mM NaCl, and 50 mM imidazole.
10. HIS-Select™ nickel affinity gel (Sigma) (see Note 1) and empty 2.5 × 30 cm Econo-Column (BioRad) for affinity gel packing.

11. Cary Eclipse spectrofluorometer (Varian, Inc.) integrated with Cary Eclipse software.
12. Cell homogenizor/sonicator (Biologics, Inc., Model: 300 V/T, Ultrasonic homogenizer) fitted with a microtip.
13. Refrigerated floor centrifuge with rotors, centrifuge tubes, bottle, and adaptors.

2.4. Protein Estimation, Denaturing Gel Electrophoresis, and Fluorescence Detection

1. Bradford protein assay kit (BioRad) – store at 4°C.
2. Other components needed for protein assay: clean 13×100 mm test tubes, test tube racks, vortex mixer, Whatman no. 1 filter paper, 1 cm path length cuvettes.
3. Bovine serum albumin (BSA) standard stock solution at 2 mg/ml in water – store at 4°C for up to a month or at −20°C for long-time storage.
4. Sodium dodecyl sulfate–polyacrylamide gel electrophoresis (SDS–PAGE) procedure is described assuming the use of BioRad protein mini gel apparatus setup.
5. BioRad precast 10% SDS–PAGE (see Note 7), electrode, see blue protein standard (Invitrogen), pipettes, and 1–200 μl gel loading tips.
6. SDS–PAGE Tris–glycine electrophoresis buffer (1×): 25 mM Tris, 250 mM glycine, 0.1% SDS (pH 8.3) in water. A 5× stock of the above buffer can be prepared by dissolving 15.1 g/l of Tris base and 94.0 g/l of glycine in 950 ml water and 50 ml of 10% (w/v) of SDS. This can be stored at 4°C until use.
7. In water, 1.0 M Tris–HCl (pH 6.8), 1.0 M dithiothreitol is prepared for making 2× protein gel loading dye: 100 mM Tris–HCl (pH 6.8), 200 mM dithiothreitol, 4% SDS, 0.2% bromophenol blue, 20% glycerol. 100 μl aliquots can be made and stored at −20°C.
8. For preparing protein samples for gel electrophoresis, boiling water bath or heat block capable of reaching temperatures of 100°C.
9. For SDS–PAGE Coomassie brilliant blue staining (detection range 100–1,000 ng):
   a. Coomassie brilliant blue stain solution: 40% methanol/50% water/10% acetic acid/0.025% Coomassie brilliant blue R-250. This solution can be stored at room temperature.
   b. Destain solution: 40% methanol/50% water/10% acetic acid, store at room temperature.
   c. Other requirements include a rocking shaker and a gel imager.

10. Cary eclipse spectrofluorometer (Varian, Inc.) integrated with Cary Eclipse software, four-side clear quartz cuvette or disposable plastic cuvettes for fluorescence measurements.

2.5. Boron-Depleted Media and Conditioned Media Preparation

1. Amberlite® IRA743 borate-specific chelating resin (Sigma).
2. Column 5 × 10 cm Econo-Column (BioRad).
3. Following stocks in water: 3.0 M ammonium hydroxide, 1.0 M hydrochloric acid, 0.16 M nitric acid, and 10.0 M potassium hydroxide for pH adjustment.
4. 3.0 kDa membrane cut-off filter with omega membrane – 3K Microsep™ centrifugal device (Pall Life Science) and 0.2 μm HT Tuffryn® membrane syringe filter (Pall Life Science).

3. Methods

For best results, the use of borosilicate glassware should be avoided for the entire procedure. BAI-2 is formed from the cyclization of DPD in the presence of borate. It has been previously determined that a DPD:borate ratio of 1:4 leads to a yield of 10% BAI-2 and potentially other borate derivatives of AI-2 (not LuxP ligand) (16). All the buffer and media preparations should be stored in clean polystyrene/polypropylene containers. Bacterial culturing is done using sterile polystyrene 14 ml culture tubes (BD Biosciences). To avoid changes in the DPD:borate ratio that potentially alter the detected BAI-2 concentrations, the media should be cleaned of any contaminating borate prior to use, following the procedure described in Subheading 3.4.

3.1. Biosensor Overexpression and Purification

1. An overnight starter culture of BL21 (*luxS*$^-$) cells transformed with pQE30-CLPY or LuxP-mutant constructs are started as single colony inoculum in 14 ml sterile tubes containing 3.0 ml of LB broth supplemented with 100 mg/l ampicillin. The cultures are grown in a roller drum or shaker for overnight (O/N) at 37°C (see Note 8).
2. A 500 ml conical flask with 75.0 ml LB broth supplemented with 100 mg/l ampicillin is inoculated with 1% of above O/N culture (750 μl) to begin second overnight shake cultures (250 rpm) at 28°C for 16 h.
3. The O/N culture is then transferred to 1.5 l of LB in 4.0 l volume Erlenmeyer flask (5% innoculum) supplemented with 100 mg/l ampicillin.
4. The culture is grown at 28°C, shaking at 200 rpm. At about 3–4 h, the optical density at 600 nm (OD_{600}) is measured using spectrophotometer. A small volume of culture is then

aliquoted into 1 cm path length cuvette and OD_{600} is measured with fresh LB media as blank control. When the OD_{600} is 0.6, 1 ml of culture is removed and cell pellet collected by centrifugation (benchtop centrifuge – 13,000 rpm/3 min) and stored at −20°C for subsequent SDS–PAGE analysis. The remaining culture is induced for CLPY expression by adding 0.3 mM IPTG and grown for an additional 6 h.

5. After 6.0 h growth, 300 μl of culture is removed to collect the cell pellet by centrifugation for use with SDS–PAGE analysis. The rest of the bacterial cells are then harvested by centrifugation using a floortop centrifuge at 8,000×*g* for 10 min and protein purifications carried out at 4°C (see Notes 9 and 10).
6. Cell pellet is resuspended in 35 ml of 25 mM NaH_2PO_4–Na_2HPO_4 (pH 8.0), 35 mM NaCl, 10 mM imidazole, 15 mM 2-mercaptoethanol, and 1.0 mM phenylmethyl sulfonyl fluoride (buffer B), placed in an ice bath and lysed by sonication as follows. A sonicator fitted with microtip is used for six rounds of sonication with power set at 40 and pulsed ten times with a 1 min pause between each round to allow for cooling.
7. Cell lysate supernatant is separated from cell debris and unlysed cells by centrifugation at 12,000×*g* for 20 min. About 200 μl of clarified lysate is removed and stored at −20°C for SDS–PAGE analysis (Subheading 3.2). The rest of the clarified cell lysate is then loaded onto a 2.5×30 cm column containing 7.5 ml bed volume of His-Select™-HC Nickel affinity gel (Sigma) equilibrated with five column volumes of buffer A.
8. The protein bound resin is washed with five column volumes of buffer A and eluted by adding three column volumes of 25 mM NaH_2PO_4–Na_2HPO_4 (pH 8.0), 35 mM NaCl, and 50 mM imidazole (buffer B). In clean tubes, eluate containing biosensor protein (characteristics pale to brighter yellow color) is collected as 3.0 ml fractions.
9. The same column is now ready to be equilibrated and reused or can be stored in 70:30 ethanol:water (v/v) for later use (see Note 1).

3.2. Biosensor Quantification and SDS–PAGE Analysis

1. The purified CLPY biosensor fractions are quantified using BioRad protein assay method using BSA as standard. The assay is carried out as detailed in the user's manual with some modifications.
2. The dye reagent is prepared by diluting one part dye reagent concentrate with four parts Millipore™ water. The solution is filtered through Whatman no. 1 filter (or equivalent) to remove particulates. This diluted dye reagent can be used for approximately 2 weeks when stored at room temperature.

3. Prepare six dilutions of a protein BSA standard (0.2, 0.4, 0.5, 0.6, 0.8, and 0.9 mg/ml) – store at 4°C or –20°C for long-term storage.
4. Add 1.0 ml of diluted dye reagent to clean-dry test tubes; add 20 μl of each standard and CLPY fraction solution and vortex for 5 s.
5. Incubate at room temperature for 5 min and measure absorbance at 595 nm using spectrophotometer (see Note 11).
6. The protein concentration is determined using the standard curve. Typically the protein yields in visibly colored fractions range from 0.1 to 0.3 mg/ml.
7. For SDS–PAGE analysis, a precast 10% acrylamide gel is assembled in the minigel apparatus as per the user's manual. Immediately add 1× Tris–glycine electrophoresis buffer to fill the gel reservoir to limit idling and drying of the gel. Using both hands carefully remove the comb in a single vertical motion (see Note 16).
8. The samples for SDS–PAGE are prepared as follows. Bacterial cell pellet, cell lysate, and purified protein (Subheading 3.1) are left on ice to provide sufficient time for frozen samples to thaw entirely.
9. Resuspend the cell pellets in 50 μl of water. 50 μl of clarified cell lysate and purified protein (say concentration ~50 ng/μl) are aliquoted in 1.5 ml centrifuge tubes. To all the samples, add 50 μl of 2× SDS–PAGE loading dye and gently mix with the pipette tip. Place the tubes in boiling water bath for 10 min. Remove the tubes and leave at RT to cool. In a benchtop centrifuge spin down (13,000 rpm/1 min), the sample tubes that contain bacterial cells. Carefully remove 30 μl of sample using the gel loading tip and load onto designated wells. Add 10–15 μl of protein standard for use as molecular weight marker.
10. Once the sample is loaded, connect the apparatus to the power supply and initially run at 20 mA (2+ hours to allow sample through stacking gel) and then increase to 40 mA until the bromophenol blue starts running out of the gel. At this point, remove the assembly and carefully remove the gel for staining.
11. Place the gel in Coomassie staining solution so it immerses the gel fully. The gel is left on a rocking shaker for 2 h to overnight in the staining solution.
12. The staining solution is transferred to a new container and can be reused later. Rinse the gel with water and destain by adding sufficient volume of destain solution and let incubate for additional 2–3 h. Following destaining remove two-thirds

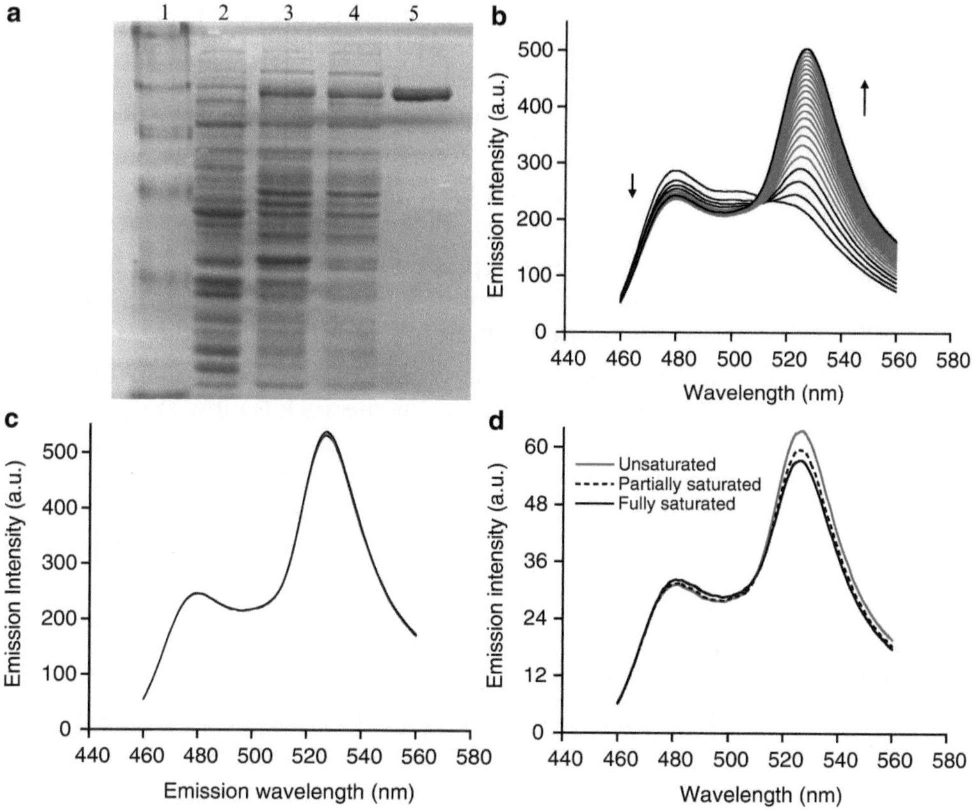

Fig. 2. Characterization of biosensor. (**a**) SDS–PAGE analysis of CLPY purification. Lanes: (1) see blue protein standard, (2) uninduced BL21(*luxS*⁻)-pQE30-CLPY, (3) 6 h induced BL21(*luxS*⁻)-pQE30-CLPY, (4) cell-lysate supernatant, (5) purified CLPY (~98 kDa). (**b**) Time-dependent change in FRET associated with YFP maturation. CLPY protein immediately after purification when monitored at room temperature [λ_{ex} 440 nm (slit 5 nm) and λ_{em} 460–560 nm (slit 5 nm) shows time-dependent FRET increase]. (**c**) Mature FRET sensor. The purified CLPY stored at room temperature for 5 h or at 4°C for 48 h showed no time-dependent FRET changes. Shown here is an overlay of several emission spectra scans recorded for mature CLPY protein over 30 min. (**d**) CLPY responses to BAI-2 ligand. Concentrations: unsaturating (*gray line*), partially saturating (*dotted black line*), and fully saturating (*black line*) concentrations of BAI-2 ligand.

of the destain solution and replace the volume with water, which allows the shrunk gel (during staining and destaining) to swell back to its original size.

13. The gel image can be captured using a gel imager or document scanner layered with saran™ wrap (Fig. 2a).

3.3. Biosensor Characterization and Flurophore Maturation

Fluorescence measurements are carried out at room temperature using Cary Eclipse spectrofluorometer (Varian, Inc.) set in a scanning mode. CLPY and LuxP mutants are monitored by CFP excitation (λ_{ex} 440 nm/slit 5 nm) and the emission spectrum (λ_{em}) is measured from 460 to 560 nm using a 5 nm slit width. Alternatively, the FRET ratio (527/485 nm) is recorded using an excitation wavelength of 440 nm.

1. A ligand-binding-independent increase in FRET is observed with recently purified biosensor (Fig. 2b). This is due to the delayed maturation of YFP over CFP, which is characterized by faster fluorophore maturation.
2. To eliminate this problem, immediately after purification, the biosensor should be left for maturation at RT (5 h) or 4°C (2 days) before beginning ligand-binding studies.
3. Following this step, the FRET change does not happen in the absence of the ligand (Fig. 2c). At this time, the biosensor is ready to use for BAI-2 ligand-binding studies.
4. The biosensor response to the detection of BAI-2 is characterized by a decrease in the FRET ratio (YFP/CFP ratio). This decrease is proportional to the amount of BAI-2 ligand concentration present in a given test sample. As reported elsewhere (9), Fig. 2d shows a representative spectrum of biosensor responses to unsaturating, partially saturating, and fully saturating BAI-2 concentrations in test sample.

3.4. Boron-Depleted Media Preparation

1. The protocol is adapted from Bennett et al. (17) with some modifications.
2. 30 ml of Amberlite® IRA743 resin is used per liter of media. Here we use AB media as an example for describing the procedure.
3. The column is treated with 150 ml 3.0 M ammonium hydroxide, 600 ml distilled water, 180 ml of 1.0 M hydrochloric acid, 150 ml distilled water, 180 ml of 0.16 M nitric acid, 300 ml distilled water.
4. Following these treatments, a liter of AB medium is passed through the column and the pH adjusted to 6.75 using 10 M potassium hydroxide.

3.5. BAI-2 Ligand Detection and Quantification from V. harveyi Cultures

1. As an example, we show here the determination of *V. harveyi* BAI-2 ligand concentrations as a function of culture age. This procedure can be suitably followed for the determination of BAI-2 from other Vibrio strains and other bacteria that synthesize and use BAI-2 as a signal molecule.
2. *V. harveyi* BB120 (wild-type) grown overnight at 28°C for 16 h is used to make 2% (v/v) inoculum in fresh AB media (2.0 ml, 0.5 mM borate) in round bottom polystyrene tubes (BD Biosciences, USA). Set up 36 tubes of 2.0 ml cultures for using three tubes at a given time point.
3. For this experiment, every 2.5 h (including the 0 h time point) the cell density is determined by plating serial dilutions of *V. harveyi* cultures on LM agar plates. Before removing cultures for plating, the culture tubes are vortexed for 10–15 s to eliminate any visible cell clumps. The dilutions are made in

fresh LM media and different volumes are plated out on LM agar (see Note 12). The agar plates are left at 30°C for 24 h before counting the colonies.

4. At the same time, the BB120 cell-free supernatant is collected using a refrigerated benchtop centrifuge – 13,200 rpm/5 min.
5. The supernatant is collected and to remove proteins (proteases) and cell debris, it is passed through a 3 kDa membrane cut-off filter (3 K Microsep™ centrifugal device, Pall Life Science) pretreated with 2.0 ml water by centrifugation at 5,000 × *g* for 20 min (see Notes 4 and 14).
6. The flow through of the culture supernatant containing the BAI-2 signal is collected after centrifugation at 5,000 × *g* for 30 min.
7. Working stock of 0.015 mg/ml of CLPY is made in buffer C. For triplicate measurements, three different batches of CLPY preparations are used.
8. In a 1.5 ml centrifuge tube aliquot 1.0 ml of 0.015 mg/ml of CLPY and leave at room temperature for 15 min to equilibrate.
9. The YFP/CFP fluorescence or FRET ratio (527/485 nm) response of CLPY for the given sample is measured at room temperature after 5 min incubation of the biosensor with the sample. The sample is mixed with 1.0 ml of CLPY by gently inverting the tube 4–5 times and incubating before transferring the contents to a cuvette for FRET measurements (see Note 13).
10. For instance, 150 μl of *V. harveyi* BB120 culture filtrate is mixed with 1.0 ml of CLPY and used for generating a graph representative of FRET response versus time and cell number. This assay provides an indication of the BAI-2 concentration in the media as a function of culture age (Fig. 3a).

3.6. BAI-2 Quantification from *V. harveyi* Cultures

1. For a given culture age, a fixed volume of diluted culture filtrate is added to the CLPY biosensor (0.015 mg/ml).
2. For instance to determine the concentration of BAI-2 from 7.5 h old 2% *V. harveyi* in AB media (0.5 mM Borate), 200 μl of various dilutions of culture supernatant (prepared as detailed in the previous section) is made in borate-free AB media (Subheading 3.3) and added to CLPY (1.0 ml, 0.015 mg/ml) to generate a biosensor response saturation curve. The following volumes 5, 10, 20, 40, 60, 80, 100, 120, 150, and 200 μl are made to 200 μl with borate-free AB media and used for generating the CLPY biosensor saturation curve.
3. The biosensor response curve is plotted as a function of dilution of cell supernatant versus the FRET ratio response (Fig. 3b, *x*-axis shown in log10 scale). The dilution volume for each data point is determined by dividing the volume

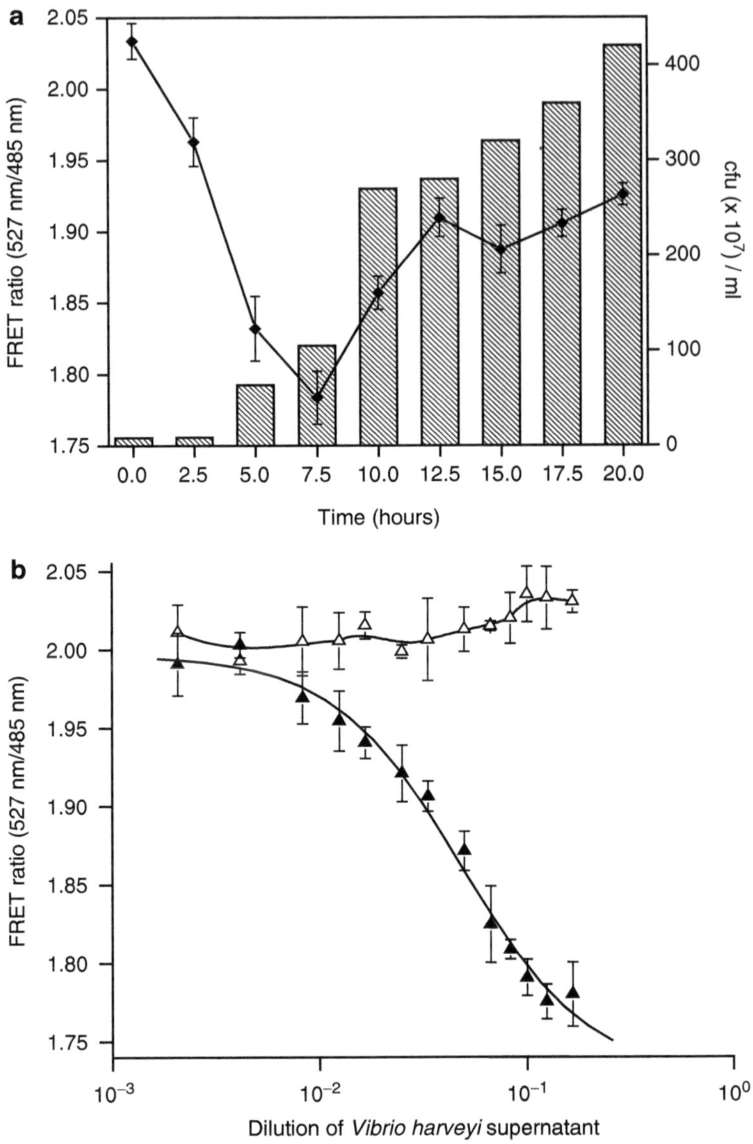

Fig. 3. Quantification of BAI-2 from *Vibrio harveyi* cultures using the CLPY biosensor. (**a**) Wild-type *V. harveyi* BAI-2 levels monitored as a function of time and culture density. Cell-free culture filtrates were prepared by passing the bacterial cell culture media through a 3 kDa MW cut-off filter. The filtrate (150 μl) was added to CLPY (0.015 mg/ml). The YFP/CFP FRET ratio (*closed diamonds*) is plotted as a function of cell density (CFU × 10^7/ml; bars). The error bars represent the standard deviations from three independent experiments. (**b**) CLPY fluorescence response to culture filtrate from wild-type (BB120) *V. harveyi* (*closed triangles*) and *luxS*$^-$ mutant (MM30, *open triangles*) culture filtrate. The error bars represent the standard deviations from three independent experiments. (**a**) and (**b**) Adapted with permission from ref. (9). Copyright© 2007 American Chemical Society.

of conditioned media used (do not include the volume of fresh borate-free AB media used for diluting) by the total volume of the FRET assay mix (this volume will remain the same when calculating all the dilutions).

4. Dilutions in borate-free AB media help maintain the ligand DPD and boron ratios the same in the bacterial growth media. This is important since changes in DPD and boron ratio might affect the equilibrium of BAI-2 ligand.
5. Cell-free supernatant from 7.5 h old *V. harveyi* MM32 (a *luxS*$^-$ mutant devoid of BAI-2 synthesis) culture is used as a control (Fig. 3b) (see Note 15).
6. Using a software program (such as "OriginLab") capable of fitting a nonlinear regression relationship shown in Eq. 1, the CLPY response to various dilutions of *V. harveyi* supernatant "*v*" can be used for determining half saturation volume "*h*". Where "*R*" is the observed CLPY FRET ratio response for different dilutions of *V. harveyi* supernatant "*v*," with "R_{max}" and "R_{min}" represent the ligand-free and ligand-saturated FRET ratios, respectively.

$$R = R_{max} - \left(\frac{(R_{max} - R_{min}) \times n \times v}{(h + v)} \right). \qquad (1)$$

7. By fitting the above equation, the "*h*" value is determined and used for calculating the unknown BAI-2 concentration "*M*" using the following relationship given in Eq. 2. As reported previously, the BAI-2 and LuxP binding affinity "K_d" – 270 nM (9) is used for determining "*M*" in a given culture supernatant.

$$M = Kd/h \qquad (2)$$

8. Other controls for these experiments include using M2CLPY and M3CLPY biosensors to confirm if the CLPY FRET response is specifically due to LuxP-BAI-2 binding (see Note 2).

4. Notes

1. There are other commercial suppliers for the same Ni-NTA affinity resin (e.g., Qiagen, Fisher Sci., etc.). Once obtained these gels can be reused 3–4 times depending on the column condition. Once the column looks faded (initially bright nickel blue), the resin can be cleaned and recharged with nickel for reuse. Following is a modified procedure from Qiagen that is used regularly for obtaining good protein

purification (refer to their procedure for more stringent treatment). Our modified procedure includes the following steps: wash resin with one column volume of water, followed by five column volumes of 100 mM EDTA (pH 8.0) to remove nickel and bound impurities and three column volumes of water and recharge the resin with two column volumes of 100 mM $NiSO_4$, wash with two column volumes of water and equilibrate with three column volumes of buffer A before use. Always store the column in small volumes of buffer for short-time storage or with 70% ethanol for long-time storage.

2. Chloride ions (Cl^-) can affect YFP fluorescence (18). You will note that the FRET ratios are lower in the presence of chloride due to YFP fluorescence (CFP remains unaffected). So care must be taken to minimize buffer and media concentrations between samples.
3. FRET ratio of mature CLPY with the choice of bacterial media can be determined to use as initial FRET ratio for your culture supernatant experiments and remember to maintain the same chloride ion concentrations (by diluting your used media in fresh media).
4. Pall centrifugal devices fitted with Omega™ membrane were found to have less interference with BAI-2 ligand sticking to membrane. So care must be taken with the choice of filtration device and the accompanying membrane.
5. In case the *E. coli* strain BL21(*luxS*⁻) is not available, DH5α can be used. However, appropriate protease inhibitors should be added as DH5α is not protease deficient strain unlike BL21 derivatives.
6. PMSF has a shore half-life in aqueous solutions (~30 min at pH 8.0), PMSF should be added to the lysis buffer immediately before initiation of cell lysis to avoid loss of potency.
7. The precast gels are recommended to be used within certain period of time. Also, in case of need for reducing expenses, SDS–PAGE gels can be conveniently prepared in the lab following published procedures. We recommend you refer to standard laboratory manuals such as molecular cloning: A laboratory manual by E.F. Fritsch, J. Sambrook, T. Maniatis (1989) for detailed procedure.
8. A single colony culture can be used to make 15% glycerol stock and stored at −80°C. For starting the cultures, scrape out small amount of frozen stock into fresh media. Care must be taken not to freeze–thaw stock vials that may result in loss of cell viability.
9. At this point, the culture flasks can be moved to 4°C and left standing O/N before harvesting. Next day, the cells can be noticed to have settled to the bottom. The culture flask can

be gently removed and ~500 ml of the culture can be decanted. The rest of the culture can be resuspended in the remaining media and the cells harvested using a refrigerated centrifuge set at 4°C.

10. The cell pellet can be stored at −80°C for prolonged periods of time without thawing. To thaw the cells, remove the cell pellet from −80°C and leave it O/N at −20°C. Transfer to ice next day to enable slow thawing of cells.
11. It is important to make the protein assay reads within minutes of each other. Over time the response intensifies, so care must be taken not to incubate for prolonged time. It is also recommended to go through user's manual for troubleshooting.
12. Under the given growth conditions for *V. harveyi* BB120, dilutions ranged from 10^6 at 0 h to 5×10^7 at 10 h and later. It is recommended that a pilot experiment is conducted to determine the dilution series required for the strain of bacteria you are working with before you take up the biosensor quantification.
13. Care must be taken not to vortex the contents; this may result in CLPY protein denaturation and loss of biosensor functionality.
14. If later time point cultures have cell debris that clogs the 3 kDa membrane, try including an additional step of passing the cell supernatant through 0.2 μm HT Tuffryn® membrane syringe filter (Pall Life Science) before 3 kDa membrane filtration.
15. In case the bacterial strain you are working with has no *luxS* deletion derivative, try using borate-free AB media grown culture filtrate as control.
16. To have good sample runs clean the wells carefully. Using a pipette set at 100 μl and fitted with 1–200 μl gel loading tip, pipette in electrophoresis buffer into wells (3–5 times) to remove any residual polymerized acrylamide on the loading well walls that might stick to sample and create a drag during sample run.

References

1. Lilley, B. N., and Bassler, B. L. (2000) Regulation of quorum sensing in *Vibrio harveyi* by LuxO and sigma-54, *Mol Microbiol* **36**, 940.
2. Bassler, B. L., Wright, M., Showalter, R. E., and Silverman, M. R. (1993) Intercellular signalling in *Vibrio harveyi*: sequence and function of genes regulating expression of luminescence, *Mol Microbiol* **9**, 773–786.
3. Henke, J. M., and Bassler, B. L. (2004) Quorum sensing regulates type III secretion in *Vibrio harveyi* and *Vibrio parahaemolyticus*, *J Bacteriol* **186**, 3794.
4. Mok, K. C., Wingreen, N. S., and Bassler, B. L. (2003) *Vibrio harveyi* quorum sensing: a coincidence detector for two autoinducers controls gene expression, *EMBO J* **22**, 870–881.

5. DeKeersmaecker, S. C. J., and Vanderleyden, J. (2003) Constraints on detection of autoinducer-2 (AI-2) signalling molecules using *Vibrio harveyi* as a reporter, *Microbiology* **149**, 1953–1956.
6. Turovskiy, Y., and Chikindas, M. L. (2006) Autoinducer-2 bioassay is a qualitative, not quantitative method influenced by glucose, *J Microbiol Methods* **66**, 407–503.
7. de Lorimier, R. M., Smith, J. J., Dwyer, M. A., Looger, L. L., Sali, K. M., Paavola, C. D., Rizk, S. S., Sadigov, S., Conrad, D. W., Loew, L., and Hellinga, H. W. (2002) Construction of a fluorescent biosensor family, *Protein Sci* **11**, 2655–2675.
8. Felder, C. B., Graul, R. C., Lee, A. Y., Merkle, H. P., and Sadee, W. (1999) The Venus flytrap of periplasmic binding proteins: an ancient protein module present in multiple drug receptors, *AAPS PharmSci* **1**, E2.
9. Rajamani, S., Zhu, J., Pei, D., and Sayre, R. (2007) A LuxP-FRET-based reporter for the detection and quantification of AI-2 bacterial quorum-sensing signal compounds, *Biochemistry* **46**, 3990–3997.
10. Fehr, M., Frommer, W. B., and Lalonde, S. (2002) Visualization of maltose uptake in living yeast cells by fluorescent nanosensors, *Proc Natl Acad Sci USA* **99**, 9846–9851.
11. Fehr, M., Lalonde, S., Lager, I., Wolff, M. W., and Frommer, W. B. (2003) In vivo imaging of the dynamics of glucose uptake in the cytosol of COS-7 cells by fluorescent nanosensors, *J Biol Chem* **278**, 19127–19133.
12. Shilton, B. H., Flocco, M. M., Nilsson, M., and Mowbray, S. L. (1996) Conformational changes of three periplasmic receptors for bacterial chemotaxis and transport: the maltose-, glucose/galactose- and ribose-binding proteins, *J Mol Biol* **264**, 350–363.
13. Zukin, R. S., Hartig, P. R., and Koshland, D. E., Jr. (1979) Effect of an induced conformational change on the physical properties of two chemotactic receptor molecules, *Biochemistry* **18**, 5599–5605.
14. Zukin, R. S., Hartig, P. R., and Koshland, D. E., Jr. (1977) Use of a distant reporter group as evidence for a conformational change in a sensory receptor, *Proc Natl Acad Sci USA* **74**, 1932–1936.
15. Bassler, B. L., Wright, M., and Silverman, M. R. (1994) Multiple signalling systems controlling expression of luminescence in *Vibrio harveyi*: sequence and function of genes encoding a second sensory pathway, *Mol Microbiol* **13**, 273–286.
16. Semmelhack, M. F., Campagna, S. R., Federle, M. J., and Bassler, B. L. (2005) An expeditious synthesis of DPD and boron binding studies, *Org Lett* **7**, 569–572.
17. Bennett, A., Rowe, R. I., Soch, N., and Eckhert, C. D. (1999) Boron stimulates yeast (*Saccharomyces cerevisiae*) growth, *J Nutr* **129**, 2236–2238.
18. Nagai, T., Ibata, K., Park, E. S., Kubota, M., Mikoshiba, K., and Miyawaki, A. (2002) A variant of yellow fluorescent protein with fast and efficient maturation for cell-biological applications, *Nat Biotechnol* **20**, 87–90.

Chapter 4

Isolation of *agr* Quorum Sensing Autoinducers

Naomi Balaban

Abstract

Autoregulation of genes is often associated with quorum sensing systems where bacteria produce and secrete molecules that allow the cells to communicate with one another, leading to the activation of certain genes at certain population densities. Here we describe the identification of the *agr* as a quorum sensing system in *Staphylococcus aureus* and the isolation of *agr* autoinducers and inhibitors by northern blotting, real-time RT-PCR, and β-lactamase reporter cells assays.

Key words: Quorum sensing, Autoinduction, β-Lactamase reporter cell assay, Temporal regulation, RNAIII, *Agr*, *Staphylococcus aureus*

1. Introduction

Bacterial cell-to-cell communication systems, or quorum sensing, regulate the expression of genes necessary for cellular functions at a certain population density (1). To determine if a gene is part of a quorum sensing system and test if it is autoinduced by a secreted molecule, it is necessary first to test if the expression of the gene of interest is temporally regulated and thus changes during the growth of the culture (2). Gene expression can be monitored by northern blotting, by real-time RT-PCR and if available, by gene reporter assays.

If the gene is in fact temporally regulated, it is possible to test if it can be autoinduced by culture supernatants. For example, if a gene is active only in the postexponential phase, postexponential supernatants should be added to cells in their early exponential phase and gene activation tested shortly after (2). If the gene can be induced by the addition of culture supernatants, the potential autoinducer can be isolated from the culture supernatant by column chromatography.

Kendra P. Rumbaugh (ed.), *Quorum Sensing: Methods and Protocols*, Methods in Molecular Biology, vol. 692,
DOI 10.1007/978-1-60761-971-0_4, © Springer Science+Business Media, LLC 2011

Identification of the *agr* as a quorum sensing system and isolation of its autoinducers will be used here as an example. *Agr* encodes a typical autoactivation circuit. The gene locus is expressed from the mid-exponential phase of growth and contains two divergent promoters, P2 and P3. P2 transcript contains *agrB, D, C, A*. AgrA and AgrC constitute a classical two-component signaling system, AgrC being the sensor and AgrA a response regulator. AgrD is the precursor of the *agr* autoinducing peptide (AIP). AgrB is a membrane endopeptidase that removes the AgrD carboxy tail. AIP is a 7–9 amino acid peptide that usually contains a thiolactone ring. The AIP binds to and activates the phosphorylation of AgrC, which activates the phosphorylation of AgrA, which upregulates its own promoter (P2) as well as that of the adjacent and divergent RNAIII promoter, P3 (3–6). RNAIII is a bifunctional molecule that encodes the δ-hemolysin protein in its 5′ end (*hld*) while it also acts as a noncoding regulatory RNA. With a total size of 514 nt folded into 14 stem-loops, RNAIII is one of the largest known regulatory RNAs (7–10). RNAIII is expressed from the mid-exponential phase of growth and is translated into δ-hemolysin at the postexponential phase of growth (10), when most other exotoxins are also produced (11).

Among isolates of *Staphylococcus aureus*, there are four different classes of *agr* systems, each recognizing a unique AIP signal. All of the AIP signals retain the basic thiolactone structure, but besides the fixed cysteine, the other residues vary among the four classes. Two of the AIP structures, type I and IV, differ by only one amino acid and function interchangeably, while AIP type II and type III are divergent, giving three different groupings of AIP signals (6). These three AIP groups (type I/IV, II, and III) cross inhibit each other's functions in a mechanism called bacterial interference. Non-*S. aureus* strains generally inhibit the *agr* response of *S. aureus* strains from each of the other groups (4, 12).

The expression of RNAIII can also be regulated in a yet unknown manner by RNAIII-activating protein (RAP) and by RNAIII-inhibiting peptide (RIP). RAP is a 277 amino acid protein found in culture supernatants of *S. aureus* (2, 13–15). From its sequence, RAP is predicted to be an ortholog of the 50 S ribosomal protein L2, which is encoded by the gene *rplB* found in eubacterial genomes (14). Inhibiting RAP by anti-RAP antibodies or by RAP-binding peptides suppressed infections in vivo (13, 15).

The RNAIII inhibitor most studied in vivo is RIP. This peptide was initially isolated from culture supernatants of a staphylococcal strain RN833. RIP is commonly used to inhibit staphylococcal infections as a linear synthetic peptide in its amide form (YSPWTNF-NH2) (16–18). The methods used in identifying the *agr* as a quorum sensing system that can be autoinduced or inhibited are described below.

2. Materials

2.1. Cell Culture and Lysis

1. Tryptic Soy Broth (TSB) (e.g., from Difco).
2. TES buffer: 100 mM Tris-HCL pH 7.2, 1 mM EDTA, 20% sucrose.
3. Lysostaphin in TES buffer: 100 μg/ml.
4. 2% SDS.
5. Proteinase K: 100 μg/ml.

2.2. Northern Blotting

1. Deionized glyoxal.
2. 16 mM Phosphate buffer, pH 7.0.
3. DMSO.
4. RNA loading buffer (Applied Biosystems).
5. 1% Agarose gel in 10 mM phosphate buffer pH 7.0.
6. Iodoacetic acid (Sigma-Aldrich).
7. 0.04% Methylene blue in 0.5 M sodium acetate pH 5.2.
8. Hybond-N+ membrane (Amersham Biosciences).
9. 1% SDS in SSPE: 10 mM sodium phosphate buffer pH 7.4/150 mM NaCl/1 mM EDTA.
10. Saline-sodium citrate (SSC) buffer: A 20× stock solution consists of 3 M sodium chloride and 300 mM trisodium citrate, adjusted to pH 7.0 with HCl.
11. Rapid-hyb (Amersham Biosciences).
12. Washing solution I: 2× SSC, 0.05% SDS.
13. Washing solution II: 0.1× SSC, 0.1% SDS.

2.3. Real-Time RT-PCR Analysis

1. TRIzol (Sigma-Aldrich).
2. DNAse I (Applied Biosystems).
3. ImProm-II™ Reverse Transcription System (Promega).
4. Random hexamers (Invitrogen).
5. LightCycler fast start DNA master SYBR Green Kit (Roche).

2.4. Preparation of Spent Culture Supernatants

1. 0.22 mm Filter (e.g., Fisher).
2. 10 kDa Cutoff membrane (Centriprep, Amicon).
3. Phosphate-buffered saline (PBS).
4. FPLC-gel-filtration column.
5. 50 mM Tris–HCl, pH 8.0.

2.5. Purification of AIP

1. 3 kDa Cutoff membrane (Centriprep, Amicon).
2. Acetonitrile.

3. Trifluoroacetic acid.
4. HPLC CI8 column.
5. 20 mM Tris buffer, pH 7.5.

2.6. Reporter Plate Assay (β-Lactamase Assay)

1. Reporter cells.
2. Microtiter 96-well plates.
3. 0.02% Azide.
4. β-Lactamase substrate: 40 µL 132 µg/ml Nitrocefin in 0.1 M sodium phosphate, pH 5.8.

3. Methods

3.1. Monitoring RNAIII Transcription During Growth

1. Grow *S. aureus* overnight. Typically 5 ml of TSB in 50 ml tubes. Start a new culture by diluting the overnight culture 1:100 in TSB. It is important to note that the genes expressed at the moment of dilution are still those of the late exponential phase. It is important to wait a few generations (at least 20–40 min) before the culture is considered to be early exponential, as far as gene expression goes.
2. Grow the cells with shaking (220 rpm) at 37°C for 6 h (to the late exponential phase of growth) and monitor growth spectroscopically at 600 nm.
3. Collect equal number of cells (e.g., 5×10^8) at time intervals (e.g., 1 ml of OD 0.1 and 100 µl of OD 1.0) and harvest the cells by centrifugation (e.g., 2 min 12,000 × *g*).
4. To prepare RNA, resuspend the cell pellet in 20 µl lysostaphin in TES buffer and incubate for 10 min at room temperature. Then add 20 µl of 2% SDS containing Proteinase K. Vigorously mix by vortexing for 10 min at room temperature. This sample now contains total RNA and DNA (2, 14).

Below is an example of methods used for the detection of RNAIII (2, 14, 19) but any protocol for RNA isolation, northern blotting and real-time PCR can be used instead.

3.1.1. Detection of RNAIII by Northern Blotting

1. To prepare the sample for northern blotting, mix 15 µl of the 40 µl total RNA/DNA sample prepared as described above with 11% deionized glyoxal, 16 mM phosphate buffer pH 7.0, and 50% DMSO, and incubate for 1 h at 65°C. Place on ice. Add RNA loading buffer and load the sample onto a 1% agarose gel in 10 mM phosphate, buffer pH 7.0 supplemented with 5 mM iodoacetic acid. Run the gel at 80 V until bromophenol blue reaches the middle of the gel.

2. Northern blot the gel onto a positively charged nylon membrane, e.g., by dry transfer.
3. Crosslink the transferred RNA to the membrane by UV irradiation for 1 min.
4. Briefly stain the membrane in 0.04% methylene blue in 0.5 M sodium acetate, pH 5.2 to ensure proper transfer and equal sample loading. Remove stain with 1% SDS in SSPE. Briefly soak the membrane in 2× SSC.
5. Prehybridize the membrane by incubating it in rapid-hyb at 37°C for 60 min with gentle shaking.
6. Place the membrane in hybridization bottle, add fresh solution of rapid-hyb, add the radiolabeled RNAIII-specific DNA probe (prepared as described in (20)), and further incubate for 4–12 h at 42°C.
7. Transfer the membrane to a glass tray filled with 200 ml washing solution I at 37–45°C. Shake at for 30 min. Repeat washing with washing solution I. Wash with washing solution II for 15 min, twice. Remove residual liquid and wrap the membrane in plastic wrap and expose to an X-ray film at −80°C.
8. As a control for the amount of RNA loaded on the gel, membrane can also be hybridized with a radiolabeled probe to *traP* (21) or any other constitutively expressed gene.

Using this method, *agr* was demonstrated as being temporally regulated, where the production of RNAIII was most evident from the mid-exponential phase of growth (Fig. 1) (2).

3.1.2. Detection of RNAIII by Real-Time RT-PCR Analysis

1. For real-time RT-PCR analysis, isolate RNA by TRIzol according to manufacturer's instructions, followed by treatment with DNAse I at 37°C for 20 min according to manufacturer's instructions.
2. Verify the absence of DNA by PCR, using the DNAse I-treated RNA samples as a template, and specific gene primers below.

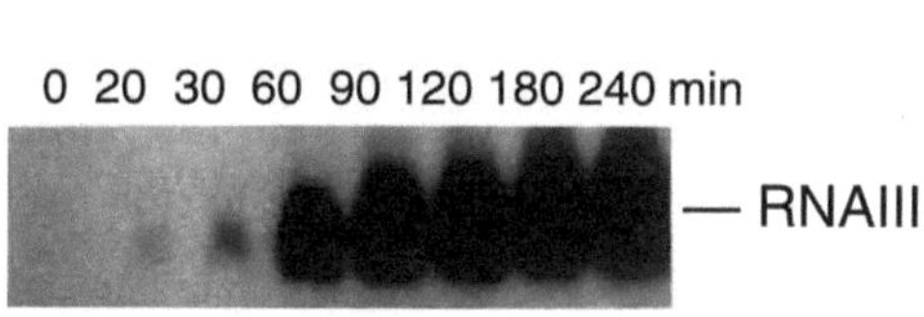

Fig. 1. Agr activity (RNAIII production) during culture growth of *S. aureus*. Cells were grown from early exponential to postexponential phase and *agr* activity was measured by analyzing RNAIII production over time. The autoradiogram of the northern blot is presented (2). Copyright (2009) National Academy of Sciences, USA.

3. For cDNA synthesis, 2 μg of each RNA sample should be used with the ImProm-II™ Reverse Transcription System according to manufacturer's instructions.
4. To prime the reaction, random hexamers should be used. 1 μl of resulting cDNA reaction should be used to set up the real-time PCR, using the LightCycler fast start DNA master SYBR Green Kit, according to manufacturer's instructions.
5. The transcripts for *hld* (which is part of RNAIII (10)) can be amplified using the *hld* primers 5′ ATGATCACAGAGATGGTA 3′ and 5′ CTGAGTCCTAGGAAACTAACT 3′. To monitor for specificity, the PCR products should be analyzed by melting curves and agarose gel electrophoresis. The values can be normalized with respect to *gyrB* expression using primers *gyrB* 5′ TTATGGTGCTGGGCAAATACA 3′ and 5′ CACCATGTAAACCACCAGATA 3′ (22).

3.2. Quorum Sensing Autoactivation

Once the gene of interest is shown to be temporally regulated, one can test if an autoinducer is present in the spent culture supernatant. In our case, the *agr* was active from the mid-exponential phase of growth, suggesting that an autoinducer may be present in the spent culture media.

3.2.1. Preparation of Spent Culture Broth

To prepare the spent culture media to be tested for the presence of an autoinducer, grow *S. aureus* cells from early exponential phase of growth for 6 h (to late postexponential phase). Harvest the cells by centrifugation at 5,000 × *g*, filter the culture supernatant on a 0.22 mm filter to remove all remaining cells, and dry the filtered supernatant by lyophilization. Resuspend the sample in water to a tenth of the original volume (10× sup).

3.2.2. Preparation of "Test" Cells in Their Early Exponential Phase

To prepare the cells to be tested for autoinduction of the *agr* and prepare cells in the early exponential phase of growth before *agr* is normally activated, grow *S. aureus* overnight at 37°C with shaking. Start a new culture by diluting the overnight culture 1:100 in TSB. It is important to note that the genes expressed at the moment of dilution are still those of the late exponential phase. It is important to wait a few generations (at least 20–40 min) before the culture is considered to be early exponential, as far as gene expression goes (see Note 1).

3.2.3. Test for Autoregulation

To test for regulation of RNAIII production by spent culture supernatants, first adjust the pH of the material in question (e.g., postexponential culture broth) to the pH of the growing cells to prevent an effect of the pH instead of an actual autoinducer. Then add to the test early exponential cells the 10× sup to a final 1× concentration, and continue growing the cells for 30–40 min to test for the activation or 2–2.5 h for inhibition of the *agr* (Fig. 2)

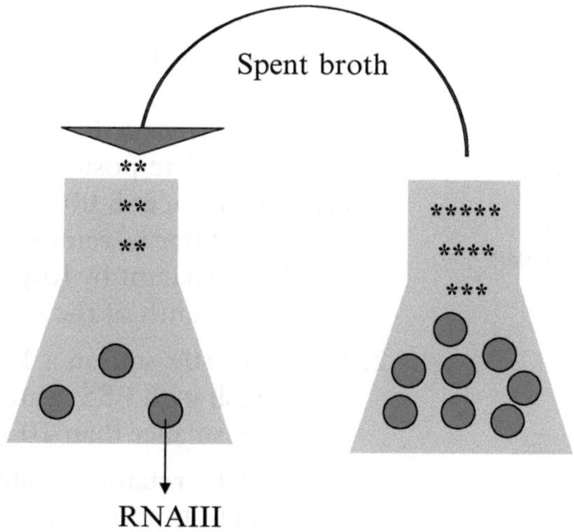

Fig. 2. Test for the presence of an autoregulator: Spent culture broth isolated from a postexponential culture is applied to cells in their early exponential phase of growth and activation of RNAIII production is tested.

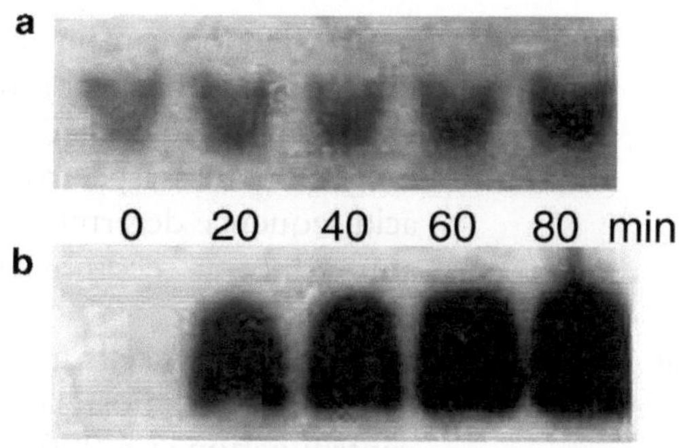

Fig. 3. Activation of *agr* (*production of RNAIII*) *by spent culture broth:* Supernatant from a 6-h culture of *S. aureus* was added to early exponential wild-type cells and the activity of *agr* was monitored during the subsequent growth of the bacteria by northern blotting with RNAIII-specific DNA as a probe. (**a**) CY broth was added. (**b**) Concentrated (I0×) supernatant was added (final dilution, 1×) (2). Copyright (2009) National Academy of Sciences, USA.

(see Note 1). Analyze RNAIII production as described above. As shown in Fig. 3, The production of RNAIII could be induced by spent culture supernatants (2), suggesting a presence of an autoinducer.

To test if the autoinducer is proteinaceous, concentrated culture supernatant should be treated with proteinase K as recommended by the manufacturer (Novagen). Briefly, adjust the pH of the culture supernatant in question to pH 7.2. Add increasing amounts of proteinase K and incubate for 1 h at 40°C. Stop the activity of the proteinase K with serine protease inhibitors such as phenylmethylsulfonyl fluoride (PMSF) or diisopropylfluorophosphate

(DFP). If activity is eliminated, one can assume that the autoinducer is a protein or a peptide.

3.3. Purification of an Autoinducer

3.3.1. Purification of the Autoinducer RAP

1. Grow *S. aureus* cells from early exponential phase of growth for 6 h (to late postexponential phase). Harvest the cells by centrifugation at 5,000 × *g*, filter the culture supernatant on a 0.22 mm filter to remove all remaining cells, and dry the filtered supernatant by lyophilization. Resuspend the sample in water to a tenth of the original volume (10× sup).
2. Filter the 10× sup on a 10 kDa cutoff membrane and extensively wash with PBS, water or Tris saline buffer, to remove all materials smaller than 10 kDa.
3. Fractionate retained material greater than 10 kDa on an FPLC-gel-filtration S-300 column in PBS or 50 mM Tris–HCl (pH 8.0) at a flow rate of 0.8 ml/min. Concentrate 1 ml fractions to one-tenth of their original volume by lyophilization and test each fraction for activation of RNAIII synthesis by northern blotting or by RNAIII reporter plate assay (2, 13, 15) (see below).
4. Active fraction can be analyzed by SDS–PAGE and visualize by Coomassie or silver staining (Fig. 4) (2). Suspected protein band can be identified by matrix-assisted laser desorption/ionization mass spectrometry (MALDI-MS) and its amino acid sequence determined. To confirm identity of the protein autoinducer, a recombinant protein can be made (14).

3.3.2. Purification of the Autoinducer AIP

1. Grow *S. aureus* cells from early exponential phase of growth for 6 h (to late postexponential phase). Harvest the cells by centrifugation at 5,000 × *g*, filter the culture supernatant on a 0.22 mm filter to remove all remaining cells, and boil material for 10 min. Centrifuge the sample, collect soluble material, and apply to a 3 kDa cutoff membrane. Collect the flow through, lyophilize, and resuspend in 2.5% acetonitrile/0.1% trifluoroacetic acid (1/40 volume of the culture supernatant).
2. Load this material onto an HPLC CI8 column in 2.5% acetonitrile/0.1% trifluoroacetic acid and elute with an acetonitrile gradient (16–48%) at 0.27% acetonitrile per minute. Lyophilize collected fractions (1.5 ml per fraction) and resuspend in 0.1 ml of 20 mM Tris buffer (pH 7.5). Test fractions for activity, e.g., by northern blotting or reporter cell assay. Fractions containing activity can be pooled and rerun on the HPLC CI8 column and eluted with an acetonitrile gradient at 0.2% acetonitrile per minute over the interest range (20–32%). AIP, eluting at an acetonitrile concentration of about 28.5%, was analyzed by MALDI-MS and its amino acid sequence was determined (12, 23). See the purification of RIP as an example (Fig. 5) (2).

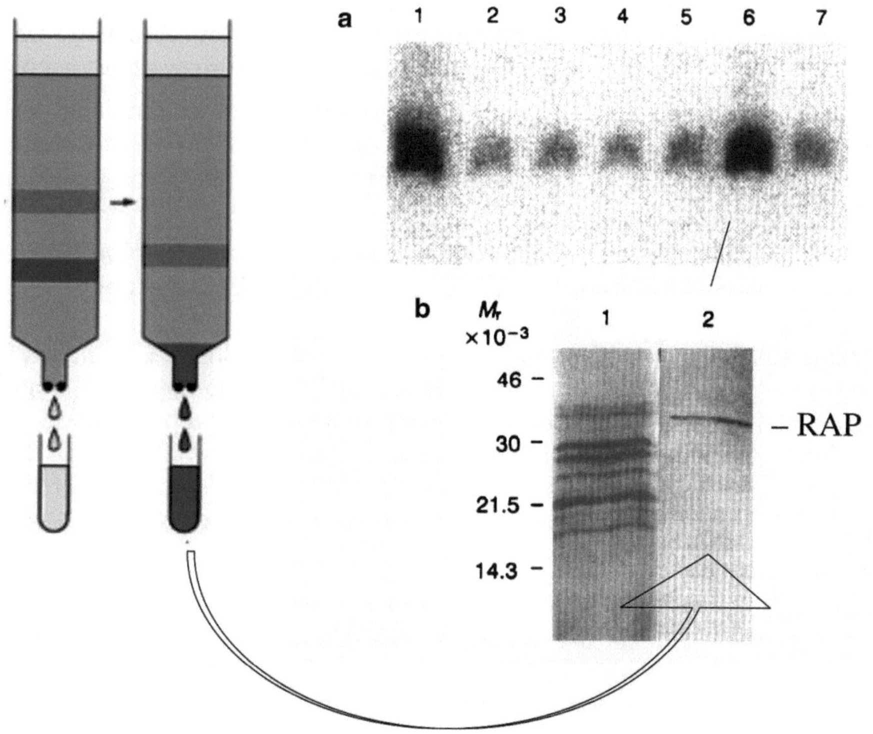

Fig. 4. Purification of RAP. (**a**) Spent postexponential culture broth was applied to an HPLC gel filtration column, and fractions were collected, lyophilized, and added to cells in the early exponential phase of growth. Cells were analyzed for RNAIII by northern blotting 20 min after the addition of total spent culture supernatant (lane 1), fresh broth (lane 2), or column fractions (lanes 3–7). (**b**) The column fractions were separated by SDS 12.5% PAGE and the gel was silver stained (Bio-Rad). Lanes: 1, column fractions having no *agr*-upregulating activity but maximum protein concentration; 2, column fraction corresponding to elution volumes which contained the *agr*-upregulating activity (2). Copyright (2009) National Academy of Sciences, USA.

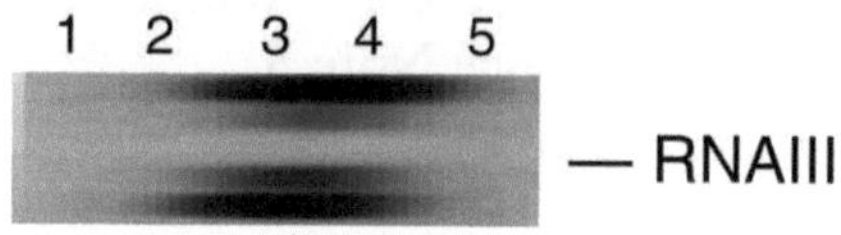

Fig. 5. Purification of RIP, an *agr* inhibitor. (**a**) Postexponential culture supernatant of RN833 was passed through a 3-kDa cutoff membrane and the flow through was applied to a C18 reverse-phase HPLC column and eluted with increasing amounts of acetonitrile in 0.1% trifluoroacetic acid. Eluted fractions were lyophilized, redissolved in water, and added to early exponential *Staphylococcus aureus* cells. The activation of *agr* was tested by northern blotting after 40 min. Inhibitor was identified by peptide sequencing.

3.4. Regulation of RNAIII Production

3.4.1. Regulation of RNAIII Production by Northern Blotting

1. Grow *S. aureus* overnight. Dilute the overnight culture 1:100 in TSB. Continue growing the cells at 37°C with shaking until they reach OD600 of about 0.2 (early exponential phase).
2. Add the sample in question (e.g., recombinant RAP, AIP, 10× supernatants or column fractions) and continue incubation for 30–40 min to test for activation or 2–2.5 h for inhibition.
3. Analyze RNAIII production by northern blotting using, e.g., radiolabeled RNAIII-specific DNA as a probe (2).

3.4.2. Regulation of RNAIII Production by β-Lactamase Reporter Cell Assay

Agr activation and inhibition can be studied in *S. aureus* strains containing plasmid-carried *agr*P3::β-lactamase (*blaZ*) reporter. See Note 2 for other reporter assays. *S. aureus* β-lactamase is encoded on the pI258 multidrug resistance plasmid and has been used extensively as a reporter (24, 25). Multiple BlaZ reporter strains are available, like *S. aureus agr* Group I tester: RN9365; Group II tester: RN9372 and Group III tester: RN9369 (26, 27).

1. Grow *S. aureus* reporter cells overnight. Dilute the overnight culture 1:100 in TSB. Continue growing the cells at 37°C with shaking until they reach OD600 of about 0.2 (early exponential phase).
2. To test when in your system RNAIII is made, place 30–50 μl early exponential reporter cells (2×10^7 CFU) into wells of microtiter plates. Grow the cells with shaking for 0.5–4 h at 37°C.
3. Stop cell growth at time intervals by the addition of 0.02% sodium azide.
4. Measure β-lactamase activity by adding a substrate for β-lactamase, nitrocefin (see Note 3).
5. Determine OD at 490/650 nm every 5 min or follow kinetics of color development. The initial slope should be considered (see Note 3).

4. Notes

1. When preparing the cells to be tested for gene activation or repression, it is important to remember that merely diluting the cells to a new OD does not mean that immediately the cells are in a new growth phase. The genes expressed at the moment of dilution are still those of, e.g., the late exponential phase. It is important to wait a few generations (at least 20–40 min), before the culture is considered to be in the early exponential phase of growth.

2. Quorum sensing can also be monitored in *S. aureus* strains using reporter plasmids with the P3 promoter driving GFP expression as described (28, 29).
3. Nitrocefin is yellow, and development of a pink color indicates RNAIII production. Assuming for example that using the above protocol, wells containing cells grown for 90 min turn pink, suggesting that RNAIII was made only after 90 min. In that case, to test for the presence of an activator, cells should be grown with the material in question (e.g., growth culture) for a period of time that is shorter than 90 min (e.g., for 30–60 min only). In that case, the wells containing an activator will turn pink, while the controls will remain yellow.

To test for the presence of an inhibitor, cells should be incubated with the material in question for longer than 90 min (e.g., 90–110 min). In that case, the wells containing an inhibitor will remain yellow, while the controls will turn pink (Fig. 6).

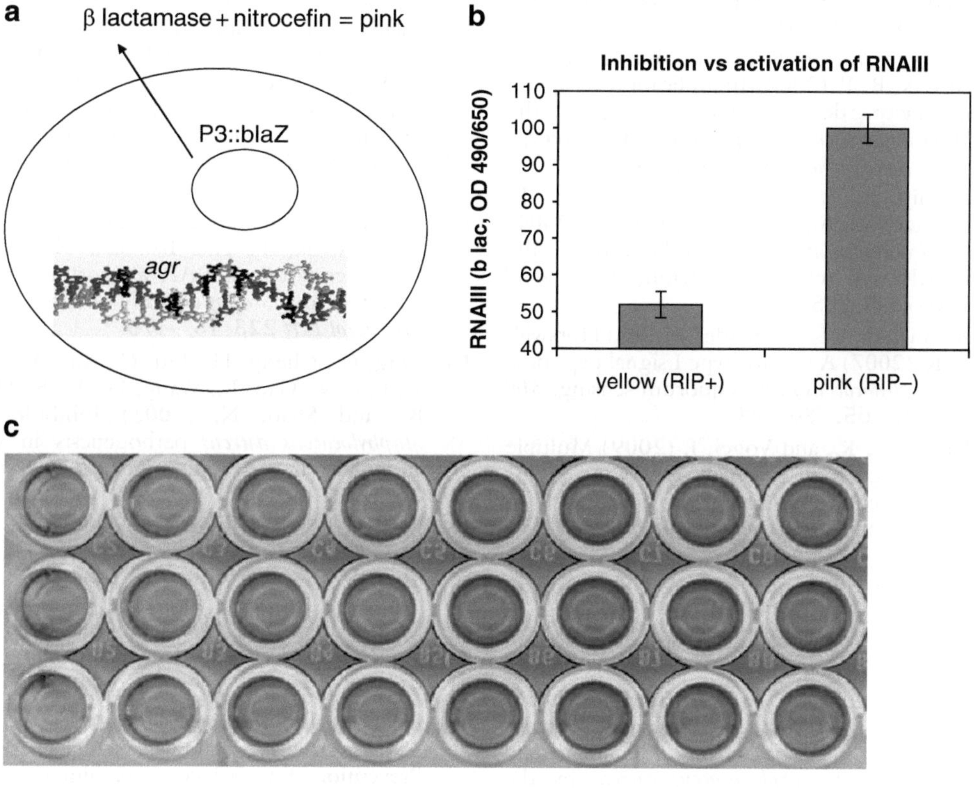

Fig. 6. Reporter RNAIII plate assay: RIP inhibits RNAIII production: 2×10^7 early exponential *Staphylococcus aureus* cells containing RNAIII::blaZ fusion construct (**a**) were grown for 2.5 h with RIP. β-Lactamase activity was determined by adding nitrocefin, a substrate for β-lactamase. When *agr* is active, β-lactamase is made and the sample turns (**b**, *right panels*). When RIP is added, *agr* is inhibited and the sample remains (**b**, *left panels*). OD490/650 is shown in (**c**).

The β-lactamase reporter assay is useful when the difference between experimental and control groups is high (at least twice) and one group turns pink, while the other remains yellow. Over time, however, all samples will turn pink. So it is important to read the plate from the time the substrate is added.

To ensure that the added material does not by itself contain β-lactamase activity and thus cause a change of color without actually activating RNAIII synthesis, material in question should be incubated with buffer only (without cells) as a negative control.

To prevent sample evaporation, it is recommended to place the microtiter plate in a humid chamber during incubations.

References

1. Greenberg, P. E. (2003) Tiny Teamwork, *Nature* **424**, 134.
2. Balaban, N., and Novick, R. P. (1995) Autocrine regulation of toxin synthesis by *Staphylococcus aureus*, *Proc Natl Acad Sci USA* **92**, 1619–1623.
3. Novick, R. P., and Geisinger, E. (2008) Quorum sensing in staphylococci, *Annu Rev Genet* **42**, 541–564.
4. Geisinger, E., George, E. A., Muir, T. W., and Novick, R. P. (2008) Identification of ligand specificity determinants in AgrC, the *Staphylococcus aureus* quorum-sensing receptor, *J Biol Chem* **283**, 8930–8938.
5. Koenig, R. L., Ray, J. L., Maleki, S. J., Smeltzer, M. S., and Hurlburt, B. K. (2004) *Staphylococcus aureus* AgrA binding to the RNAIII-agr regulatory region, *J Bacteriol* **186**, 7549–7555.
6. Kavanaugh, J. S., Thoendel, M., and Horswill, A. R. (2007) A role for type I signal peptidase in *Staphylococcus aureus* quorum sensing, *Mol Microbiol* **65**, 780–798.
7. Papenfort, K., and Vogel, J. (2009) Multiple target regulation by small noncoding RNAs rewires gene expression at the post-transcriptional level, *Res Microbiol* **160**, 278–287.
8. Benito, Y., Kolb, F. A., Romby, P., Lina, G., Etienne, J., and Vandenesch, F. (2000) Probing the structure of RNAIII, the *Staphylococcus aureus* agr regulatory RNA, and identification of the RNA domain involved in repression of protein A expression, *RNA* **6**, 668–679.
9. Morfeldt, E., Taylor, D., von Gabain, A., and Arvidson, S. (1995) Activation of alpha-toxin translation in *Staphylococcus aureus* by the trans-encoded antisense RNA, RNAIII, *EMBO J* **14**, 4569–4577.
10. Balaban, N., and Novick, R. P. (1995) Translation of RNAIII, the *Staphylococcus aureus* agr regulatory RNA molecule, can be activated by a 3′-end deletion, *FEMS Microbiol Lett* **133**, 155–161.
11. Lowy, F. D. (1998) *Staphylococcus aureus* infections, *N Engl J Med* **339**, 520–532.
12. Ji, G., Beavis, R., and Novick, R. P. (1997) Bacterial interference caused by autoinducing peptide variants, *Science* **276**, 2027–2030.
13. Balaban, N., Goldkorn, T., Nhan, R. T., Dang, L. B., Scott, S., Ridgley, R. M., Rasooly, A., Wright, S. C., Larrick, J. W., Rasooly, R., and Carlson, J. R. (1998) Autoinducer of virulence as a target for vaccine and therapy against *Staphylococcus aureus*, *Science* **280**, 438–440.
14. Korem, M., Sheoran, A. S., Gov, Y., Tzipori, S., Borovok, I., and Balaban, N. (2003) Characterization of RAP, a quorum sensing activator of *Staphylococcus aureus*, *FEMS Microbiol Lett* **223**, 167–175.
15. Yang, G., Cheng, H., Liu, C., Xue, Y., Gao, Y., Liu, N., Gao, B., Wang, D., Li, S., Shen, B., and Shao, N. (2003) Inhibition of *Staphylococcus aureus* pathogenesis in vitro and in vivo by RAP-binding peptides, *Peptides* **24**, 1823–1828.
16. Balaban, N., Cirioni, O., Giacometti, A., Ghiselli, R., Braunstein, J. B., Silvestri, C., Mocchegiani, F., Saba, V., and Scalise, G. (2007) Treatment of *Staphylococcus aureus* biofilm infection by the quorum-sensing inhibitor RIP, *Antimicrob Agents Chemother* **51**, 2226–2229.
17. Balaban, N., Stoodley, P., Fux, C. A., Wilson, S., Costerton, J. W., and Dell'Acqua, G. (2005) Prevention of staphylococcal biofilm-associated infections by the quorum sensing inhibitor RIP, *Clin Orthop Relat Res* **437**, 48–54.
18. Kiran, M. D., Giacometti, A., Cirioni, O., and Balaban, N. (2008) Suppression of biofilm

related, device-associated infections by staphylococcal quorum sensing inhibitors, *Int J Artif Organs* **31**, 761–770.

19. Korem, M., Gov, Y., Kiran, M. D., and Balaban, N. (2005) Transcriptional profiling of target of RNAIII-activating protein, a master regulator of staphylococcal virulence, *Infect Immun* **73**, 6220–6228.
20. Novick, R. P., Ross, H. F., Projan, S. J., Kornblum, J., Kreiswirth, B., and Moghazeh, S. (1993) Synthesis of staphylococcal virulence factors is controlled by a regulatory RNA molecule, *EMBO J* **12**, 3967–3975.
21. Balaban, N., Goldkorn, T., Gov, Y., Hirshberg, M., Koyfman, N., Matthews, H. R., Nhan, R. T., Singh, B., and Uziel, O. (2001) Regulation of *Staphylococcus aureus* pathogenesis via target of RNAIII-activating protein (TRAP), *J Biol Chem* **276**, 2658–2667.
22. Livak, K. J., and Schmittgen, T. D. (2001) Analysis of relative gene expression data using real-time quantitative PCR and the 2(-Delta Delta C(T)) method, *Methods* **25**, 402–408.
23. Ji, G., Beavis, R. C., and Novick, R. P. (1995) Cell density control of staphylococcal virulence mediated by an octapeptide pheromone, *Proc Natl Acad Sci USA* **92**, 12055–12059.
24. Novick, R. P. (1991) Genetic systems in staphylococci, *Methods Enzymol* **204**, 587–636.
25. Wang, P. Z., Projan, S. J., Leason, K. R., and Novick, R. P. (1987) Translational fusion with a secretory enzyme as an indicator, *J Bacteriol* **169**, 3082–3087.
26. Lyon, G. J., Mayville, P., Muir, T. W., and Novick, R. P. (2000) Rational design of a global inhibitor of the virulence response in *Staphylococcus aureus*, based in part on localization of the site of inhibition to the receptor-histidine kinase, AgrC, *Proc Natl Acad Sci USA* **97**, 13330–13335.
27. Geisinger, E., Muir, T. W., and Novick, R. P. (2009) agr receptor mutants reveal distinct modes of inhibition by staphylococcal autoinducing peptides, *Proc Natl Acad Sci USA* **106**, 1216–1221.
28. Yarwood, J. M., Bartels, D. J., Volper, E. M., and Greenberg, E. P. (2004) Quorum sensing in *Staphylococcus aureus* biofilms, *J Bacteriol* **186**, 1838–1850.
29. Boles, B. R., and Horswill, A. R. (2008) Agr-mediated dispersal of *Staphylococcus aureus* biofilms, *PLoS Pathog* **4**, e1000052.

Chapter 5

Liquid Chromatography/Mass Spectrometry for the Detection and Quantification of *N*-Acyl-L-Homoserine Lactones and 4-Hydroxy-2-Alkylquinolines

François Lépine and Eric Déziel

Abstract

High-performance liquid chromatography (HPLC) coupled in-line with mass spectrometry (MS) permits rapid and specific identification and quantification of *N*-acyl-L-homoserine lactones (AHLs) and 4-hydroxy-2-alkylquinolines (HAQs). We are presenting here methods for the analysis of these molecules directly from biological samples using LC/MS.

Key words: Bacteria, Quorum sensing, Acyl homoserine lactone, 4-Hydroxy-2-alkylquinoline, Analysis, Quantification, Mass spectrometry, Liquid chromatography

1. Introduction

Most Gram-negative bacteria produce *N*-acyl-L-homoserine lactones (AHLs), such as *N*-(3-oxododecanoyl)-L-homoserine lactone (3-oxo-C_{12}-HSL) and *N*-butanoyl-L-homoserine lactone (C_4-HSL) for *Pseudomonas aeruginosa* (1, 2). A few species, such as *P. aeruginosa*, *Burkholderia pseudomallei*, and *B. thailandensis*, release a number of 4-hydroxy-2-alkylquinolines (HAQs), also known as 2-alkyl-4-quinolones (3–6) and derivatives, such as 3-methyl analogues (HMAQs).

Although bioassays are essential for screening and for rapid qualitative determination of quorum sensing signalling molecules, more precise and sensitive methods are required to provide definitive confirmation of their presence and allow formal structural confirmation and accurate quantification. Because of the very large number of structurally related signalling molecules,

Kendra P. Rumbaugh (ed.), *Quorum Sensing: Methods and Protocols*, Methods in Molecular Biology, vol. 692,
DOI 10.1007/978-1-60761-971-0_5, © Springer Science+Business Media, LLC 2011

a preliminary separation is required using high performance liquid chromatography (HPLC). Identification and quantification are then performed by mass spectrometry coupled in-line with the HPLC.

2. Materials

2.1. Bacterial Cell Cultures

1. Bacterial strain: the methods presented here use cultures of the Gram-negative bacterium *P. aeruginosa*, strain PA14 (7).
2. Tryptic soy broth (TSB) medium (Difco).
3. 18×150 mm borosilicate glass tubes.
4. TC-7 roller drum (New Brunswick).

2.2. LC/MS Analysis

1. Quattro II (Waters) triple quadrupole mass spectrometer (MS) equipped with a Z-spray interface.
2. Nitrogen is used for drying and argon is used as collision gas in multiple reactions monitoring (MRM) mode.
3. 1100 HP HPLC equipped with a 4.6×150 mm Eclipse XDB C8 column (Agilent): the MS is connected to the HPLC through a T splitter (Valco). The third output of the splitter is fitted with a tube of internal diameter and length such that only 10% of the initial flow goes to the electrospray probe.
4. Solvent A: distilled water containing 1% ACS grade acetic acid.
5. Solvent B: acetonitrile (or 2-propanol) HPLC grade, containing 1% ACS grade acetic acid.

2.3. Internal Standards

1. 5,6,7,8-Tetradeutero-4-hydroxy-2-heptylquinoline (HHQ-d_4) and 5,6,7,8-tetradeutero-3,4-dihydroxy-2-heptylquinoline (PQS-d_4) are synthesized as described (8).
2. Standard stock solution: 10 mg/l HHQ-d_4 and 20 mg/l PQS-d_4 are prepared in ACS grade methanol and kept at -20°C.

3. Methods

Mass spectrometric analysis of the samples is performed under positive electrospray ionization conditions, a process that adds a proton to the analytes, thus producing pseudomolecular ions that correspond to the mass of the neutral molecule with the addition of one proton.

The methods presented here mostly deal with the quantification of 4-hydroxy-2-heptylquinoline (HHQ), 3,4-dihydroxy-2-heptylquinoline (*Pseudomonas* Quinolone Signal; PQS), and 4-hydroxy-2-heptylquinoline *N*-oxide (HQNO), each the best studied HAQ congener of its family (3), and of C_4-HSL and 3-oxo-C_{12}-HSL as the most representative AHLs of *P. aeruginosa* (Fig. 1). Within each family, the various congeners only differ by the length of their alkyl side chain. These molecules can thus be analysed by the same methods by taking into account the mass difference due to the successive addition (or subtraction) of one methylene unit (14 Da) going from one congener to the next. However, the most abundant HAQs, contrary to AHLs, contain an odd number of carbons on their side chain, so going from the most abundant congener to the next most abundant ones generally entails successive addition (or subtraction) of 28 Da. These methods can also be used to quantify the open form of the AHLs (*o*-AHLs) (Fig. 1), in which the lactone ring is hydrolysed, which corresponds to the addition of one molecule of water (18 Da) to the unmodified AHLs. Depending on the growth conditions and age of the cultures, these *o*-AHLs can be found in abundance. The methods presented can be adapted to perform the desired analysis. For instance, some *Burkholderia* species produce modified HAQs carrying an additional methyl substitution (6); detection and quantification of these 4-hydroxy-3-methyl-2-alkylquinolines

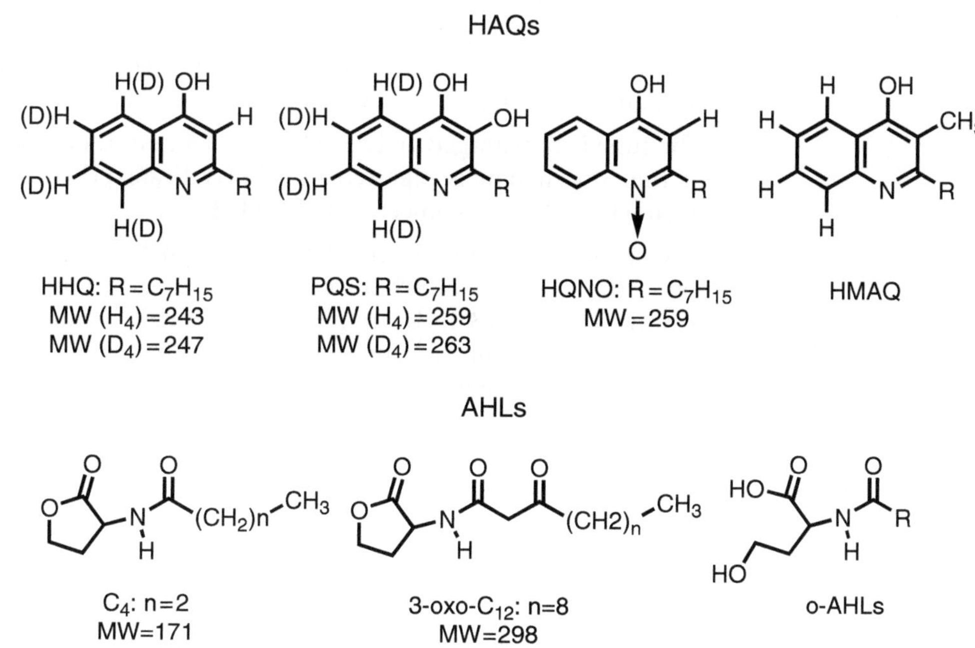

Fig. 1. Chemical structure of HAQs and AHLs.

(HMAQs) are possible by taking into account the additional mass of 14 Da resulting from the replacement of a proton by a methyl group on the quinoline ring (Fig. 1).

Various scanning modes can be used with a triple quadrupole to acquire quantitative data, each mode having his advantages and limitations. Hence, in full scan mode, a wide range of masses can be acquired, allowing for subsequent analysis of other compounds of interest aside from the intended HAQs and AHLs. In this scanning mode, the sensitivity is not maximal if only quantification of a few molecules is needed. On the other hand, it is very useful for detecting other members of the various families of HAQs or AHLs or other metabolites directly in culture broth. In MRM mode, one takes advantages of the ability of the instrument to select and to fragment specific pseudomolecular ions, and monitor the intensity of one specific resulting fragment ion. This scanning mode is limited to a preset series of fragmentation reactions (called transitions), but due to the specificity of the fragmentation, it provides a much better signal-to-noise ratio than the full scan mode, thus increasing the sensitivity of the analysis. The MRM mode is especially interesting for the analysis of AHLs, or for low concentrations of HAQs produced in complex matrices, such as from infected animal tissue samples.

3.1. Direct Quantification of HAQs from Culture Broth in Full Scan Mode

1. *P. aeruginosa* PA14 cultivated at 37°C and 200 rpm overnight (see Note 1) is used to inoculate 3 ml of fresh TSB medium at a starting OD_{600} of 0.05 (see Note 2). The cultures are then incubated under the same conditions, typically until they reach an OD_{600} of 3.0 (see Note 3). A 300 µl culture sample is then transferred to a microcentrifuge tube and 300 µl methanol containing the internal standards is added (see Note 4).
2. The tube is centrifuged at 13,000 × *g* for 15 min, then 500 µl of the supernatant is pipetted in a borosilicate HPLC vial, from which 20 µl are injected in the HPLC.
3. The solvent gradient for the chromatographic run is as follows: from 0 to 1 min 70% solvent A; from 1 to 13 min 100% solvent B; from 13 to 23 min 100% solvent B; from 23 to 25 min 70% solvent A; from 25 to 28 min 70% solvent A (see Note 5). Flow rate is set at 400 µl/min split to 40 µl/min by the T splitter.
4. The MS parameters are: positive mode; needle voltage 3.0 kV; cone 30 V; block temperature 120°C and drying gas 150°C; nebulising gas 20 l/min and drying gas 200 l/min.
5. In full scan mode, the scanning range is set to m/z 100–400. The chromatogram of all the ions monitored (total ion chromatogram or TIC) is presented in Fig. 2.
6. Figure 3 shows the chromatogram of the pseudomolecular ions $(M+H)^+$ of HHQ, PQS, and HQNO at 244, 260, and

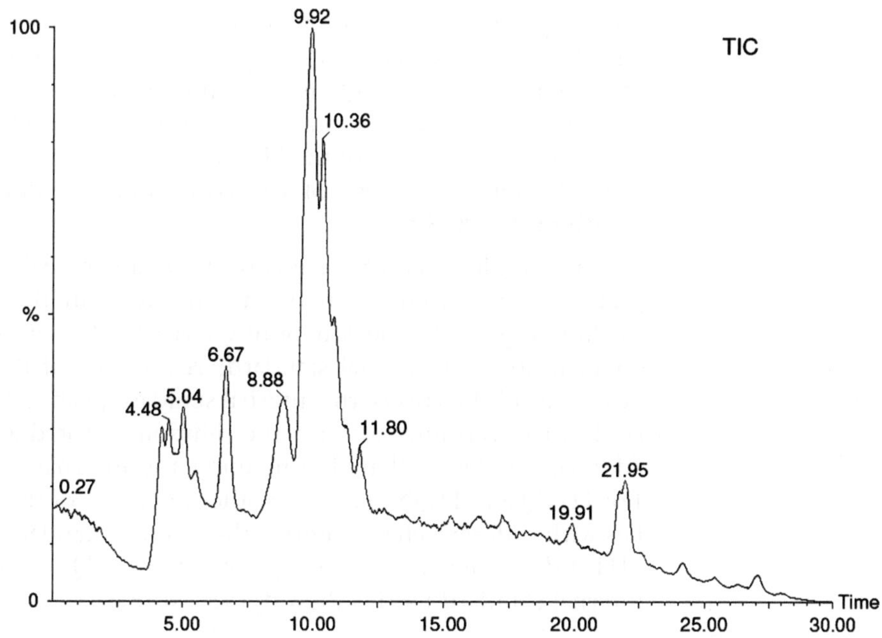

Fig. 2. Total ion chromatogram of the supernatant of a *Pseudomonas aeruginosa* strain PA14 culture grown to an OD_{600} of 3.0.

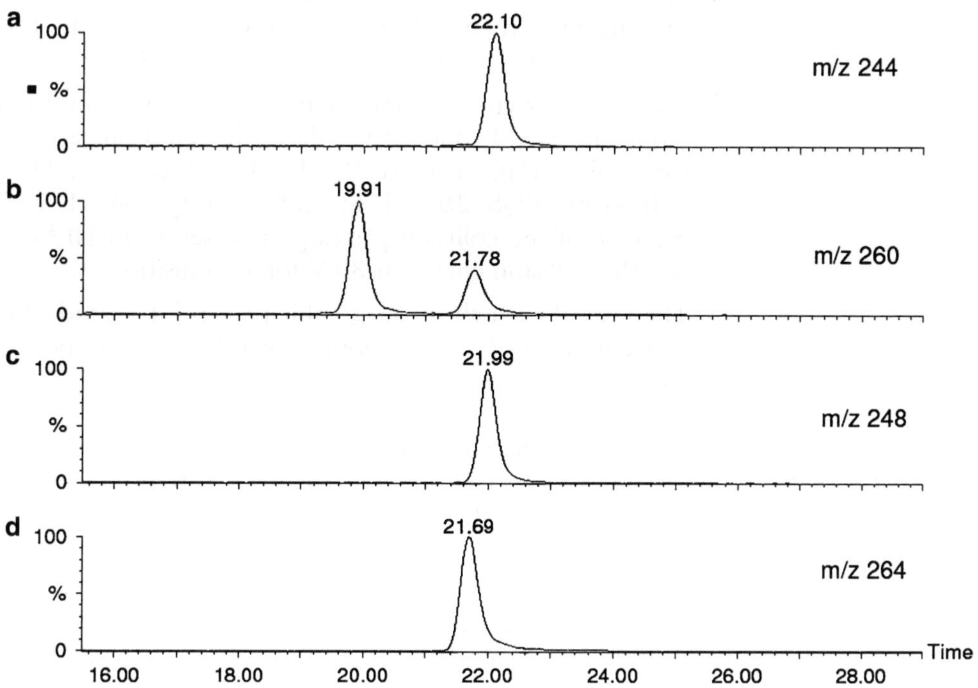

Fig. 3. Ion chromatograms of the supernatant of a *Pseudomonas aeruginosa* strain PA14 culture grown to an OD_{600} of 3.0. (**a**) HHQ, (**b**) HQNO and PQS, (**c**) HHQ-d_4, and (**d**) PQS-d_4.

260, respectively, and those of the internal standards HHQ-d_4 and PQS-d_4 at 248 and 264, respectively. With this column, the retention time of PQS is 21.8 min, while the one of the isobaric (an ion with the same nominal *m/z* value) HQNO is 19.9 min. All members of the HQNO family have a retention time shorter than those of the corresponding alkyl chain length in the PQS family.

7. The area under each of these chromatographic peaks is integrated. The concentration of the analyte A in the culture medium is given by the equation: $C \times A \times 2/H$,where C is the concentration of internal standard; A the area of the HAQ peak; and H the area of the internal standard peak. The factor of 2 is to keep into account the dilution factor due to the addition of the methanol containing the internal standard. For HHQ and HQNO and the other members of these families with different chain lengths, the internal standard used is HHQ-d_4, while for the congeners of the PQS family, the internal standard used is PQS-d_4 (see Note 6).

3.2. Quantification of HAQs from a Complex Matrix in MRM Mode

1. HAQs can be analysed from a sample of muscle tissue from a mouse infected with *P. aeruginosa* (*9*). One hundred milligrams of muscle tissues are put in a 2 ml microcentrifuge tube to which 500 μl of methanol containing 0.2 mg/l of HHQ-d_4 and PQS-d_4 is added. The tissues are homogenized with a Polytron then centrifuged at 13,000×*g* for 15 min. The supernatant is then collected and 80 μl injected in the HPLC under the same conditions as in Subheading 3.1.
2. The source operating parameters are the same as in full scan mode. In MRM mode, the following transitions are monitored: for HHQ 244→159; HHQ-d_4 248→163; HQNO 260→159; PQS 260→175; and PQS-d_4 264→179. The pressure of the collision gas (argon) is set at 2×10^{-3} mTorr and the collision energy at 30 V for all transitions.
3. The area of each chromatographic peak is integrated and the concentration of each compound is calculated as above (see Note 7).

3.3. Direct Quantification of AHLs from Culture Broth in MRM Mode

1. The bacteria are cultivated as in Subheading 3.1 (see Note 8). To 500 μl of culture is added 500 μl of a 20 mg/l methanolic solution of HHQ-d_4 and the mixture is centrifuged at 13,000×*g* for 15 min. A 500 μl aliquot is collected in a HPLC vial and 15 μl are injected (see Note 9).
2. The solvent gradient is as follows: from 0 to 1 min 100% solvent A; from 1 to 5 min 50% solvent B; from 5 to 13 min 100% solvent B; from 13 to 23 min 100% solvent B; from 23 to 25 min 100% solvent A; from 25 to 28 min 100% solvent A. Flow rate is set at 400 μl/min split to 40 μl/min by the splitter.

Under these conditions, the open form of each AHL always has a shorter retention time than the corresponding closed form.

3. The following transitions are monitored: C_4-HSL 172→102; *o*-C_4-HSL 190→120; HHQ-d_4 248→163; 3-oxo-C_{12}-HSL 298→102; and *o*-3-oxo-C_{12}-HSL 316→120. The pressure of the collision gas is set at 2×10^{-3} mTorr and the collision energy is 15 V for the AHLs and 33 V for HHQ-d_4.
4. A calibration curve is performed to obtain the response factor of C_4-HSL and 3-oxo-C_{12}-HSL relative to HHQ-d_4. To do so, a series of solutions with increasing concentration of the AHL from 0 to 15 mg/l are made with a constant concentration of 10 mg/l of HHQ-d_4. The same response factor is used for the open and closed forms of each AHLs.
5. The concentration of the AHLs in the solution is calculated as follows: $R \times C \times A \times 2/H$, where, R is the response factor, C the concentration of internal standard (HHQ-d4), A the area of the AHL peak, H the area of the internal standard (HHQ-d4) peak.

4. Notes

1. HAQs have been detected in all *P. aeruginosa* wild-type strains we have tested, except PAK and PA7.
2. The cultivation can also be performed in larger volumes, such as 50 ml of TSB in 250 ml flasks. However, we found that upon scaling up, the concentration of HAQs tends to decrease significantly with larger culture volumes for the same final cell density.
3. HAQs such as PQS or HQNO accumulate up to concentrations in excess of 15 mg/l in the culture medium under the described culture conditions and are thus easily detected in normal scanning mode. However, HHQ and the other members of the same family of compounds are the precursors of PQS and its congeners and thus show a decrease in their concentration following an initial increase. It is thus important to select the proper cultivation time, if HHQ and its congeners are to be measured.
4. HAQs such as PQS and HHQ have a limited solubility in water. For example, the solubility of PQS is less than 5 mg/l in TSB. However, concentrations in excess of 15 mg/l have often been measured in whole cultures. This is an indication that these compounds are either pseudosolubilized by bacterial exoproducts or are adsorbed on the cell surface. As bacteria

must be removed from the medium prior to analysis in order to avoid plugging the HPLC column, this can have an important effect on the concentration of HAQs detected in the samples. To dissociate HAQs from the bacteria cell surface, methanol is added to the culture at a 50% v/v concentration prior to removing the cells by centrifugation (8). In fact, we measure typically 50% less HAQs if the cells are centrifuged prior to adding methanol. Thus addition of methanol (which contains the internal standards) to the culture samples serves two purposes: to release bound HAQs and to act as carrier for the internal standards for quantification purposes.

5. Because of worldwide uncertainties in the supply of acetonitrile, an alternative gradient method can be used with 2-propanol as solvent according to the following steps: at the time of injection 90% solvent A; from 0 to 3 min 40% solvent B (2-propanol containing 1% acetic acid); from 3 to 10 min 65% solvent B; from 10 to 21 min 70% solvent B; from 21 to 23 min 100% solvent B; from 23 to 25 min 100% solvent B; from 25 to 26 min 90% solvent A; and from 26 to 30 min 90% solvent A.

6. One problem with PQS quantification is that when the HPLC column ages, the shape of the PQS peaks tends to present tailing and the area of the corresponding peak often vary considerably from one injection to another, which makes the quantification more difficult without using an appropriate deuterated internal standard. The PQS-d_4 internal standard fluctuates in the same manner, thus correcting for these variations.

7. The most abundant members of each families of compound can be quantified using their respective internal standard (PQS-d_4 for the PQS family and HHQ-d_4 for all the other HAQs) by adding (or subtracting) 28 Da to the weight of the corresponding pseudomolecular ion of the internal standard while monitoring the same fragment ion. Under these conditions, fragmentation of all HAQs produces a fragment ion common to all the congeners of a given family (10).

8. Because the Las system, which mediates 3-oxo-C_{12}-HSL, is expressed before the Rhl system that produces the C_4-HSL, the maximal concentration of 3-oxo-C_{12}-HSL in cultures is achieved prior the one of C_4-HSL. Thus timing of sampling will critically affect the concentrations of AHLs obtained.

9. Because the lactone ring of AHLs can spontaneously open at high pH or the open form can close at low pH, care should be taken to avoid these conditions if the samples are stored for a certain period of time prior to analysis.

Acknowledgments

The important contribution of Sylvain Milot to the development of these methods is gratefully acknowledged. We also thank Émilie Gauthier for the help in the development of the AHLs quantification methods.

References

1. Pearson, J. P., Gray, K. M., Passador, L., Tucker, K. D., Eberhard, A., Iglewski, B. H., and Greenberg, E. P. (1994) Structure of the autoinducer required for expression of *Pseudomonas aeruginosa* virulence genes, *Proc Natl Acad Sci U S A 91*, 197–201.
2. Pearson, J. P., Passador, L., Iglewski, B. H., and Greenberg, E. P. (1995) A second *N*-acylhomoserine lactone signal produced by *Pseudomonas aeruginosa*, *Proc Natl Acad Sci U S A 92*, 1490–1494.
3. Déziel, E., Lépine, F., Milot, S., He, J., Mindrinos, M. N., Tompkins, R. G., and Rahme, L. G. (2004) Analysis of *Pseudomonas aeruginosa* 4-hydroxy-2- alkylquinolines (HAQs) reveals a role for 4-hydroxy-2-heptylquinoline in cell-to-cell communication, *Proc Natl Acad Sci U S A 101*, 1339–1344.
4. Diggle, S. P., Lumjiaktase, P., Dipilato, F., Winzer, K., Kunakorn, M., Barrett, D. A., Chhabra, S. R., Camara, M., and Williams, P. (2006) Functional genetic analysis reveals a 2-alkyl-4-quinolone signaling system in the human pathogen *Burkholderia pseudomallei* and related bacteria, *Chem Biol 13*, 701–710.
5. Pesci, E. C., Milbank, J. B. J., Pearson, J. P., McKnight, S., Kende, A. S., Greenberg, E. P., and Iglewski, B. H. (1999) Quinolone signaling in the cell-to-cell communication system of *Pseudomonas aeruginosa*, *Proc Natl Acad Sci U S A 96*, 11229–11234.
6. Vial, L., Lépine, F., Milot, S., Groleau, M. C., Dekimpe, V., Woods, D. E., and Déziel, E. (2008) *Burkholderia pseudomallei, B. thailandensis*, and *B. ambifaria* produce 4-hydroxy-2-alkylquinoline analogues with a methyl group at the 3 position that is required for quorum-sensing regulation, *J Bacteriol 190*, 5339–5352.
7. Rahme, L. G., Stevens, E. J., Wolfort, S. F., Shao, J., Tompkins, R. G., and Ausubel, F. M. (1995) Common virulence factors for bacterial pathogenicity in plants and animals, *Science 268*, 1899–1902.
8. Lépine, F., Déziel, E., Milot, S., and Rahme, L. G. (2003) A stable isotope dilution assay for the quantification of the *Pseudomonas* quinolone signal in *Pseudomonas aeruginosa* cultures, *Biochim Biophys Acta 1622*, 36–41.
9. Xiao, G., Déziel, E., He, J., Lépine, F., Lesic, B., Castonguay, M. H., Milot, S., Tampakaki, A. P., Stachel, S. E., and Rahme, L. G. (2006) MvfR, a key *Pseudomonas aeruginosa* pathogenicity LTTR-class regulatory protein, has dual ligands, *Mol Microbiol 62*, 1689–1699.
10. Lépine, F., Milot, S., Déziel, E., He, J., and Rahme, L. G. (2004) Electrospray/mass spectrometric identification and analysis of 4-hydroxy-2-alkylquinolines (HAQs) produced by *Pseudomonas aeruginosa*, *J Am Soc Mass Spectrom 15*, 862–869.

Chapter 6

Detection of Autoinducer (AI-2)-Like Activity in Food Samples

Kirthiram K. Sivakumar, Palmy R. Jesudhasan, and Suresh D. Pillai

Abstract

The contamination, survival, and possible foodborne disease outbreaks are major issues confronting the food industry. However, from a microbial perspective, any food whether natural or processed is just another environmental niche that is available for colonization. Quorum sensing or cell–cell communication is a process by which microorganisms are thought to communicate with each other using a variety of small molecules termed autoinducers. The autoinducer AI-2 is thought to be a universal signaling molecule due to its ability to modulate the gene expression of a number of different bacterial species and genera. Pathogens such as *Pseudomonas aeruginosa*, *Aeromonas hydrophila*, *Vibrio anguillarum*, *Streptococcus* sp., and *Burkholderia cepacia* form biofilms on a variety of man-made and natural surfaces using cell–cell mechanisms. It is important to detect and study autoinducers and their activities in foods, since a better understanding of these molecules in food and food ingredients may help in designing new approaches to thwart microbial persistence and biofilm formation. The autoinducer AI-2 is thought to be involved in microbial attachment and biofilm formation leading to food spoilage. To better understand microbial cell–cell signaling in foods especially as it relates to pathogen persistence, biofilm formation, and food spoilage, methods to process, extract, and purify autoinducer molecules need to be developed. This chapter details methods to process food samples to obtain cell-free supernatants (CFS), which could subsequently be tested for the presence of AI-2 or "AI-2-like activity" in the extracted CFS using autoinducer bioassays. Additionally, the method of synthesizing AI-2 in the laboratory is also provided. The methods that are presented in this chapter are based on previously published research articles from the authors' laboratory.

Key words: Quorum sensing, Autoinducer, Prokaryotic, Biofilm, Cell-free supernatant

1. Introduction

The autoinducer-2 is particularly important as it is believed to be used for both intra and inter species communication (1). AI-2 is known to be produced by more than 50 bacterial species and homologs of the AI-2 synthase gene, *luxS* is widely distributed

Kendra P. Rumbaugh (ed.), *Quorum Sensing: Methods and Protocols*, Methods in Molecular Biology, vol. 692, DOI 10.1007/978-1-60761-971-0_6,

across different microbial species and genera (2). The AI-2 molecule has been reported to influence the expression of a wide variety of genes responsible for virulence factors in *Streptococcus pneumonia*, *Streptococcus pyogenes*, motility in *Campylobacter jejuni*, and biofilm formation in *Streptococcus godonii*, *Bacillus cereus*, *Escherichia coli* K-12, *Streptococcus mutans*, and *Klebsiella pneumonia* (3). Studies in our laboratory have shown the presence of AI-2-like activity on the surface of frozen fish, in tofu, in milk samples, vegetable produce such as carrots, tomatoes, and cantaloupe (4). Interestingly, certain foods such as raw uncooked turkey patties, uncooked chicken breast, homemade cheeses, uncooked beef steak, and uncooked frozen beef patties did not exhibit AI-2-like activity (5). Later studies showed that there were inhibitory compounds primarily medium and long-chain fatty acids in meat and poultry products that inhibited the AI-2 bioassay (6, 7). Since the autoinducer bioassays are used for detecting autoinducers, we have used the convention of referring to such activity as autoinducer-like activity, for example, AI-2-like activity (5). The AI-2-like activity in food samples is detected by means of the *Vibrio harveyi* reporter strain BB170 originally designed by Bassler et al. (8). *V. harveyi* produces bioluminescence when the bacterium senses either AI-1 or AI-2 via their sensors LuxN and LuxPQ, respectively. In the reporter strain *V. harveyi* BB170, the sensor LuxN for recognizing AI-1 is mutated. Thus, BB170 is able to respond specifically only to AI-2 (2).

Briefly, the assay involves the collection of cell-free supernatant (CFS) of the test samples and the addition of the collected CFS to *V. harveyi* reporter strain (usually in 96-well microtiter plates). The light (bioluminescence) produced as a result of the presence of AI-2 in the different sample is measured using a luminometer plate reader. The appropriate positive and negative controls are included to verify the bioluminescence results. This is a relatively inexpensive assay that can be used to screen and quantify the relative levels of AI-2-like activity in different food and food ingredient samples. This is also a general protocol that can be used to quantify the amount of AI-2 or the AI-2 that is present in the CFS of different bacterial cultures. By including appropriate controls, it is also possible to identify foods and food ingredients that may be inhibiting the AI-2 assay. Understanding AI-2 assay inhibition can provide insight into possible AI-2 inhibitors that could be used to control microbial development.

2. Materials

2.1. Preparation of Food Samples

1. Food samples to be tested.
2. Cotton swab.

3. Whirl-Pak bags.
4. Sterile syringe filters, 0.2 μm pore size.
5. 10-ml Syringes.
6. Sterile 1.5-ml microcentrifuge tubes.
7. Luminometer, or equivalent device for measuring bioluminescence (490 nm).

2.2. Preparation of AI-2 Positive Control

2.2.1. In Vitro Synthesis of AI-2 Molecules

1. LB medium: 10 g/l tryptone, 5 g/l yeast extract, 10 g/l NaCl.
2. IPTG (1 M solution): 238 mg/ml in H_2O, filter sterilize and store in aliquots at −20°C.
3. Lysis buffer (1 l): 50 mM NaH_2PO_4–6.9 g $NaH_2PO_4{\cdot}H_2O$, 300 mM NaCl (17.54 g), 10 mM imidazole (0.68 g) into 800 ml distilled H_2O, pH to 8.0.
4. Wash buffer (1 l): 50 mM $NaH_2PO_4{\cdot}H_2O$ (6.9 g), 300 mM NaCl (17.54 g), 20 mM imidazole (0.68 g), pH 8.0.
5. Elution buffer (1 l): 50 mM $NaH_2PO_4{\cdot}H_2O$ (6.9 g), 300 mM NaCl (17.54 g), 250 mM imidazole (0.68 g), pH to 8.0.
6. Ni-NTA slurry (Qiagen, Hilden, Germany) – mix well before use.
7. *S*-adenosyl-L-homocysteine-10 mM (Sigma-Aldrich, USA) – prepare, filter sterilize, and store in aliquots at −20°C.
8. Lysozyme.
9. Ampicillin.
10. Conical flasks (100 ml and 2 l).
11. 50-ml Conical tubes.
12. Sterile spatulas.
13. Sonicator.
14. Magnetic stirrer.
15. Polypropylene column tubes.
16. Dialysis cassette.
17. 5 kDa Amicon centrifugal filter (Millipore, Billerica, Massachusetts).

2.2.2. Preparation of Positive and Negative Controls Using V. harveyi CFS

1. *V. harveyi* BB120 (ATCC-BAA-1117).
2. *V. harveyi* BB152 (ATCC-BAA-1119).
3. *V. harveyi* MM32 (ATCC-BAA-1121).
4. Autoinducer bioassay (AB) liquid medium (1 l): 17.5 g sodium chloride (0.3 M final), 12.3 g magnesium sulfate heptahydrate (50 mM final), 2.0 g casamino acids, vitamin-free (0.2% w/v final), pH to 7.5, and autoclaving at 121°C for 15 min. Cool down completely, and add 10 ml of 1 M potassium phosphate buffer (pH adjusted to 7.0 using 10 N sodium

hydroxide, sterilized by autoclaving), 10 ml of 0.1 M L-arginine (prepare it fresh and filter through 0.2-μm sterile filter), and 20 ml of 50% glycerol (sterilized by autoclaving).

5. Sterile culture tubes.
6. Inoculating loop.
7. Gas burner or laminar air flow chamber.
8. Sterile syringe filters, 0.2 μm pore size.
9. 10-ml Syringes.

2.3. Preparation of the Working Sample of the Reporter Strain V. harveyi BB170

1. Reporter strain *V. harveyi* BB170 (ATCC-BAA-1117).
2. Inoculating loop.
3. Sterile syringe filters, 0.2 μm pore size.
4. 10-ml Syringes.
5. 96-Well microtiter plate – white polystyrene flat bottomed plate for luminescence measurement (Whatman 7701-3350).
6. Luminometer or equivalent instrument for measuring light (Perkin Elmer-Wallac Victor 2 plate reader).
7. Petri plates.
8. 50-ml Conical tube.
9. 100-ml Erlenmeyer flask.

3. Methods

3.1. Processing of Food Samples: Vegetable Produce

1. Take a cotton swab and wet it thoroughly using AB medium, under aseptic conditions.
2. Swab the entire surface of test produce thoroughly with the AB medium-wetted cotton swab (see Note 1).
3. Resuspend the produce-swabbed cotton swab in tubes containing 3 ml of AB medium.
4. Aliquot 1 ml of the resuspended AB medium into three 1.5-ml centrifuge tubes.
5. Centrifuge 1 ml of aliquoted sample for 5 min (6,000×*g* at 4°C).
6. Carefully remove the supernatant and filter through 0.2 μm pore size syringe filter to obtain a CFS.
7. Store the CFS sample at –20°C until the AI-2 activity bioassay is performed.

3.2. Meat Products

1. Measure 50 g of meat aseptically and place it in sterile Whirl-Pak bags (see Note 2).
2. Add 10 ml of cold (4°C) AB medium to the sample bag.

3. Shake the sample gently for 1 min by hand.
4. Using a sterile 1-ml micro-pipette, pipette 1 ml of the sample and place it in sterile 1.5-ml microcentrifuge tubes.
5. Centrifuge the sample immediately for 5 min (6,000 × *g* at 4°C).
6. Carefully remove the supernatant and filter through 0.2 μm pore size syringe filter to obtain the CFS.
7. Store the CFS of samples at −20°C until the AI-2 activity bioassay is performed.

3.3. Liquid Food Samples, for Example, Milk

1. One milliliter of the liquid sample is placed directly into sterile 1.5-ml microcentrifuge tube.
2. Immediately centrifuge the samples for 5 min (6,000 × *g* at 4°C).
3. Carefully remove the supernatant and filter through 0.2 μm pore size syringe filter to obtain the CFS.
4. Store the CFS of samples at −20°C until the AI-2 activity bioassay is performed.

3.4. Preparation of the AI-2 Positive Control

3.4.1. In Vitro AI-2 Synthesis

The synthesis of the cell-to-cell signaling molecule AI-2 is catalyzed by two enzymes, LuxS and Pfs. Pfs is a nucleosidase that catalyzes the conversion of the substrate, *S*-adenosylhomocysteine (SAH) to *S*-ribosylhomocysteine (SRH) by cleaving the substrate's adenine base. Lux-S is the autoinducer synthase that converts *S*-ribosylhomocysteine into homocysteine and 4,5-dihydroxy-2,3-pentanedione (DPD). DPD is a highly unstable compound that spontaneously rearranges by reacting with water and produces the reactive AI-2 molecules. Here we describe the method to express and purify LuxS and Pfs in *E. coli*, carrying plasmids pVS212 and pVS214, respectively, by Ni-NTA chromatography and dialysis (9, 10). Under the following conditions: 1 mM SAH, 1 mg/ml of Pfs and LuxS for 15 min at 37°C, approximately 0.4 mM AI-2 is produced. The AI-2 is then purified using 5 kDa centrifugal filters and its activity is confirmed using a *V. harveyi* bioluminescence assay.

3.4.2. Protein Expression

1. Innoculate an *E. coli* strains containing the expression vectors pVS212 and pVS214 separately in 30 ml of LB broth containing 100 μg/ml ampicillin. Grow at 37°C overnight with vigorous shaking.
2. Innoculate a 1-l flask of LB broth containing 100 μg/ml ampicillin with 1:50 overnight culture (i.e., 20 ml overnight culture in 1 l of LB broth).
3. Incubate at 37°C with vigorous shaking until an OD_{600} of 0.6 is reached.

4. Take a 1 ml sample immediately before inducing with IPTG. This sample is the “non-induced control,” pellet the cells by centrifuging at 4,000 × *g* for 5 min in a tabletop centrifuge and resuspend in 50 μl of 2× SDS–PAGE sample buffer. Freeze until use at −20°C.
5. Induce the protein expression by adding IPTG to a final concentration of 1 mM.
6. Incubate the culture for 4–5 h at 37°C with vigorous shaking. Collect a second 1 ml sample. This is the “induced control,” pellet the cells by centrifuging at 4,000 × *g* for 5 min in a tabletop centrifuge and resuspend in 100 μl of 2× PAGE sample buffer. Freeze until use at −20°C.
7. Harvest the cells by centrifugation at 4,000 × *g* for 20 min at 4°C. Care should be taken that the culture is always maintained at 4°C during these steps.
8. Freeze the cells in dry ice–ethanol or liquid nitrogen, or store the cell pellet overnight at −20°C.

3.4.3. Protein Purification

1. Thaw the cell pellet for 15 min by keeping it on ice. Scrape the cell pellet using a spatula and place it in a 50-ml conical tube. Add lysis buffer at 2–5 ml/g wet weight of cell pellet and vortex the cells thoroughly (see Note 3).
2. Add 1 mg/ml lysozyme to the cell pellet-lysis buffer suspension and incubate on ice for 30 min. Vortex the cells thoroughly every 10 min.
3. After incubation, keep the conical tube with cell suspension on ice and sonicate the cells using a sonicator equipped with a microtip. Before sonication, make sure the microtip is surface sterilized with 70% isopropanol.
4. Sonicate six times with 10 s bursts at 200–300 W with a 10-s cooling period between each burst. Surface sterilize the microtip with 70% isopropanol before sonicating the second culture (*E. coli* containing pVS214).
5. This is an optional step. Add RNase A (10 μg/ml) and DNase I (5 μg/ml) if the lysate is very viscous and incubate on ice for 10–15 min.
6. Centrifuge the cell lysate at 10,000 × *g* for 20–30 min at 4°C to pellet the cellular debris. Save the supernatant at 4°C (see Note 4).
7. Add 1 ml of the 50% Ni-NTA slurry to 4 ml cleared lysate and mix gently by shaking (200 rpm on a rotary shaker) at 4°C for 1 h.
8. Add the lysate–Ni–NTA mixture into the 15-ml polypropylene column with the bottom outlet capped.

9. Remove bottom cap, apply mild pressure at top by using the column cap and collect the column flow through. Save flow through for SDS–PAGE analysis.
10. Wash the column twice with 4 ml wash buffer, collect wash fractions for SDS–PAGE analysis.
11. Elute the protein four times from the column with 0.5 ml elution buffer. Collect the elute in four tubes and analyze by SDS–PAGE.
12. Dialyze the two proteins separately overnight at 4°C in PBS using a dialysis cassette kept in a magnetic stirrer.
13. After dialysis, carefully remove the protein and keep it in ice.

3.4.4. In Vitro Synthesis of AI-2

1. Measure the amount of protein. Add 1 mg each of the purified enzymes to 10 mM of substrate, *S*-adenosyl-L-homocysteine (SAH) in a 50-ml conical tube.
2. Incubate the conical tube at 37°C water bath for 1 h with shaking.
3. After incubation, transfer the contents to a 5-kDa centrifugal filter (Amicon, Billerica, Massachusetts) and centrifuge it at 3,500 × *g* for 20 min.
4. Repeat the centrifugation if necessary, collect the flow through and store it at 4°C. Check its activity by bioluminescence assay before further use.

3.4.5. Preparation of Positive and Negative Control Using V. harveyi's CFS

1. Transfer 10 ml of AB media into a sterile culture tube.
2. Inoculate with a single bacterial colony of *V. harveyi* strain BB120 or *V. harveyi* BB152 on an inoculating loop, by dipping and shaking the loop in the medium (see Note 5).
3. Cap the tube and grow at 30°C to a mid-log phase (typically 3–4 h).
4. Centrifuge the culture at 10,000 × *g* for 5 min. Carefully remove the supernatant and pass it through a 0.2-μm filter using 10-ml syringe.
5. Collect the CFS and store this sample at −20°C. Use this CFS as the positive control for autoinducer bioluminescence assay (see Note 6).
6. Repeat the same procedure described as above to prepare negative control by inoculating *V. harveyi* MM32 and collect the CFS.

3.5. Preparation of the Working Sample of the Reporter Strain V. harveyi BB170

1. Inoculate a loopful or 10 μl of *V. harveyi* strain BB170 from a frozen glycerol stock into 5 ml of AB medium (see Note 7).
2. Incubate and grow overnight at 30°C with shaking until the culture reaches its late log phase and turbid (OD_{600} = 0.7–1.2) (see Note 8).

3. Dilute the *V. harveyi* overnight culture 1:5,000 into fresh AB medium by transferring 5 μl of the overnight culture into 25 ml of fresh AB medium (see Note 9).
4. Transfer 90 μl of diluted reporter cells (1:5,000) into each used well of the 96-well plate. Use 3–5 wells per sample to be tested.
5. Add 10 μl of the sample (CFS of food samples) to be tested into the 96-well plate containing 90 μl of diluted reporter cells (see Note 10).
6. Include the following control samples in triplicate:
   (a) Positive control: 10% CFS (10 μl of CFS from *V. harveyi* BB152 + 90 μl of diluted reporter cells).
   (b) Negative control: 10% AB medium (10 μl of AB medium + 90 μl of diluted reporter cells) (see Note 11).
7. Incubate the plate at 30°C with shaking at 100 rpm.
8. Measure the luminescence produced by each sample using a luminometer. Take measurements at every 30 min after 3–3:30 h of incubation until the average values of the negative control reaches 100.
9. After 3–3:30 h of incubation, measure the luminescence produced by each sample at 30 min intervals using a luminometer until the average value of the negative control reaches 100 (see Note 12).
10. The relative AI-2 activity is calculated as follows:

    Relative AI-2 activity = average value of sample/average value of negative control.

    If the AI-2 inhibition assay is performed, its relative inhibition can be calculated as: [1/(average value of sample/average value of positive control)].
11. For trouble-shooting, see Table 1 in Subheading 4.

4. Notes

1. Swab only the desired area of the produce sample if the objective is to detect the presence of AI-2-like activity at a specific location.
2. For larger sample volumes or amounts such as whole beef patties, turkey patties, and cheese blocks, add 20 ml of cold AB medium instead of 10 ml and follow the procedure same as processing of meat products. Note: the exact volume of AB

Table 1
Trouble shooting BB170 assay

Problem	Cause	Solution
After incubation, luminescence is not detected	The BB170 culture is turbid with no luminescence. Possible cross contamination of the culture	If the culture is turbid and no light is detected, begin the assay with the freshly prepared AB medium. Ensure that the inoculating loop, glass wares, and other materials are sterile before inoculating the culture and inoculate it in the laminar air flow chamber to prevent contamination
	The AB media is not prepared properly	Prepare fresh AB medium with all the necessary ingredients and adjust the pH
	The incubator temperature is not set to 30°C	Ensure that the incubator temperature is set to exactly 30°C as *V. harveyi* does not produce light at temperature higher than 30°C
	The culture from the glycerol stock is not active with viable cells	Before starting the autoinducer assay, confirm that the *V. harveyi* inoculated from the glycerol stock produces light by examining under dark
All cell-free supernatant samples including the negative control and the medium control produced light in the bioassay	The luminometer or the 96-well microtiter plate reader is not functioning properly	The culture produces luminescence when examined in the dark. But if no luminescence is detected using the luminometer, check the settings of the luminometer or contact the manufacturer
	The AB media and all samples could be contaminated with AI-2 producing bacteria	Prepare fresh sterile AB medium and perform the bioassay, as the AI-2 producing bacterial contaminants present in the media can induce *V. harveyi* bioluminescence
	The duration of the bioassay has exceeded the required incubation time	3 h after the bioassay has started, take the readings every 30 min and stop the assay when the values of negative control shows around 100
The bacterial medium controls the production of luminescence, whereas negative control does not	Too many cells were present in the initial dilution of *V. harveyi* BB170	Do not grow the initial *V. harveyi* BB170 culture longer than 14 h. The assay can be performed alternatively with 1:10,000 initial dilution of a turbid culture

(continued)

Table 1 (continued)

Problem	Cause	Solution
The cell-free supernatant did not contain AI-2, i.e., the bacteria did not produce AI-2 molecules	Wrong plates are being used	Always use flat-bottomed white 96-well microtiter plate for the bioluminescence assay
	Bioluminescence in *V. harveyi* is induced when the bacterial growth medium has boron or any other component that induces its bioluminescence	If the bioluminescence is induced even in the absence of boric acid in media, use the cell-free supernatant of bacteria grown in a different medium. If boric acid is a necessary ingredient in the bacterial growth medium then reduce the amount of boric acid to less than 10 μM
The sample in triplicates did not produce consistent light output as the luminescence values are widely varied	The cell-free supernatant did not contain AI-2, i.e., the bacteria did not produce AI-2 molecules	Alter bacterial growth parameters such as growth medium, temperature, growth factors, etc.
	The bacterial medium has the component, which inhibits the bioluminescence or hinders the growth of *V. harveyi*	Check the medium control. If it does not produce luminescence then grow the bacterium in a different medium and collect the cell-free supernatant
	The bacteria secrete a component that inhibits the bioluminescence or hinders the growth of *V. harveyi*	Collect the cell-free supernatant under the conditions the bacteria does not produce the inhibiting component. Reduce the amount of CFS in bioassay to 1%
	The *V. harveyi* culture is not grown under proper aerated conditions	Aeration is an important parameter for the growth and luminescence of *V. harveyi*. The luciferase enzymatic reaction requires aeration to produce bioluminescence. Grow *V. harveyi* culture with proper aeration
	The assay is unstable if lower concentrations of AI-2 is present in the sample and will not be detected properly	To improve the reproducibility and reduced variance of the bioassay, perform the modified protocol of preparing fresh working solution (AB-Fe medium with 1.2 mM of iron)
	The 96-well plate reader or the luminometer is not optimized properly to measure bioluminescence accurately	Before measuring the bioluminescence, examine that the luminometer's settings are adjusted (490 nm) and optimized properly according to the manufacturer's instructions

medium that should be used will depend on the sample amount. This volume may require optimization depending on the experimental objectives.

3. Wet weight can be calculated by the difference in weight of conical tube with cell pellet and an empty conical tube. Wet weight of cell pellet = weight of conical tube with cell pellet − weight of an empty conical tube.
4. Add 5 μl 2× SDS–PAGE sample buffer to 5 μl supernatant and store at −20°C for SDS–PAGE analysis, if so desired.
5. Inoculate the culture aseptically in a laminar air flow chamber or in front of Bunsen burner.
6. The same CFS can be used as the "preformed AI-2 CFS" for studying the AI-2 inhibitory activity.
7. It is preferred to inoculate the culture from frozen glycerol stock instead of inoculating by picking up a colony or from liquid culture. It is considered that cultures from glycerol stock produce maximum light. Always store *V. harveyi* culture at −20°C as storage in a cold room or refrigerator may drastically reduce cell viability.
8. Before starting the bioassay, ensure that the overnight culture is brightly luminescent by examining it in a dark room or check its luminescence using a luminometer.
9. The required volume of diluted reporter cells depends on the number of samples to be analyzed.
10. For detecting the presence of any "AI-2 inhibitory activity," add 5 μl of the sample to 5 μl of preformed AI-2 CFS (CFS obtained from *V. harveyi* BB152) or the in vitro synthesized AI-2. Mix well and add this sample mixture to 90 μl of diluted reporter cells.
11. Include a media control (10 μl of bacteria growth medium + 90 μl of diluted reporter cells), if the assay is performed to determine the growth conditions under which AI-2 is produced.
12. It is extremely important to monitor the readings of the negative control. After 3 h of incubation, the luminescence has to be measured at 30-min intervals. Once the value of the negative control starts increasing, it is better to take readings at 10–15-min intervals until the average of the negative control reaches 100. The assay must be stopped once the average value reaches 100. If the average value is above 100, use the luminescence values from the previous reading for data analysis (Fig. 1).

65	84	78	489	410	388	338	337	419	359	294	209
10295	9232	8275	600	547	518	391	475	472	330	302	221
617	653	642	913	858	704	1450	794	771	588	545	525
1334	1151	770	1184	1130	872	877	719	834	155	160	112
61	616	633	621	516	454	397	402	361	147	123	149
1207	1195	56	363	101	739	315	516	457	472	405	477
1630	1006	373	2318	1968	1840	1579	900	1413	653	605	563
2555	389	2203	2424	2778	2512	1241	1272	989	46	66	56

Average value of sample = (10295+9232+8275)/3 =9267.33

Average value of Negative control = (65+84+78)/3 =75.66

Relative AI-2 activity = Average value of sample/Average value of negative control.

= 9267.33/75.66

= 122.48 RLU

Fig. 1. Example luminescence values and calculations.

References

1. Lonn-Stensrud, J., Petersen, F. C., Benneche, T., and Scheie, A. A. (2007) Synthetic bromated furanone inhibits autoinducer-2-mediated communication and biofilm formation in oral streptococci, *Oral Microbiol Immunol* **22**, 340–346.
2. Taga, M. E. (2005) Methods for analysis of bacterial autoinducer-2 production, *Curr Prot Microbiol, Chapter 1*, 1C. 1.
3. Gonzalez Barrios, A. F., Zuo, R., Hashimoto, Y., Yang, L., Bentley, W. E., and Wood, T. K. (2006) Autoinducer 2 controls biofilm formation in *Escherichia coli* through a novel motility quorum-sensing regulator (MqsR, B3022), *J Bacteriol* **188**, 305–316.
4. Lu, L., Hume, M. E., and Pillai, S. D. (2005) Autoinducer-2-like activity on vegetable produce and its potential involvement in bacterial biofilm formation on tomatoes, *Foodborne Pathog Dis* **2**, 242–249.
5. Lu, L., Hume, M. E., and Pillai, S. D. (2004) Autoinducer-2-like activity associated with foods and its interaction with food additives, *J Food Prot* **67**, 1457–1462.
6. Widmer, K. W., Soni, K. A., Hume, M. E., Beier, R. C., Jesudhasan, P., and Pillai, S. D. (2007) Identification of poultry meat-derived fatty acids functioning as quorum sensing signal inhibitors to autoinducer-2 (AI-2), *J Food Sci* **72**, M363–M368.
7. Soni, K. A., Jesudhasan, P., Cepeda, M., Widmer, K., Jayaprakasha, G. K., Patil, B. S., Hume, M. E., and Pillai, S. D. (2008) Identification of ground beef-derived fatty acid inhibitors of autoinducer-2-based cell signaling, *J Food Prot* **71**, 134–138.
8. Bassler, B. L., Wright, M., Showalter, R. E., and Silverman, M. R. (1993) Intercellular signalling in *Vibrio harveyi*: sequence and function of genes regulating expression of luminescence, *Mol Microbiol* **9**, 773–786.
9. Schauder, S., Shokat, K., Surette, M. G., and Bassler, B. L. (2001) The LuxS family of bacterial autoinducers: biosynthesis of a novel quorum-sensing signal molecule, *Mol Microbiol* **41**, 463–476.
10. Sperandio, V., Torres, A. G., Jarvis, B., Nataro, J. P., and Kaper, J. B. (2003) Bacteria-host communication: the language of hormones, *Proc Natl Acad Sci USA* **100**, 8951–8956.

Chapter 7

Detection of Bacterial Signaling Molecules in Liquid or Gaseous Environments

Peter Edmonson, Desmond Stubbs, and William Hunt

Abstract

The detection of bacterial signaling molecules in liquid or gaseous environments has been occurring in nature for billions of years. More recently, man-made materials and systems has also allowed for the detection of small molecules in liquid or gaseous environments. This chapter will outline some examples of these man-made detection systems by detailing several acoustic-wave sensor systems applicable to quorum sensing. More importantly though, a comparison will be made between existing bacterial quorum sensing signaling systems, such as the *Vibrio harveyi* two-component system and that of man-made detection systems, such as acoustic-wave sensor systems and digital communication receivers similar to those used in simple cell phone technology.

It will be demonstrated that the system block diagrams for either bacterial quorum sensing systems or man-made detection systems are all very similar, and that the established modeling techniques for digital communications and acoustic-wave sensors can also be transformed to quorum sensing systems.

Key words: Acoustic wave biosensors, State-space mapping, RFID/biosensors, Chemically orthogonal antibodies, Antibody promiscuity, *Vibrio harveyi* two-component model

1. Introduction

This chapter details several techniques based on acoustic wave devices for the non-invasive, detection of microorganisms in both the liquid and vapor phases. This is a real-time detection method, which is reliable, specific and easy to use. It is a detection method that takes a radically different and innovative approach than most currently established techniques. Rather than detect the presence of the microbe as is done in such techniques as PCR or immunocapture, our approach is to identify the microbes and their activities by detecting the signaling molecules being secreted by microbes (1). These so-called quorum sensing molecules represent

Kendra P. Rumbaugh (ed.), *Quorum Sensing: Methods and Protocols*, Methods in Molecular Biology, vol. 692,
DOI 10.1007/978-1-60761-971-0_7, © Springer Science+Business Media, LLC 2011

the communication signals within specific microbial communities. In this chapter, both the acoustic wave detection systems and the microbes themselves will be modeled with digital radio communication techniques whereby several input stimuli signals are presented to the detector or microbe simultaneously, but only certain selected stimuli signals are accepted to generate a response.

The basis of this model stems from the ability of a digital radio system to identify and differentiate from the many analogous inputs presented to the system. For example, if your cell phone uses a CDMA (code division multiple access) communication protocol, the signals arriving at the antenna all look very similar as they are sequences of ones and zeroes, but only the signals with the properly coded ones and zeros can be decoded by your cell phone. The other signals just look like noise.

Further, we propose that the analogous inputs can then be processed and positioned within a state-space map. The structures of the state-space map, which are populated with these signals, are seen to indicate differences in the type of inputs present. This method of demodulating and classifying input data has been well studied within the area of digital communications (2, 3). This chapter presents an expansion of this concept to include acoustic wave biosensor detection systems and the bioluminescent marine bacterium *Vibrio harveyi.*

1.1. Acoustic Wave Biosensors

Almost every biomolecular event in living systems involves the following three principle components.

1. *Molecular recognition* – the lock and key interaction whereby one biomolecule or receptor (e.g., a protein) recognizes with a high degree of specificity another molecule. In the case of electrophysiology, this extends to the recognition of an ion, say Na^+, by a channel protein which has been incorporated into the plasma membrane.
2. *Conformational change* – the change in the molecular structure of the receptor molecule. At times it helps to think of this as the chemical phase change of the molecule. No additional chemical groups have been added to the molecule, but the internal structure of the molecule has changed. Condensed matter physics is replete with examples of crystal structure radically affecting macroscopic physical characteristics (e.g. crystalline silicon vs. polysilicon).
3. The *hydrolysis of nucleotide triphosphates* (ATP, GTP, UTP, and CTP) as an energy source.

Acoustic wave biosensors are a sensor technology well suited for the translation of the first two principles of the canon into electrical signals (4, 5). Combined, these principles manifest themselves as mass attachment to the sensor surface and stiffness changes in the biological receptor layer. These in turn will shift the resonant frequency of the device. Variations in the affinity of the receptor molecule to a collection of analytes within a class of biochemicals (e.g. estrogens) will alter the time course of the frequency signature. When the affinity is large as is the case for monoclonal antibody-antigen interactions, the analyte will bind tightly to the immobilized antibody resulting in a baseline frequency shift for the sensor. Dissociation constants, for these antibody and antigen interactions tend to be in the picomolar (pM) range. When the analyte is a chemical analog of the original antigen against which the monoclonal antibody was generated, the affinity is not so high. In immunology, this concept is referred to as antibody promiscuity (6–10). Frequency-offset biosensors based on acoustic wave devices are known to provide extremely high sensitivity and selectivity where the target is detected and identified based on the amount of frequency shift. Typically these acoustic wave biosensors are in the form of an oscillator based detection system. However, acoustic wave detection systems can also be constructed based on time and phase shifts in a return signal or by incorporating communications radar technology such as signal interrogation and correlation techniques (11).

1.2. Digital Radio Communications Techniques and Methods

The details of the similarities between a multiple-channel acoustic-wave biosensor, a two-component quorum sensing system, and a digital radio receiver are herein described. All three systems accept multiple analogous types of signal inputs, yet identify and differentiate amongst specific conditions and responses that these signals impose. The concept of state-space mapping will also be introduced where the multiple analogous types of signal inputs are identified and differentiated into functional data clusters such that each cluster has a specific role and outcome. One of the key mechanisms of state-space mapping system is the development of an orthogonal channel separation system that can separate the input signals into their orthogonal *x* and *y* components.

Digital radio has utilized this concept to increase the data content of its transmitted signals in order to effectively map the data into distinct clusters (3). One such system is quadrature amplitude modulation (QAM). The mapping generates a so-called constellation diagram. Digital communications receivers selectively detect various groups of communication signals. These groups can be regionalized depending on their chosen method of modulation. A typical digital communication receiver system is illustrated in Fig. 1. A communication input signal that could contain a multitude of modulation schemes and noise is presented to the digital communication receiver system. The specific artificial intelligence embedded within the hardware and software of the digital communication receiver system differentiates and identifies the desired group of signals using the in-phase (I) channel detector and the quadrature-phase (Q) channel detector.

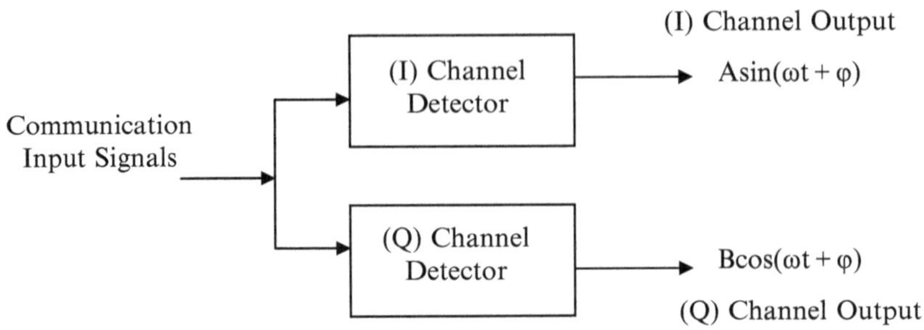

Fig. 1. A typical digital communication receiver.

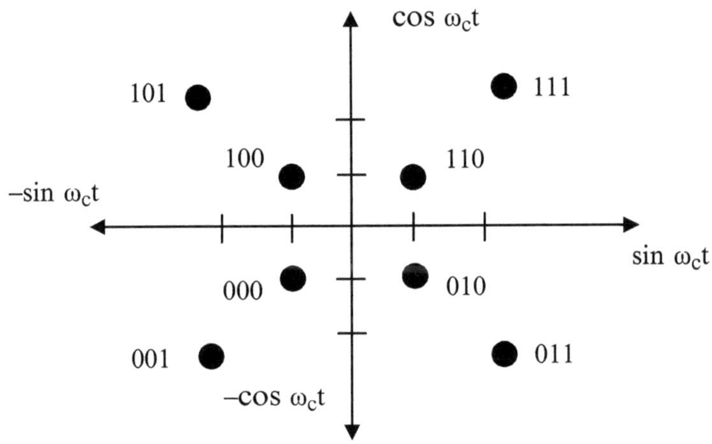

Fig. 2. A digital communications constellation diagram for an 8-QAM communication detection system.

The (I) channel output would have a signal comparable to $A\sin(\omega t+\varphi)$ and the (Q) channel output would have a signal comparable to $B\cos(\omega t+\varphi)$. These two orthogonal outputs would then be used as the values mapped to the coordinates of the magnitude-phase constellation plot. Figure 2 illustrates a complex communication system constellation diagram of an 8-level quadrature amplitude M-ary (8-QAM) encoding scheme. Here, the digital information is contained in the amplitude (A), frequency (ω) and phase (φ) of the detected signals with only the peak values being shown as filled-in circles.

A similar method is also used to exploit an acoustic wave biosensor system that incorporates chemically orthogonal antibodies as the biolayer detection components within multi-channel system configurations. For a two-dimensional detection system, input substances are simultaneously presented to both the X channel detector and Y channel detector. The X channel detector has a biolayer with X-type antibodies, and the Y channel detector has a biolayer with Y-type antibodies. The X-channel output signal $A\sin(\omega_X t)$, will depend on the binding action between the input substances and antibody X within the X channel detector and similarly, the Y-channel output signal $B\sin(\omega_Y t)$, will depend on the binding action between the input substances and antibody Y within the Y channel detector. The outputs of both channels are then mapped.

A model of a two-component system, such as that found in the *V. harveyi* quorum sensing, will also be presented. This model incorporates both the cross-reactivity of analogous autoinducers with that of the mathematical vector function called the cross product, to describe how varying amounts of different autoinducers can alter the output gene response. Autoinducers have been identified as extra- and intracellular signaling molecules, that play an important role in controlling complex processes including multicellularity, biofilm formation, and virulence (12).

Cross-reactivity will explain how the utilization of a common moiety with side chain variations can assist during detection, in the identification and differentiation of various autoinducer-signaling molecules. Further studies by Rumbaugh have shown that autoinducers that exhibit similar structures can influence mechanisms within the organism that allow them to sense and respond to each other's signaling molecules (13–16). The model implies that the cross-reactivity events occur within the separate LuxN and LuxQ channels and the cross-product function occurs within the LuxU region.

2. Materials

In this section, we describe a variety of acoustic wave biosensor system configurations. At the core of all of these approaches is the transduction of molecular recognition and conformational change into an electrical signal. These various approaches all include a high frequency acoustic wave device constructed on a piezoelectric material (e.g. ST-Quartz) with receptor molecules immobilized onto its surface. The transduction process and mapping to an electrical signal varies then by how many acoustic wave biosensor elements are in the particular system, how they are configured and how the electrical signal is extracted. The following is a description of a selected group of these configurations.

2.1. Oscillator Based Systems

Frequency-offset biosensors based on acoustic wave devices are known to provide extremely high sensitivity and selectivity, where the target is detected and identified based on the amount of frequency shift. The signal output of an acoustic wave oscillator follows Eq. 1,

$$a(t) = Asin(2\pi f_o t) \qquad (1)$$

where, A is the amplitude of the output that is determined by the combination of the oscillator loop amplifier, and losses and f_0 is the free-running frequency of the oscillator loop that is primarily determined by the frequency response of the acoustic wave device (see Note 1).

For the application of an acoustic wave oscillator sensor, the acoustic wave device is injected with an input stimuli that can be physical (temperature or pressure), chemical (explosives or cocaine), or biological (autoinducer signaling molecules), that the specific sensor is design to detect. As the injected input stimuli interfaces with the acoustic wave device, the acoustic wave that propagates within the acoustic wave device is subjected to a modification of its acoustic velocity. This change in velocity transcribes into a frequency change as shown in the modified Sauerbrey equation 2 as included in the publication by Hunt et al. (5),

$$\Delta f = -\frac{2f_u^2 h_f}{\sqrt{\rho_q \mu_q}}\left[\Delta\rho - \frac{\Delta\mu}{V_s^2}\right] \quad (2)$$

where V_s is the acoustic velocity; ρ is the density of the film; h_f is the thickness of the film; μ_q and ρ_q are the shear stiffness and density of the piezoelectric crystal, respectively; μ is the stiffness of the film; Δ is the difference between perturbed and unperturbed (denoted by subscript u) quantities. The stiffness of the film, μ, is affected by the conformational change of the recognition molecules (*see* **Note 2**).

Oscillator based sensor detection systems also present some operational concerns. The first concern involves the stability of the oscillator due to the thermal drift and load pulling of the amplifier portion of the circuit. The second concern is the instability due to possible coupling of modes between adjacent oscillators that would introduce injection-locking phenomena from stray coupling within the oscillator circuits. The largest concern that an oscillator detection system has is the loss of possible information of the detected substances due to the averaging effect of the oscillator (17).

2.2. Ladder Based Systems

This section addresses the issue of implementing multiple arrays of biosensors in a simple fashion. Each element of the biosensor array would consist of an independent measuring biolayer, therefore allowing for the whole array to measure a multitude of biomolecules or to improve the statistical analysis and measure duplicate biomolecules. Each ladder-based structure is passive to eliminate any instability found in active circuits, eliminates any averaging effects found in oscillator sensor circuits, and introduces a means to include sensed information obtained over a swept frequency range. The composition of this structure includes

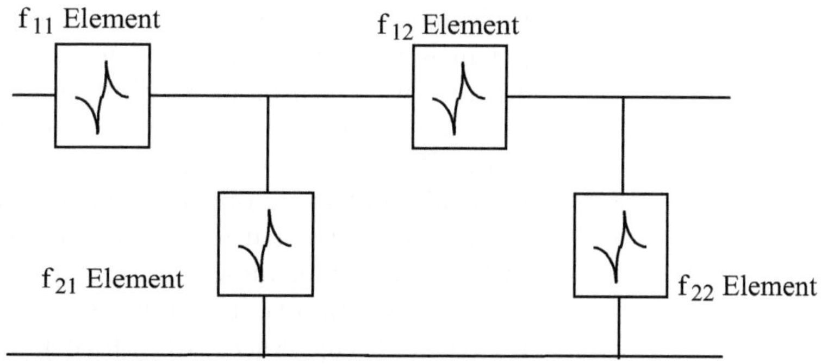

Fig. 3. A schematic of a 4-element ladder structure.

cascading certain resonant structures as illustrated in Fig. 3, which includes micro-electrical-mechanical-systems (MEMS), such as thin film bulk acoustic resonators (FBARs), surface acoustic wave (SAW) resonators, and other acoustic wave resonators such as bulk acoustic wave (BAW), leaky surface acoustic wave (LSAW) and other known acoustic modes of propagation (11, 18).

Experimental data from ladder type structures including up to a 9-element ZnO FBAR based ladder sensor have confirmed that output response parameters such as magnitude, phase and frequency changes derived from a swept frequency response is enhanced when compared to an oscillator based detection process (19–22). Several of these ladder sensors can also be multiplexed to produce large sensor arrays of 2^n sensors where $n \approx 8$.

2.3. Neural-Network Based Systems

Another variation of an acoustic wave oscillator sensor is a Neural Network (NN) type of configuration (23). Neurons typically consist of axons feeding dendrites through synapses. The operation of such neurons is highly parallel, with each network element performing independently. The neuron is a simple element consisting of nodes and links that is part of a more complex network with each simple element performing as an independent processor. Within a simple neuron physiology model, dendrites convey input stimuli to the cell body, and the axon conducts impulses away from the cell body. The neuron has a distribution of ions both on its inside and outside. An action potential is a very rapid change in the distribution of these ions, resulting when the neuron is stimulated. Neurons typically adhere to the "All-or-None Law" in which action potentials occur maximally or not at all. The input stimulus either activates the action potential, or it is not achieved, and no action potential occurs.

This very low-cost biosensor NN system is based on a hardware acoustic wave structure, and contains simple electronic components that are no more complicated than an amplifier. There is no need for digital signal processing (DSP) to generate a detection event, as the network is self-organized, and the signaling molecules convey input stimuli to the acoustic wave sensor, and perturbs the operating frequency as outlined in Eq. 2 of the acoustic wave sensor in a fashion such that the oscillator then oscillates and conducts a detection signal. An acoustic wave biosensor NN system will detect, in real-time, a specific signaling molecule and can easily be expanded to include several more acoustic wave resonators, all cascaded in series to detect several different signaling molecules. Our prior work on acoustic wave based NN systems indicates an effective processing performance of 1 Gigaflop/Watt, which greatly exceeds most supercomputers.

2.4. RFID Based Systems

Acoustic wave biosensors can also be configured as small transponders, similar to the radio frequency identification devices (RFIDs), which are located within merchandise or credit and debit cards (24).

The major advantage of these acoustic wave RFID/biosensors are that they are wireless, and therefore can be easily interrogated within distances ranging from a few centimeters to kilometers when properly configured. Figure 4 illustrates a schematic of a simple reflective type of acoustic wave RFID/biosensor. An antenna receives a radio frequency (RF) interrogation signal (f_o), and the input/output transducer transforms the RF signal into an acoustic wave signal. Since the input/output transducer is bi-directional, incident acoustic waves propagate out from either end. A reference reflector array, located on the left side of the input/output transducer, and then reflects the incident acoustic waves back towards the input/output transducer. These reflected waves from the reference reflector array retain the frequency characteristics of the original interrogation signal (f_o),

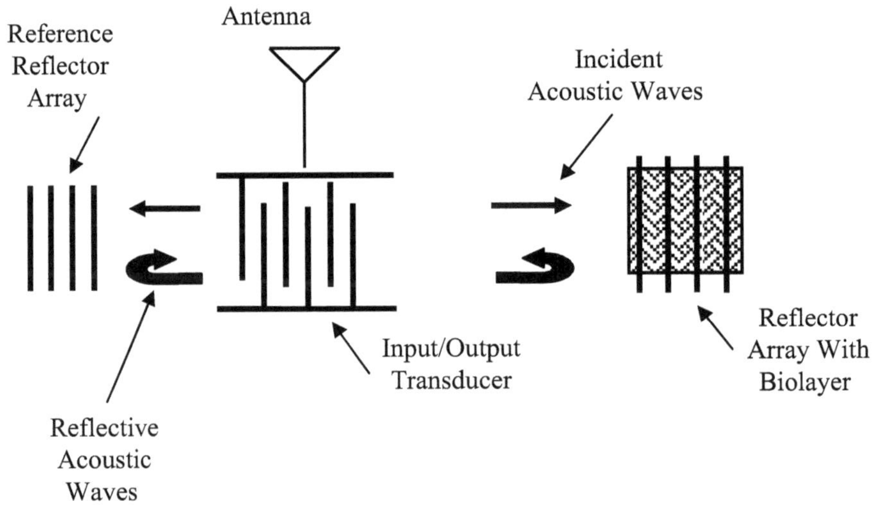

Fig. 4. A schematic of a simple reflective type of acoustic wave RFID/biosensor.

and are transformed back to an RF signal and retransmitted back to the interrogation unit via the antenna. Meanwhile, the incident acoustic waves propagating from the right side of the input/output transducer will then reflect from the reflector array containing the biolayer. Again, an effect will take place following the relationship outlined in Eq. 2 and the reflective wave will have a frequency of $f_r = f_0 - \Delta f$, where Δf is a function of the molecular binding taking affect within the biolayer. Since the distance is greater to the reflector array containing the biolayer than the reference reflector array, there will be no "collision" of waves as they reach the input/output transducer. Therefore, the interrogation unit will actually see multiple signals returning where the first set of signals are due to the reference reflector array at f_0 and the next set of signals will be due to the reflector array containing the biolayer at $f_r = f_0 - \Delta f$. Circuitry within the interrogation unit can then determine Δf, which corresponds to the concentration of the specific signaling molecule.

A further advantage that an RFID/biosensor has over an oscillator based biosensor is the different measurement parameters it can have. An RFID/biosensor can also detect a delta frequency (Δf) along with other parameter changes due to this change in velocity such as, change in time (Δt), change in phase ($\Delta \varphi$) or a change in the correlation pattern (Δc) (see Note 3).

3. Methods

This section will describe in detail the methods by which the molecular recognition-conformational change events are mapped into a signal space that both facilitates detection and discrimination and elucidates some of the intricacies of quorum sensing.

3.1. State-Space Mapping Techniques for Identification and Differentiation

This section will explain the similarities between a multiple-channel biosensor, a multiple-component quorum sensing system, and a digital radio receiver. All three systems accept multiple orthogonal type of signal inputs, yet identify and differentiate specific conditions that these signals impose. The concept of state-space mapping will also be introduced where the multiple orthogonal type input signals are identified and differentiated into functional data clusters such that each cluster has a specific role.

One of the key mechanisms of state-space mapping system is the development of the orthogonal channel separation system that can separate the input signals into their orthogonal x and y components. Digital radio has utilized this concept to increase the data content of its transmitted signals and effectively mapping the data into distinct clusters (2, 3). This concept is further exploited by an acoustic wave biosensor system that incorporates

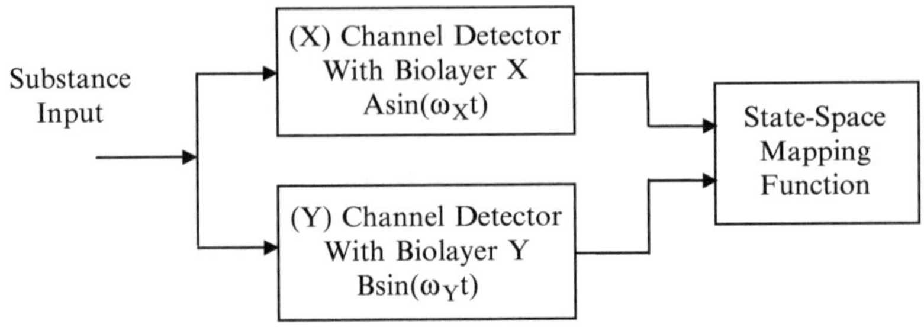

Fig. 5. A two channel biosensor state-space mapping detection system.

antibodies as the biolayer detection components as shown in a two-dimensional orthogonal biosensor state-space mapping detection system of Fig. 5 (4). Input substances are presented to the system and are simultaneously available to both the X channel detector and Y channel detector. The X channel detector has a biolayer with X-type antibodies and the Y channel detector has a biolayer with Y-type antibodies. The X-channel output signal $A\sin(\omega_X t)$, will depend on the binding action between the input substances and antibody X within the X channel detector and similarly, the Y-channel output signal $B\sin(\omega_Y t)$, will depend on the binding action between the input substances and antibody Y within the Y channel detector. The outputs of both channels are then mapped. The ability of antibody X within the X detector to cross react with multiple antigens is known as the promiscuity of the antibody. This conformational diversity allows related groups of substances to bind with the antibody. The ability of an antibody to recognize multiple epitopes allows for the binding of analogous chemical or biological groups. This binding of structural analogs evolves from variations in conformational heterogeneity of the combining site, which controls both the affinity and specificity of the site (7).

The concept of analogous substances cross-reacting with each other due to their similar structures is quite common (see Note 4). Problems are encountered at airports where conventional mobility spectrometers searching for explosives and trace levels of chemical warfare agents can't always determine the intended target signal out of the many other chemicals in the environment, such as perfumes, and may be susceptible to false positives, causing delays and passenger frustration.

Similarly, within bacterial quorum sensing systems, the bacterial autoinducers control gene expression in the bacterial cells, but also alter the gene expression in mammalian cells due to the similar structural interface between the bacterial autoinducer and the mammalian host cell. This "cross-reactivity" of analogous structures may lead to a modification of cellular activities and an increase in bacterial pathogenisis (16).

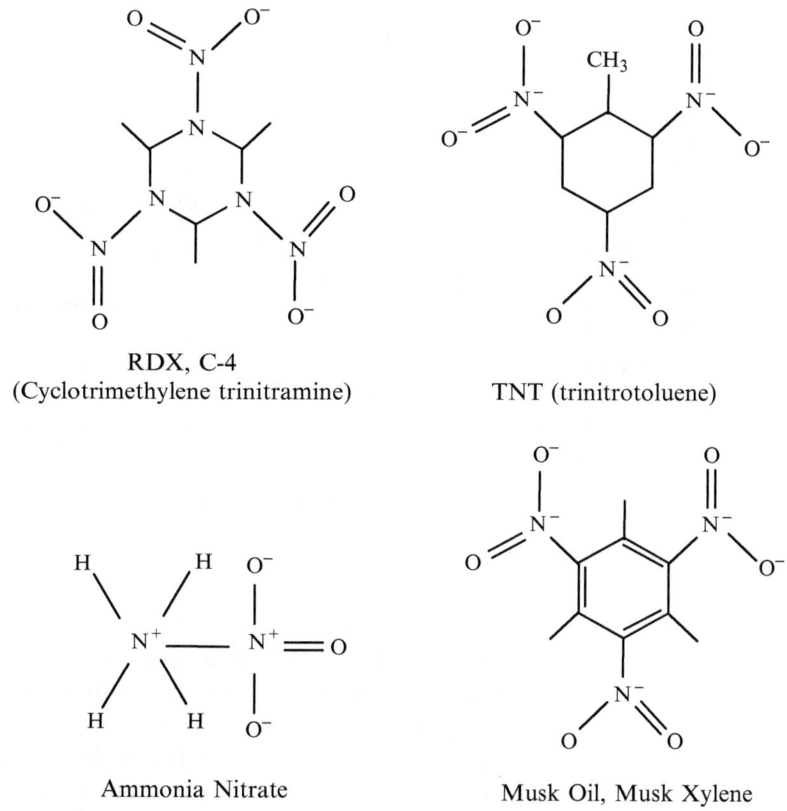

Fig. 6. Nitro based analogous substances.

The notion of state-space mapping for the identification and differentiation of analogous substances that are either orthogonal or semi-orthogonal can be explained via experimental data involving explosive substances and a common interferer. The analogous substances in this case are related via an NO_2 branch and included, Trinitrotoluene (TNT), Cyclotrimethylenetrinitramine (RDX – acronym derived from Royal Demolition Explosive), Musk Oil or Musk Xylene and ammonium nitrate (AN) and are depicted in Fig. 6. All of these nitro-based substances bind differently with respect to TNT antibodies and RDX antibodies. A two-dimensional biosensor detection system previously shown in Fig. 5 was constructed, and input substances were presented separately to the system at various distances and configurations from the biosensor input sampling head. A pneumatic system would draw through an unheated 5-micron filter the input substances into the detector system, where the X channel detector implemented the TNT antibody layer and the Y channel detector implemented the RDX antibody layer. The frequency component of the X-channel output signal $A\sin(\omega_X t)$ and the frequency component of the Y-channel output signal $B\sin(\omega_Y t)$ were then stored.

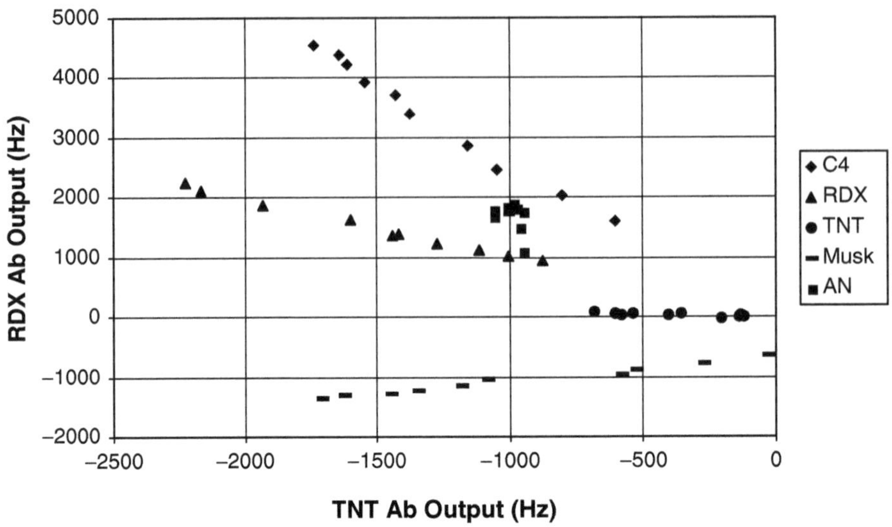

Fig. 7. State-space diagram of analogous nitro-based substances.

A nitro-based signal state-space map was constructed and is shown in Fig. 7. The state-space map *x*-axis is comprised of the frequency component of the X-channel output signal $A\sin(\omega_X t)$, and the *y*-axis is comprised of the frequency component of the Y-channel output signal $B\sin(\omega_Y t)$ of Fig. 5. It is clearly shown that each substance is distinctively mapped onto a region of the signal state-space map. This was achieved with a minimal amount of calculation and with no matrix or intricate mathematical computation. The difference in magnitude between analogous substances can also be determined from the signal state-space map of Fig. 7. The C4 substance was a larger sample (>1 g) when compared with the RDX substance, which contained 50.3 pg. This is illustrated within Fig. 7 by the C4 data having higher coordinates values with respect to the RDX substance. Even though the RDX and C4 explosive illustrated in Fig. 6 are depicted as the same, in the real world, these two explosives could vary slightly and that is why the two clusters identifying RDX and C4 in the state-space map of Fig. 7 are similar but distinct. It should also be recognized that the signal state-space map of Fig. 7 only contains ten samples of each substance. These samples were acquired during the transient stage of the pneumatic system at 1 second intervals. Even with this short accumulation of data, clear and defined regions appear on the map that involved a very low computational effort. The sampling rate can range from milliseconds to tens or hundreds of seconds depending on the application.

3.2. The Cross-Reactivity and Cross-Product Model of a V. harveyi Quorum Sensing System

Previous studies have shown that within quorum sensing, bacteria communicate with one another by the exchange of chemical signals called autoinducers. In the bioluminescent marine bacterium *V. harveyi*, two different autoinducers (AI-1 and AI-2) regulate the light emission via a two-component system (25). A block diagram of *V. harveyi's* two-component sensing system is shown in Fig. 8. If there is an absence of the autoinducers AI-1 or AI-2 then LuxU transfers phosphate onto LuxO that in turn activates the regulation function such that an output of five regulatory small RNAs (sRNAs) called Qrrl–5 (Quorum Regulatory RNA) occurs. During the presence of AI-1 and AI-2 a dephosphorylation of LuxU and LuxO takes place, which deactivates the regulation function such that no *qrr* sRNA expression occurs.

The functionality of this two-component system strikes a remarkable resemblance to that of a two-channel biosensor state-space mapping detection system from the previous Fig. 5. The performance of this system depends upon the presence of both autoinducers AI-1 and AI-2 and will vary depending upon the ratios of the two autoinducers.

3.2.1. Analogous Signaling Molecules

Investigation of the autoinducers depicted as the input stimulus of Fig. 8, shows that the structures of the AI-1 (acyl homoserine lactones (AHLs), and the AI-2 (furanosylborate diester) share common moieties as illustrated in Fig. 9. In previous publications (4), we have demonstrated the ability to detect and differentiate analogous molecules by exploiting the intrinsic promiscuous nature of all antibodies, first introduced by Cameron and Erlanger (26).

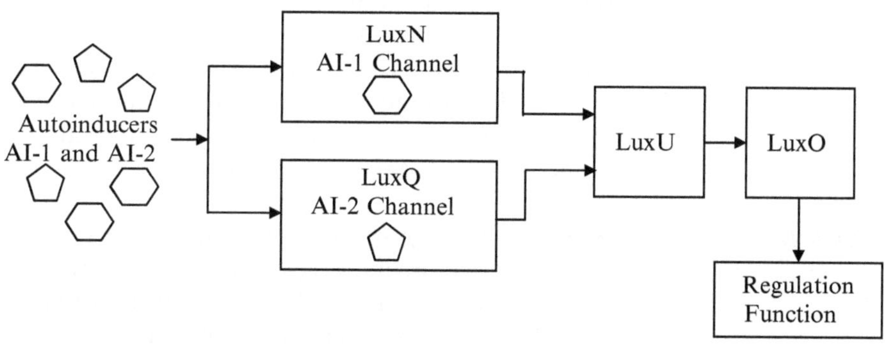

Fig. 8. A two-component sensing system within *V. harveyi*.

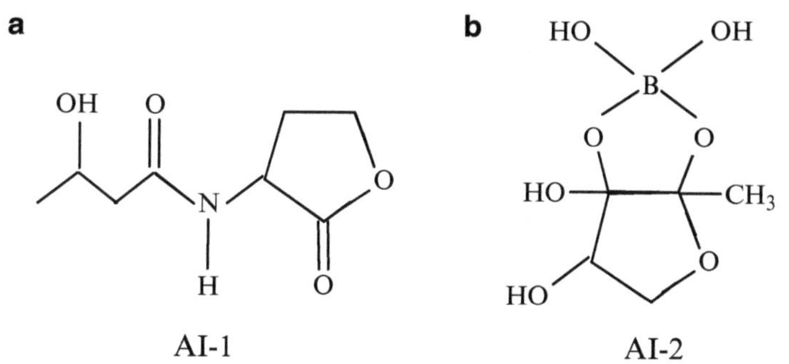

Fig. 9. Autoinducer signaling molecules, (**a**) AI-1, (acyl homoserine lactones (AHLs)) and (**b**) AI-2 (furanosylborate diester).

They discovered that the cross reactivity phenomenon between antibodies, antigens, and their structural homologues was a result of the presence of both electrostatic and hydrophobic binding interactions caused by a high presence of hydrophobic amino acid residues in the antigen binding site.

We have defined this pattern of antibody cross activity as a phenomenon that unveils a molecular signature that is unique, quantifiable, and applicable among most immuno-sensing systems. Here, we present evidence of a cross-reactive anti-lactone antibody RS2-1G9, capable of detecting and differentiating individual signaling molecules in quorum sensing known as N-acyl homoserine lactones (AHLs) among a sea of structural analogs.

Antibody RS2-1G9 was elicited against a lactam mimetic of the N-acyl homoserine lactone and represents the only reported monoclonal antibody that recognizes the naturally-occurring N-acyl homoserine lactone with high affinity (27). Surrette et al. (27) first crystallized the Fab RS2-1G9 antibody in complex with a lactam analog. This revealed a complex that showed complete encapsulation of the polar lactam moiety in the antibody-combining site. The ability of RS2-1G9 to discriminate between closely related AHLs was shown to be conferred by six hydrogen bonds. More specifically, cross-reactivity of RS2-1G9 towards the lactone ring was said to likely originate from conservation of these hydrogen bonds as well as an additional hydrogen bond to the oxygen of the lactone ring. Conversely, the crystal structure of the antibody without the bound lactam or lactone ligands revealed a considerably altered antibody-combining site with a closed binding pocket suggesting that molecular recognition events was triggered by the presence of the lactone moiety.

3.2.2. Cross-Reactivity

A simplified block diagram of a two-component sensing system within *V. harveyi* that includes cross-reactivity is shown in Fig. 10. Here, the autoinducers AI-1 and AI-2 are both presented to the AI-1 channel and the AI-2 channel simultaneously. The LuxN protein displays a promiscuous ability to also respond to AI-2

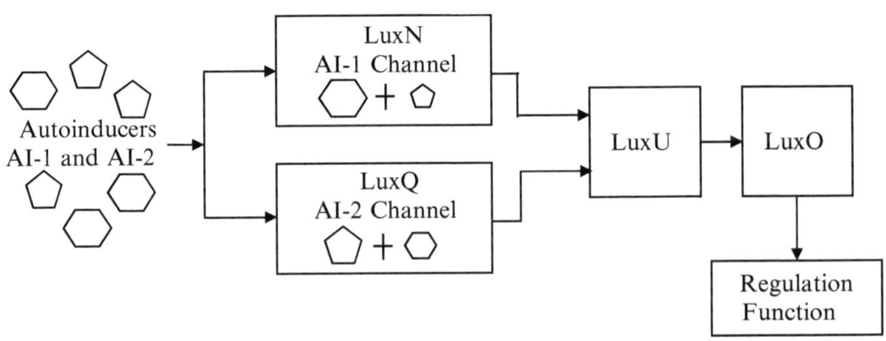

Fig. 10. A simplified block diagram of a two-component sensing system within *V. harveyi*.

autoinducers. This response is scaled differently to that of an AI-1 stimulation, but the output of the LuxN channel contains information that is a function of both autoinducers. Similarly, the output of LuxQ channel also contains information that is a function of both autoinducers. Mok et al. (25), suggested that the active response of the targeted gene when only AI-1 or AI-2 were only present corresponded to a "leakage" within the system.

3.2.3. The Cross Product of AI-1 and AI-2

Previous studies by Mok et al. (25), have suggested that the two autoinducers, AI-1 and AI-2 act synergistically and both autoinducers need to be present to produce the necessary response of the targeted gene. A similar approach is also evident within the mathematical function called the cross product where inputs must be non-zero in order that the function has any affect.

The cross product of two vectors **a** and **b** is denoted by **a**×**b**. Generally, in a three-dimensional Euclidean space, with a right-handed coordinate system, **a**×**b** is defined as a vector **c** that is perpendicular to both **a** and **b**, with a direction given by the right-hand rule and a magnitude equal to the area of the parallelogram that the vectors span. A simple example of a cross product is a propeller, in that the pitch of the propeller is broken down into an x and y component with a motion direction of the propeller perpendicular to x and y. Equation 3 illustrates the cross product mathematical function

$$a \times b = ab\sin(\theta)n \tag{3}$$

where θ is the measure of the angle between **a** and **b** ($0° \leq \theta \leq 180°$), a and b are the magnitudes of vectors **a** and **b**, and n is a unit vector perpendicular to the plane containing **a** and **b** in the direction given by the right-hand rule. If the vectors **a** and **b** are collinear (i.e., the angle θ between them is either 0° or 180°), by the above formula, the cross product of **a** and **b** is zero.

To model the *V. harveyi* quorum sensing system with the cross product function, the two vectors **a** and **b** have been replaced by the autoinducers AI-1 and AI-2. The transformation of a chemical molecule to vector form requires a magnitude component, which is accounted for in this case by the ratio of the autoinducer presented to the model and a coordinate direction. For this study, the vector's direction has been transformed to the equivalent of a molecular alignment, within the chemistry realm and is initially set as a unit vector. Equation 3 was then modified to include the magnitudes of each of the two autoinducers with the angle θ, initially set to 90° as illustrated in Eq. 4,

$$(\mathrm{AI1}) \times (\mathrm{AI2}) = K(L + (\mathrm{R1} \times \mathrm{R2})) \tag{4}$$

where K is a constant, L is a constant to adapt for the condition when there are no autoinducers present, and R1 and R2 are the ratios of AI-1 and AI-2 respectively.

To account for the cross-reactivity where the LuxN channel partially responds to the AI-2 input stimuli, and where the LuxQ channel partially responds to the AI-1 input stimuli of Fig. 10, Eq. 4 was then modified as illustrated in Eq. 5,

$$(\mathrm{AI1}) \times (\mathrm{AI2}) = K(L + (\mathrm{R1} \times \mathrm{R2})) \pm (K_2(\mathrm{R1}) \times K_1(\mathrm{R2})) \quad (5)$$

where K_1 is a constant defining the cross reactivity between the LuxN channel and AI-2 and similarly, K_2 is a constant defining the cross reactivity between the LuxQ channel and AI-1. The ± function depends upon whether the response is an activation (+) or a repression (−).

3.2.4. Simulated Results

A set of simulated results were performed that implemented both Eqs. 4, with no cross-reactivity, and 5 with cross-reactivity, and compared to the data presented in the β-galactosidase activity from of Mok et al. (25). Figure 11 illustrates the β-galactosidase activity of the fusions in the luxS, luxLM derivatives of strains KM314 and Fig. 12 illustrates the β-galactosidase activity of the fusions in the luxS, luxLM derivatives of strains KM321.

3.3. Summary and Conclusions

In this chapter we have presented acoustic wave biosensors as a platform for the electrical transduction of molecular recognition, and conformational change between an immobilized biomolecule and an analyte molecule, which, for the purposes of quorum sensing will be a small molecule. We presented various system and signal conditioning approaches for these acoustic wave biosensors and explored the very close analogy between the signal conditioning of a particular approach, state space mapping, and the signal conditioning which goes on in cells due to the incoming quorum

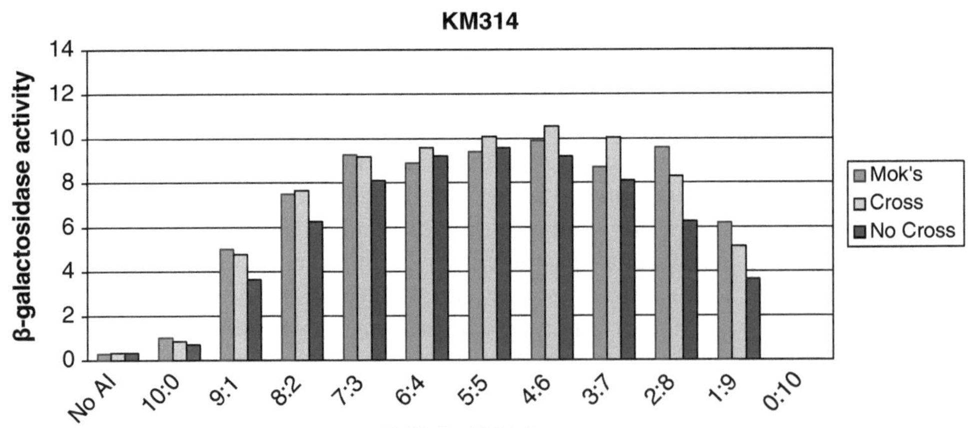

Fig. 11. Comparative plot of Wok, Wingreen and Bassler's figure 5a, illustrating β-galactosidase activity for a two-component system model implementing Eq. 4, no cross-reactivity and Eq. 5.

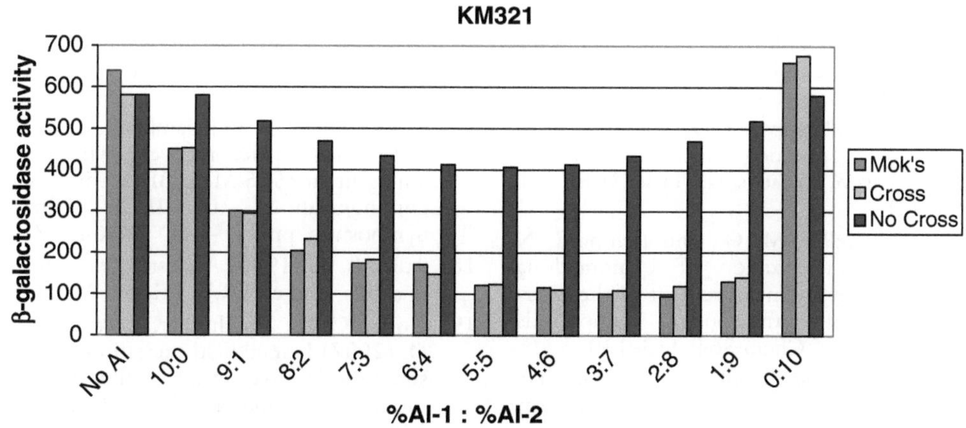

Fig. 12. Comparative plot of Wok, Wingreen and Bassler's figure 5b illustrating β-galactosidase activity for a two-component system model implementing Eq. 4, no cross-reactivity and Eq. 5.

sensing molecules. It is our hope that the tightness of fit of this analogy may elucidate some of the intricacies of the biology.

4. Notes

1. Power consumption of the oscillator increases as frequency increases especially for $f_o > 1$ GHz.
2. Since frequency change is dependent on the square of the center frequency of the oscillator, it may seem obvious to increase this center frequency as high as possible. However, the receptive area of the biosensor also decreases by the square of the frequency resulting in less bioreceptors to bind with.
3. Further information on how to extract binding information from an RFID biosensor can be found at http://www.google.com/patents/about?id=QNF3AAAAEBAJ&dq=edmonson+rfid
4. W.L. Jorgensen recognized that flexible molecules can change their conformation during binding events accounting for cross reactivity among these molecules refuting static molecular recognition models. "*Rusting of the lock and key model for protein-ligand binding*", *Science*, 1991 Nov 15;254(5034):954–5.

References

1. Stubbs, D. D., Hunt, W. D., and Edmonson, P. J. Acoustic Wave Biosensor for The Detection and Identification of Characteristic Signaling Molecules in A Biological Medium, US Patent No. US 7,651,843 B2, issued January 26, 2010.
2. Van Trees, H. L. (1968) Detection, estimation, and modulation theory, Wiley, New York.
3. Wozencraft, J. M., and Jacobs, I. M. (1965) Principles of communication engineering, Wiley, New York.

4. Hunt, W. D., Sang-Hun, L., Stubbs, D. D., and Edmonson, P. J. (2007) Clues from digital radio regarding biomolecular recognition, IEEE Trans Biomed Circuits Syst 1, 50–55.
5. Hunt, W. D., Stubbs, D. D., and Sang-Hun, L. (2003) Time-dependent signatures of acoustic wave biosensors, Proc IEEE 91, 890–901.
6. Zeck, A., Weller, M. G., and Reinhard, N. (1999) Characterization of a monoclonal TNT-antibody by measurement of the cross-reactivities of nitroaromatic compounds, Fresenius J Anal Chem 364, 113–120.
7. James, L. C., and Tawfik, D. S. (2003) The specificity of cross-reactivity: promiscuous antibody binding involves specific hydrogen bonds rather than nonspecific hydrophobic stickiness, Protein Sci 12, 2183–2193.
8. Kramer, A., Keitel, T., Winkler, K., Stocklein, W., Hohne, W., and Schneider-Mergener, J. (1997) Molecular basis for the binding promiscuity of an anti-p24 (HIV-1) monoclonal antibody, Cell 91, 799–809.
9. Ober, R. J., Radu, C. G., Ghetie, V., and Ward, E. S. (2001) Differences in promiscuity for antibody-FcRn interactions across species: implications for therapeutic antibodies, Int Immunol 13, 1551–1559.
10. Sethi, D. K., Agarwal, A., Manivel, V., Rao, K. V., and Salunke, D. M. (2006) Differential epitope positioning within the germline antibody paratope enhances promiscuity in the primary immune response, Immunity 24, 429–438.
11. Campbell, C. (1998) Surface acoustic wave devices for mobile and wireless communications, Academic Press, San Diego.
12. Camilli, A., and Bassler, B. L. (2006) Bacterial small-molecule signaling pathways, Science 311, 1113–1116.
13. Rumbaugh, K. P. (2007) Convergence of hormones and autoinducers at the host/pathogen interface, Anal Bioanal Chem 387, 425–435.
14. Rumbaugh, K. P., Diggle, S. P., Watters, C. M., Ross-Gillespie, A., Griffin, A. S., and West, S. A. (2009) Quorum sensing and the social evolution of bacterial virulence, Curr Biol 19, 341–345.
15. Rumbaugh, K. P., Griswold, J. A., and Hamood, A. N. (2000) The role of quorum sensing in the in vivo virulence of *Pseudomonas aeruginosa*, Microbes Infect 2, 1721–1731.
16. Rumbaugh, K. P., Griswold, J. A., Iglewski, B. H., and Hamood, A. N. (1999) Contribution of quorum sensing to the virulence of *Pseudomonas aeruginosa* in burn wound infections, Infect Immun 67, 5854–5862.
17. Edmonson, P. J., and Campbell, C. K. (1992) SAW-based carrier recovery without phase ambiguity for 915 MHz BPSK wireless digital communications, in IEEE Ultrasonics Symposium, pp 241–244, Tucson, AZ.
18. Auld, B. A. (1990) Acoustic fields and waves in solids, 2nd ed., R.E. Krieger, Malabar, FL.
19. Corso, C. D., Dickherber, A., and Hunt, W. D. (2007) Lateral field excitation of thickness shear mode waves in a thin film ZnO solidly mounted resonator, J Appl Phys 101, 54514–54511.
20. Edmonson, P. J., Hunt, W. D., Corso, C. D., Dickherber, A., and Csete, M. E. An acoustic wave sensor assembly utilizing a multi-element structure, United States Patent Application No.11/822045 filed July 2, 2007.
21. Corso, C. D., Dickherber, A., and Hunt, W. D. (2008) An investigation of antibody immobilization methods employing organosilanes on planar ZnO surfaces for biosensor applications, Biosens Bioelectron 24, 811–817.
22. Corso, C. D., Dickherber, A., Hunt, W. D., and Edmonson, P. J. (2008) Passive sensor networks based on multi-element ladder filter structures, pp 538–541, IEEE, Piscataway, NJ, USA.
23. Edmonson, P. J., and Hunt, W. D. Sensing systems utilizing acoustic wave devices, US Patent No. US 7,608,978 B2, issued October 27, 2009.
24. Edmonson, P. J., Campbell, C. K., and Hunt, W. D. A surface acoustic wave sensor or identification device with biosensing capability, United States Patent No. 7,053,524 B2 issued May 30, 2006.
25. Mok, K. C., Wingreen, N. S., and Bassler, B. L. (2003) *Vibrio harveyi* quorum sensing: a coincidence detector for two autoinducers controls gene expression, EMBO J 22, 870–881.
26. Cameron, D. J., and Erlanger, B. F. (1977) Evidence for multispecificity of antibody molecules, Nature 268, 763–765.
27. Surette, M. G., Miller, M. B., and Bassler, B. L. (1999) Quorum sensing in *Escherichia coli*, *Salmonella typhimurium*, and *Vibrio harveyi*: a new family of genes responsible for autoinducer production, Proc Natl Acad Sci USA 96, 1639–1644.

Chapter 8

Determination of Acyl Homoserine Lactone and Tetramic Acid Concentrations in Biological Samples

Colin A. Lowery, Gunnar F. Kaufmann, and Kim D. Janda

Abstract

Within environmental communities, there is a constant struggle for survival, as nutrients are often limited. In response, bacteria have developed elaborate methods to deal with competitors. One such mechanism is the coordination of behaviors and function via the exchange of small chemical signals in a process known as quorum sensing. This process is especially prominent in the pathogenicity of *Pseudomonas aeruginosa*, an opportunistic human pathogen that forms sessile communities known as biofilms. These biofilms play an important role in the lifestyle of *P. aeruginosa,* either in their natural environment or during establishment and maintenance of infection in human hosts; thus, they often have grievous effects on human health. As such, a method for the detection of these QS signals may provide insights into the pathogenicity and survival of *P. aeruginosa*. In this chapter, we present a method for the extraction and quantitation of the *P. aeruginosa* QS signal *N*-3-oxo-dodecanoyl-homoserine lactone, and its rearranged tetramic acid product, C_{12}-TA, which itself has implications as a survival tactic used by *P. aeruginosa*.

Key words: Quorum sensing, Biofilm, *Pseudomonas aeruginosa*, Acyl homoserine lactone, Tetramic acid

1. Introduction

Quorum sensing (QS) is the process through which individual bacteria coordinate gene expression as a function of cell density via the exchange of small chemical signals, called autoinducers. This process plays a vital role in the lifestyle of *Pseudomonas aeruginosa*, a Gram-negative bacterium that is a common environmental microorganism, but, by capitalizing on weaknesses in the host immune system, has become an opportunistic pathogen in humans. Over the last 20 years, significant progress has been made in elucidating the molecular mechanisms underlying *P. aeruginosa* pathogenicity, in which QS plays an integral role by controlling

Kendra P. Rumbaugh (ed.), *Quorum Sensing: Methods and Protocols*, Methods in Molecular Biology, vol. 692, DOI 10.1007/978-1-60761-971-0_8, © Springer Science+Business Media, LLC 2011

virulence factor production, iron acquisition, and biofilm formation (1). This process is necessary for *P. aeruginosa* to establish infections, as QS-deficient mutants exhibit aberrant biofilm formation and attenuated virulence in animal models (2, 3).

Two acyl homoserine lactones (AHLs), *N*-3-oxo-dodecanoyl homoserine lactone (3-oxo-C_{12}-HSL), synthesized by LasI, and *N*-butyrylhomoserine lactone (C_4-HSL), synthesized by RhlI, have been identified as the main autoinducers in *P. aeruginosa* (4–6). Although this AHL-based QS is involved in the development of infection, QS mutants arise during chronic *P. aeruginosa* infections that exhibit a loss of function in the gene encoding the 3-oxo-C_{12}-HSL receptor LasR, while the homoserine lactone synthase LasI remains active and continues to synthesize the autoinducer, albeit at lower levels, suggesting 3-oxo-C_{12}-HSL provides additional benefits to *P. aeruginosa* besides its role in QS signaling (1, 7–12).

Recently, we reported the conversion of 3-oxo-C_{12}-HSL to the tetramic acid 3-(1-hydroxydecylidene)-5-(2-hydroxyethyl) pyrrolidine-2,4-dione, herein called C_{12}-TA, through an irreversible, non-enzymatic Claisen-like reaction (Fig. 1) (13). Tetramic acids (TAs) are a class of compounds containing a 2,4-pyrrolidinedione ring system with a broad spectrum of biological activity including antibacterial, antiviral, cytotoxic, mycotoxic, fungicidal, as well as anti-cancer properties (14). With this in mind, we hypothesized that 3-oxo-C_{12}-HSL, via the action of C_{12}-TA, might be used by *P. aeruginosa* as a strategy to hamper encroachment by competing bacteria and in a previous study, we detailed the bactericidal activity of both 3-oxo-C_{12}-HSL and C_{12}-TA (15). Notably, C_{12}-TA exhibits significantly more potent antibacterial activity than 3-oxo-C_{12}-HSL (13).

Bacterial infections often come in the form of sessile communities known as biofilms; in fact, it has been estimated that bacterial biofilms are responsible for over 80% of human infections (16). Biofilms of *P. aeruginosa* are common causes of infections in immunocompromised patients, most prominently in lung infections of cystic fibrosis patients. In fact, once formed in the CF lung, these infections are never cleared (17). QS has been linked to *P. aeruginosa* biofilm formation under a variety of conditions, and signal accumulation likely occurs in biofilms due to the high

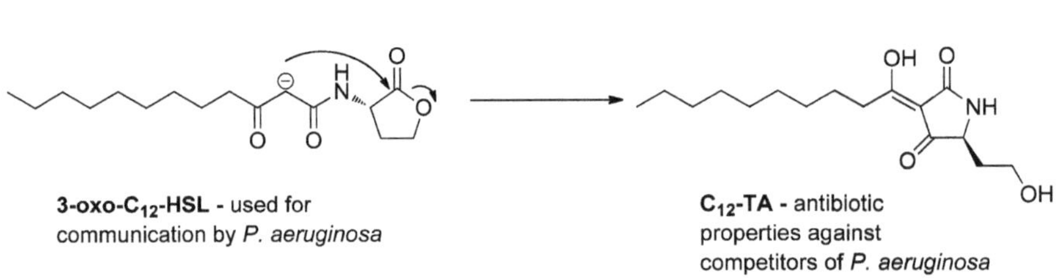

Fig. 1. Structures of 3-oxo-C_{12}-HSL and C_{12}-TA.

cell density and limited diffusion caused by the exopolysaccharide matrix (18, 19). However, the role of QS in biofilm formation in certain settings, such as in the lungs of CF patients, remains unclear (17). In this chapter, we present a reproducible method for the extraction and quantitation of both 3-oxo-C_{12}-HSL and C_{12}-TA from biofilm cultures of *P. aeruginosa*, providing a glimpse of the role of these two compounds in the biofilm lifestyle. Importantly, the methods presented herein may also be applied to the analysis of clinical and environmental samples to gain a broad overview of the behaviors of *P. aeruginosa* in a variety of settings.

2. Materials

2.1. Biofilm Growth

1. M9 minimal salts (Difco).
2. Casamino acids.
3. Peptone tryptic soy broth (PTSB): 5% peptone (Difco) and 0.25% tryptic soy broth.
4. Filter-sterilized 20% glucose solution.
5. 1 M $MgSO_4$ solution.
6. 100 mg/mL carbenicillin solution.
7. *Pseudomonas aeruginosa* strain PAO1 (wildtype) constitutively expressing gfp: this strain harbors the plasmid pTDK-gfp, resulting in constitutive gfp expression as it is under the control of the *lac* promoter (in this case, this strain was a gift generously provided by Prof. Barbara Iglewski).
8. Peristaltic pump, tubing and luer fittings. The tubing and fittings should be autoclaved at 121°C for 15 min before use.
9. Flow cell. In the example presented herein, an FC-81 model flow cell (BioSurface Technologies Corp., Bozeman, Montana), a model designed for microscopy applications, is used. One side of the flow chamber is a standard 25 × 75 × 1 mm glass microscope slide as one viewing window and a no. 2, 24 × 60 mm cover slip serves as the opposite viewing window. The flow channel (i.e. the separation between the two glass surfaces) is approximately 2 mm.

2.2. Preparation for Confocal Microscopy

1. Phosphate buffered saline (PBS): 1× stock containing 137 mM NaCl, 2.7 mM KCl, 10 mM Na_2HPO_4, 2 mM KH_2PO_4, pH 7.4.
2. Formaldehyde solution: 2% (w/v) solution in PBS buffer.
3. Counterstain solution: propidium iodide/Syto85 solution, composed of 2 μM propidium iodide and 5 μM Syto85 (Molecular Probes, Eugene, Oregon) in PBS buffer.

4. Confocal microscope apparatus: Bio-Rad (Zeiss) Radiance 2100 Rainbow laser scanning confocal microscope (LSCM) connected to a Nikon TE2000-U. Image analysis, in this example, was performed using the program Imaris (Bitplane, St. Paul, Minnesota).

2.3. Extraction of Biofilm Samples

1. Acidified ethyl acetate: solution of 0.01% glacial acetic acid in ethyl acetate.
2. 50 mL polypropylene conical tubes.
3. HPLC grade methanol, cooled to –20°C.

2.4. LC/MS Analysis of Samples

1. Synthetic standards of 3-oxo-C_{12}-HSL and C_{12}-TA (see Note 2).
2. LC/MS apparatus: In the example described, an Agilent 1100 MSD LC/MS system was used. This device is an HPLC-single quadrupole mass spectrometer which utilizes an electrospray ionization method, equipped with an Agilent Zorbax column (5 μm, 300SB-C8, 4.6 × 50 mm) and a Phenomenex Security Guard® guard column (C8, double stacked, 4.0 × 3.0 mm cartridge).
3. Autosampler vials (Wheaton).

3. Methods

3.1. Measurement of *P. aeruginosa* Biofilm Volume

1. For growing biofilms, prepare a 5× solution of M9 minimal salts growth medium, according to the manufacturer's instructions, and supplement with 2.5% w/v casamino acids. Add 200 mL sterile 5× M9 solution to 750 mL sterile water. Aseptically add 10 mL of filter-sterilized 20% glucose solution, 1 mL sterile 1 M $MgSO_4$ solution, and 2 mL of 100 mg/mL carbenicillin solution. Adjust the final volume to 1 L. Final concentrations: 0.2% glucose, 1 mM $MgSO_4$, 200 μg/mL carbenicillin.
2. Grow an overnight culture of *P. aeruginosa* at 37°C in a 15 mL culture tube containing 3 mL of PTSB medium supplemented with 200 μg/mL carbenicillin.
3. Prepare the flow cell for inoculation with the culture of *P. aeruginosa*. Attach the outlet tubing to the flow cell. Insert the end of the inlet tubing into the fresh medium reservoir, and turn on the pump to allow the medium to fill the inlet tubing. Allow the medium to run into a sterile flask to ensure the tube is full of medium (see Notes 3–5).
4. Dilute the overnight culture into fresh M9 medium to an OD_{600} of about 1.5.

5. Using a sterile needle and 3 mL syringe, withdraw 2 mL (or a slightly larger volume than the volume of the flow chamber) of the freshly diluted culture. Carefully remove the needle, invert the syringe, and pull in about 0.5 mL of air. Tap the syringe to remove any bubbles from the bacterial culture, and depress the plunger to remove excess air from the syringe (see Notes 1 and 3).
6. Slowly inject the culture into the flow cell, being careful not to introduce any air bubbles into the flow chamber. Continue injecting the culture until the flow chamber is free of bubbles.
7. After injection, remove the syringe and immediately attach the inlet tubing to the flow cell.
8. Turn off the pump, and incubate the flow cell at 37°C for 90 min under static conditions.
9. Turn the peristaltic pump on at a flow rate of 4 mL/h (shear rate 0.205/s) via peristaltic pump (see Note 6).
10. After the desired time (6 days, in this case), turn the pump off and replace the reservoir containing fresh medium with a reservoir containing a 2% formaldehyde solution in PBS buffer. Turn the pump back on, and allow the formaldehyde solution to flow over the biofilm for 2 h. Be sure to account for the volume of the tubing when calculating the time required for biofilm fixation. For example, if the tubing volume is 8 mL, then the formaldehyde solution should be pumped through the flow cell for 4 h. *If your institution has a microscopy facility equipped to handle BSL-2 agents, then this step may be omitted* (see Note 7).
11. At this time, stop the flow again, remove the inlet tubing from the flow cell, and slowly inject 3 mL of a solution of propidium iodide/Syto-85, again using a slightly larger volume that that of the flow chamber. After injection, leave the syringe in place and incubate the flow cell in the dark for 30 min (see Note 9).
12. Remove the syringe, and attach a 10 mL syringe filled with fresh PBS buffer. Every 10 min, inject 2 mL of PBS buffer. Repeat this for a total of three times.
13. Detach the outlet tubing and the syringe from the flow cell, and cap each end with a male luer fitting (see Note 13).

3.2. Confocal Microscopy to Measure Biofilm Volume

1. Place the flow cell carefully on the platform of the microscope, being careful not to disrupt the biofilm. The flow cell should be placed such that the cover slip is facing down.
2. Using the 10× objective, obtain z-series images (at 6 μm interval steps) of the fluorescent signal associated with the biofilm. Excitation at 543 nm induces red fluorescence

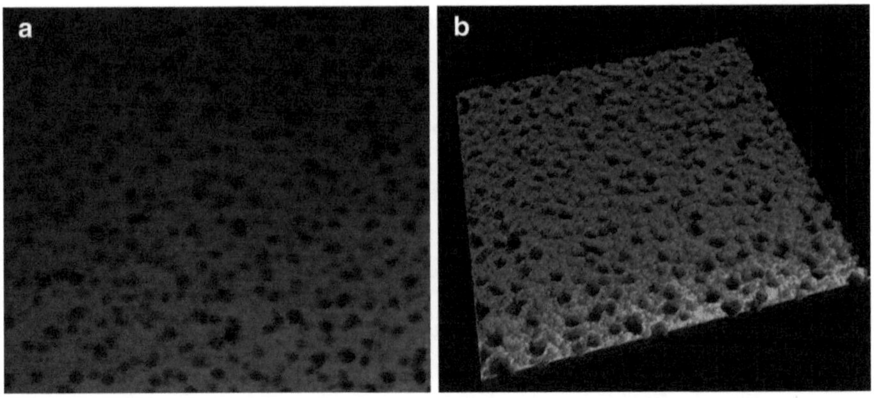

Fig. 2. Volumetric analysis for determination of [3-oxo-C_{12}-HSL] and [C_{12}-TA]. (**a**) CLSM image of a biofilm of *P. aeruginosa* PA01-gfp grown for 6 days in the presence of carbenicillin (200 μg/mL). (**b**) 3-dimensional representation (generated using the program Imaris) of the biofilm in (**a**) used in the calculation of biofilm volume.

(propidium iodide/Syto-85), and 488 nm induces green fluorescence (GFP).

3. To select an area for measurement, divide the flow cell laterally into sections and choose seven fields of view randomly within each section. Record an image at each 6 μm interval. To calculate mean biofilm volume for each image, analyze the z-series images for volume using imaging software (Fig. 2).
4. From these calculations, an average height of the biofilm may be obtained and used to calculate the maximum mean volume for the cover slip by multiplying by the surface area of the biofilm on the cover slip.
5. To gain an average biofilm volume, the processes described in Subheadings 3.1 and 3.2 should be performed at least in duplicate.

3.3. Extraction of Biofilm Samples

1. Grow biofilms as described above without the fixation steps (i.e. follow steps 1–9 in Subheading 3.1). Alternatively, the same biofilm sample may be used for both image analysis and extraction, if the institution has a microscopy facility equipped for the analysis of BSL-2 agents (see step 9 in Subheading 3.1 and Note 7).
2. After 6 days, or the desired length of time, turn off the flow and detach the inlet and outlet tubing.
3. Tilt the flow cell on to its side and allow the medium inside to flow into a 50 mL conical tube. Using a 1 mL pipette, gently rinse the flow chamber by pipetting 1 mL of water through the chamber.
4. Unscrew the baseplates of the flow cells, and collect the biofilm (primary) sample by removing the cover slip, placing into the 50 mL conical tube, and crushing it.

5. Swab each baseplate and the microscope slide with a cotton gauze, and place the cotton gauze in the same 50 mL conical tube. At this point there should be 3–4 mL of spent culture in the conical tube.
6. Add 10 vol. of acidified ethyl acetate to the tube, and mix the contents vigorously. At this point, it may be necessary to crush the cover slip further. Allow the contents to settle, using a glass pipette to push the cotton gauze to the bottom of the tube (see Note 10).
7. Remove half of the ethyl acetate layer and transfer to a dry Erlenmeyer flask, and add $MgSO_4$ to absorb any excess water. Filter the resulting slurry, and concentrate the filtrate using a rotary evaporator. Once the volume is <2 mL, transfer the solution to a 2 mL Eppendorf tube, and finish the concentration process using a GeneVac or a similar apparatus.
8. Resuspend the residue in a minimum amount of HPLC-grade methanol (≤200 μL), cooled to a temperature of −20°C, by pipetting up and down several times. There should be a small amount of precipitate formed once the residue is exposed to cold methanol.
9. Centrifuge the Eppendorf tubes to allow the formation of a pellet of the precipitated material. Transfer the supernatant to autosampler vials in preparation for LC/MS analysis (see Note 11).

3.4. Analysis of Biofilm Samples by LC/MS

1. Set up a gradient on the LC apparatus as follows: from 0 to 1 min: 5% MeCN, from 1 min to 9 min: gradient of 5% MeCN to 98% MeCN, and from 9 min to 13 min: 98% MeCN.
2. Set up the mass spectrometer to read in the single ion monitoring mode for the following ions: 298 ($M+H^+$), 316 ($M+H_2O+H^+$), and 320 ($M+Na^+$).
3. Using the synthetic 3-oxo-C_{12}-HSL and C_{12}-TA, run varying concentrations of each standard on the LC/MS to establish a standard curve for concentrations and retention times for the two analytes. Suggested concentrations range from 100 μM to 0.01 μM (see Note 2).
4. Load the MS file from the LC/MS software. From this file, extract the data from the individual ion currents to examine the 298, 316, and 320 MS signals. Integrate the peaks in these chromatograms, and create an *XY*-plot to establish a standard curve (X = integration values, Y = moles of standard).
5. Fit the data to a linear equation, and use this equation for the later determination of 3-oxo-C_{12}-HSL and C_{12}-TA concentrations in the biofilm samples.
6. Run the extracted biofilm samples on the LC/MS using the LC gradients and MS settings as established for the two standards (see Note 8).

7. Analyze the data as described in Subheading 3.3, step 4, integrating the peaks in these chromatograms corresponding to the peaks of the standard samples (based on the retention time of the standards). Using the equation generated from the standard curve, calculate the number of moles in the LC/MS sample. If it is a 10 μL injection from a 200 μL sample, it will be necessary to multiply this number by 20 to obtain the total moles in the LC/MS sample.
8. Because of the noise associated with the biofilm samples (due to the presence of a myriad of compounds), it will be necessary to perform a co-injection of a biofilm sample with the two standards to confirm the identities of the peaks. To do so, add a small volume of the standards to the biofilm samples to approximately double the number of moles in the biofilm sample. Perform the LC/MS analysis again, and monitor the increase of the peaks previously identified as 3-oxo-C_{12}-HSL and C_{12}-TA (see Note 12).
9. To perform a more rigorous quantitation, stable isotope marker compounds (i.e. ^{13}C-labeled 3-oxo-C_{12}-HSL and C_{12}-TA) should be added at known concentrations to the biofilm samples as internal standards. Thus, comparison of the non-labeled signals against the heavy labeled standards will provide another level of quantitation. It should be noted that if the heavy-labeled standards are added as a reference, it will be necessary to include additional ion settings as described in Subheading 3.4, step 2.
10. To calculate the total amount of 3-oxo-C_{12}-HSL and C_{12}-TA in the biofilm sample, take the number of moles calculated for the LC/MS samples and multiply by 2 (because half of the organic phase was taken from the extraction). Divide by the volume calculated for the biofilm to find the total concentration.

4. Notes

1. *P. aeruginosa* is a BSL-2 agent as defined by the Centers for Disease Control. Care should be taken when performing all manipulations, but extreme precautions should be taken when using needles or other sharp objects (e.g. broken cover slips).
2. Compounds must be of high purity for the generation of reliable standard curves. For more rigorous analyses, it is also advisable to include non-radioactive, isotopically-labeled (e.g. ^{13}C, ^{18}O, ^{2}H, or ^{15}N) 3-oxo-C_{12}-HSL and C_{12}-TA in the LC/MS analysis.
3. To connect the tubing to the flow chamber, we have found that it works best to attach a short (1–2 in.) piece of tubing

to the inlets of the flow chamber, with a barb to female luer at the opposite ends of the tubes. This allows for easier injection of the bacterial culture into the flow chamber, as well as quicker isolation and capping of the flow cell.

4. When assembling the flow cell, a no. 2 (0.19–0.23 mm thick) cover slip should be used. In our experience, anything thinner cracks under the pressure of the two baseplates, and the cultures leak.
5. Another consideration for assembly of the flow cell is to take care not to crack the cover slip when tightening the baseplates together. This often occurs when the cover slip moves out of the recessed area when placing the final baseplate on the apparatus.
6. Once the flow has begun, monitor the flow cell periodically over the course of 1–2 h to check for leaks. Any leaks should be fixed immediately. This can most often be accomplished by simply tightening the screws in the baseplate.
7. We have also found that fixation of the biofilm sample for CLSM results in diminished extraction efficiency. If the institution has a microscopy facility that allows for work with BSL-2 agents, the fixation steps may be skipped, and the same biofilm sample used for both CLSM and extraction, which is ideal.
8. When running the biofilm LC/MS samples, it is beneficial to run standards or blanks in between biofilm samples to ensure that there are no obstructions in the column or drift in the signals.
9. During the injection of the PI/Syto85 stain and the subsequent washing steps, it is important not to introduce any bubbles into the flow chamber, as these will disrupt the biofilm. One option is to use a bubble trap up stream of the flow cell, but we have found that this is not absolutely necessary.
10. For extraction, other organic solvents may be used, such as methylene chloride. However, we have found ethyl acetate to be the best, as it gives better extraction efficiency, does not dissolve the conical tubes, and is lighter than water, which allows for easier isolation of the organic layer.
11. Once the extraction procedure is performed, the LC/MS analysis should be performed promptly, as 3-oxo-C_{12}-HSL is a labile compound, and may undergo conversion to C_{12}-TA.
12. When performing the co-injection controls, it is necessary to add the synthetic standards in a minimal volume of solvent. Adding too much solvent will dilute the original signals from the biofilm. Another point for consideration is the addition of a modest amount (about twofold) of synthetic standard, as adding too much of the standard will overwhelm the original signals, and two peaks may then appear as one.

13. After extraction or CLSM analysis, disassemble the flow cell apparatus and soak the flow cell in 70% ethanol for 30 min. The tubing may be soaked in 10% bleach for 30 min. After 30 min, thoroughly rinse all components with water and allow to air dry. Autoclave all components (excluding the cover slip, which may be discarded in the apporopiate biohazardous waste) at 121°C for 15 min before growth of a new biofilm.

Acknowledgments

This work was supported by the National Institutes of Health (AI079503 to KDJ, AI080715 to GFK), the Skaggs Institute for Chemical Biology, and a Sanofi-Aventis Graduate Fellowship (CAL). We thank Prof. Barbara Iglewski for providing gfp-expressing *P. aeruginosa* and Dr. Victoria Wagner for helpful discussions. We would also like to thank Dr. Malcolm Wood and Dr. William Kiosses for assistance with microscopy experiments, as well as Dr. Gary Siuzdak and Bill Webb of the Scripps Center for Mass Spectrometry for assistance with the mass spectrometry experiments. This manuscript has been assigned TSRI manuscript #20556.

References

1. Heurlier, K., Denervaud, V., and Haas, D. (2006) Impact of quorum sensing on fitness of *Pseudomonas aeruginosa*. *Int J Med Microbiol* **296**, 93–102.
2. Smith, R. S., and Iglewski, B. H. (2003) *P. aeruginosa* quorum-sensing systems and virulence. *Curr Opin Microbiol* **6**, 56–60.
3. Davies, D. G., Parsek, M. R., Pearson, J. P., Iglewski, B. H., Costerton, J. W., and Greenberg, E. P. (1998) The involvement of cell-to-cell signals in the development of a bacterial biofilm. *Science* **280**, 295–8.
4. de Kievit, T. R., Kakai, Y., Register, J. K., Pesci, E. C., and Iglewski, B. H. (2002) Role of the *Pseudomonas aeruginosa* las and rhl quorum-sensing systems in rhli regulation. *FEMS Microbiol Lett* **212**, 101–6.
5. Pesci, E. C., Pearson, J. P., Seed, P. C., and Iglewski, B. H. (1997) Regulation of las and rhl quorum sensing in *Pseudomonas aeruginosa*. *J Bacteriol* **179**, 3127–32.
6. Wagner, V. E., Li, L. L., Isabella, V. M., and Iglewski, B. H. (2007) Analysis of the hierarchy of quorum-sensing regulation in *Pseudomonas aeruginosa*. *Anal Bioanal Chem* **387**, 469–79.
7. Cabrol, S., Olliver, A., Pier, G. B., Andremont, A., and Ruimy, R. (2003) Transcription of quorum-sensing system genes in clinical and environmental isolates of *Pseudomonas aeruginosa*. *J Bacteriol* **185**, 7222–30.
8. Favre-Bonte, S., Chamot, E., Kohler, T., Romand, J. A., and van Delden, C. (2007) Autoinducer production and quorum-sensing dependent phenotypes of *Pseudomonas aeruginosa* vary according to isolation site during colonization of intubated patients. *BMC Microbiol* **7**, 33.
9. Lee, B., Haagensen, J. A., Ciofu, O., Andersen, J. B., Hoiby, N., and Molin, S. (2005) Heterogeneity of biofilms formed by nonmucoid *Pseudomonas aeruginosa* isolates from patients with cystic fibrosis. *J Clin Microbiol* **43**, 5247–55.
10. Schaber, J. A., Carty, N. L., McDonald, N. A., Graham, E. D., Cheluvappa, R., Griswold, J. A., and Hamood, A. N. (2004) Analysis of quorum sensing-deficient clinical isolates of *Pseudomonas aeruginosa*. *J Med Microbiol* **53**, 841–53.
11. Sandoz, K. M., Mitzimberg, S. M., and Schuster, M. (2007) Social cheating in

Pseudomonas aeruginosa quorum sensing. *Proc Natl Acad Sci U S A* **104**, 15876–81.

12. Diggle, S. P., Griffin, A. S., Campbell, G. S., and West, S. A. (2007) Cooperation and conflict in quorum-sensing bacterial populations. *Nature* **450**, 411–4.
13. Kaufmann, G. F., Sartorio, R., Lee, S. H., Rogers, C. J., Meijler, M. M., Moss, J. A., Clapham, B., Brogan, A. P., Dickerson, T. J., and Janda, K. D. (2005) Revisiting quorum sensing: discovery of additional chemical and biological functions for 3-oxo-n-acylhomoserine lactones. *Proc Natl Acad Sci U S A* **102**, 309–14.
14. Royles, B. J. L. (1995) Naturally occurring tetramic acids: structure, isolation, and synthesis. *Chem Rev* **95**, 1981–2001.
15. Lowery, C. A., Park, J., Gloeckner, C., Meijler, M. M., Mueller, R. S., Boshoff, H. I., Ulrich, R. L., Barry, C. E., Bartlett, D. H., Kravchenko, V. V., Kaufmann, G. F., and Janda, K. D. (2009) Defining the mode of action of tetramic acid antibacterials derived from *Pseudomonas aeruginosa* quorum sensing signals. *J Am Chem Soc* **131**, 14473–79.
16. Davies, D. (2003) Understanding biofilm resistance to antibacterial agents. *Nat Rev Drug Discov* **2**, 114–22.
17. Winstanley, C., and Fothergill, J. L. (2009) The role of quorum sensing in chronic cystic fibrosis *Pseudomonas aeruginosa* infections. *FEMS Microbiol Lett* **290**, 1–9.
18. Charlton, T. S., de Nys, R., Netting, A., Kumar, N., Hentzer, M., Givskov, M., and Kjelleberg, S. (2000) A novel and sensitive method for the quantification of n-3-oxoacyl homoserine lactones using gas chromatography-mass spectrometry: application to a model bacterial biofilm. *Environ Microbiol* **2**, 530–41.
19. Kirisits, M. J., and Parsek, M. R. (2006) Does *Pseudomonas aeruginosa* use intercellular signalling to build biofilm communities? *Cell Microbiol* **8**, 1841–9.

Chapter 9

Luminescent Reporters and Their Applications for the Characterization of Signals and Signal-Mimics that Alter LasR-Mediated Quorum Sensing

Ali Alagely, Sathish Rajamani, and Max Teplitski

Abstract

In many pathogenic bacteria, quorum sensing (QS) controls expression of genes that are involved in virulence, production and resistance to antibiotics, formation and maintenance of microbial multicellular consortia on biotic and abiotic surfaces of medical and industrial importance. *N*-acyl homoserine lactones (AHL) are the best characterized quorum sensing signals in Gram-negative bacteria. Interference with AHL-mediated QS, therefore, is considered an attractive strategy for controlling virulence in pathogens. The search for AHL signals and their mimics has been facilitated by the development of sensitive bioassays, in which QS reporters luminesce in response to AHL signals. These bioassays have already led to the identification of dozens of compounds with QS modulating activities. The characterization of the mode of action of QS signals and their mimics requires follow-up biochemical studies. Here, we describe a set of luminescent reporters, which could be used in high, medium or low throughput format, for the discovery and validation of agonists or antagonists of the Las QS system of *Pseudomonas aeruginosa*. These nearly isogenic reporters contain truncations or point mutations in the AHL binding domain of the AHL receptor LasR, as well as mutations in the promoter for the gene encoding LasI AHL synthase. We also developed reporters for documenting the regulation of *lasI* and *lasR* promoters. The use of these reporters significantly streamlines identification and characterization of the Las QS signal agonists and antagonists prior to biochemical experiments. To test the usefulness of these reporters, we carried out bioassays with patulin, a known inhibitor of Las QS.

Key words: Cell-to-cell signaling, *rsaL*, Quorum sensing inhibition, Quorum sensing signal-mimic, LuxR

1. Introduction

1.1. Components of AHL-Mediated Quorum Sensing

Quorum sensing (QS) is one of the mechanisms by which microbes change global patterns of gene expression in response to increases in their population densities within a diffusion-limited environment. A discovery that QS controls virulence genes in

Kendra P. Rumbaugh (ed.), *Quorum Sensing: Methods and Protocols*, Methods in Molecular Biology, vol. 692,
DOI 10.1007/978-1-60761-971-0_9, © Springer Science+Business Media, LLC 2011

human and plant pathogens was an important impetus for developing therapies based on the inhibition of bacterial QS. The best-characterized QS signals in Gram-negative bacteria are *N*-acyl-L-homoserine lactones (AHLs). AHLs produced by different bacterial species may differ in length of their side chains, degree of saturation, and decorations at the third carbon of the side chain (1). Synthesis of AHLs from acyl-acyl carrier proteins (acyl-ACP) and S-adenosyl methionine (SAM) is typically catalyzed by the homologs of LuxI (2). Inside bacterial cells, AHLs are typically detected by the homologs of LuxR proteins (3, 4). LuxR homologues have an AHL-binding N-terminal domain, and a DNA-binding C-terminal domain. AHL binding pockets of LuxR homologues contain conserved Y56, W60, and D73 amino acid residues (Fig. 1) which are involved in hydrogen bonding to the 1-carbonyl, the ring carbonyl, and the NH group (respectively) of AHLs (3, 4). Binding of an AHL to the cognate receptor leads to the oligomerization of the protein, and allows the LuxR-AHL complex to bind its target DNA sequences, "*lux* boxes" (3, 5, 6). *lux* boxes are 20-bp inverted repeats found within promoters of genes encoding *luxI* homologues and other QS-regulated genes (6). Binding of the LuxR-AHL complex to the *lux* box upstream of *luxI* leads to the upregulation of *luxI* expression. This positive feedback is commonly called "autoinduction," and is considered to be an important feature of many QS regulatory circuits (2).

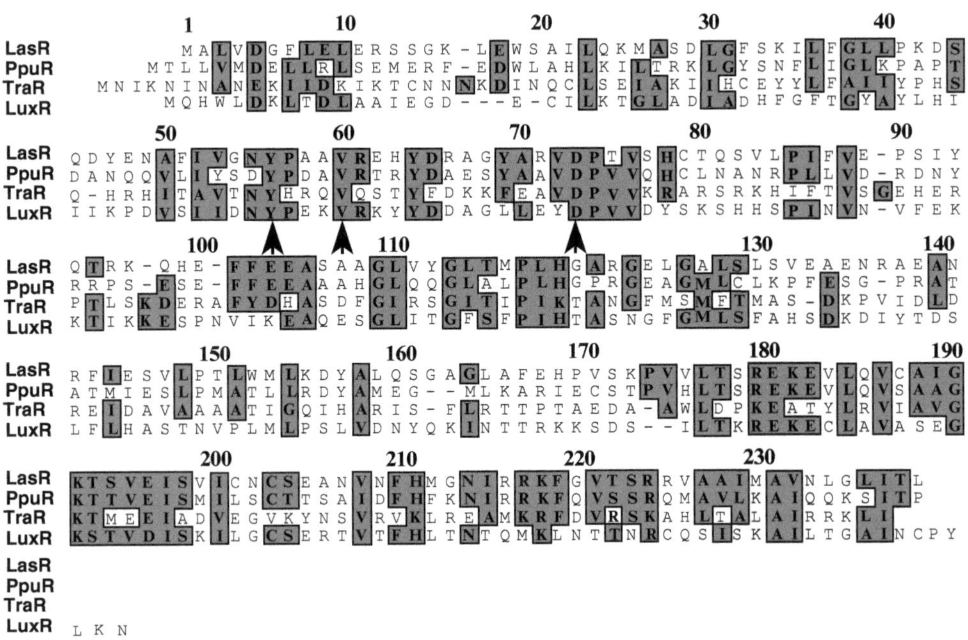

Fig. 1. CLUSTAL-W alignment of AHL receptors. LasR of *Pseudomonas aeruginosa* (accession # BAA06489), PpuR of *P. putida* (accession # AAZ80478), TraR of *Agrobacterium tumefaciens* (AAC28121) and LuxR of *Vibrio fischeri* (AAW87995). Conserved amino acid residues are *shaded gray*. *Arrows* point at conserved amino acid residues (Tyr56, Trp60 and Asp73), which were shown to interact with the AHL ligands.

1.2. QS in Pseudomonas aeruginosa

In *P. aeruginosa*, an opportunistic pathogen of humans, animals and plants (7), QS is mediated by 3-oxo-C_{12}-HSL synthesized by LasI, C_4-HSL synthesized by RhlI and by 2-alkyl-4-quinolones (the *Pseudomonas* quinolone signal PQS (2-hepty-2-hydroxy-4-quinolone) and HHQ (2-heptyl-4-quinolone)) (8). AHLs are perceived by LuxR-type AHL receptors LasR and RhlR (which bind 3-oxo-C_{12}-HSL and C_4-HSL, respectively). The *lasIR* and *rhlIR* systems are hierarchically arranged. An "orphan" AHL receptor VsqR also controls production of 3-oxo-C_{12}-HSL by regulating *lasI* expression (9). The *lasI* gene is autoinduced via LasR, and LasR activates *rhlRI*, which is in turn autoregulated (8, 10). In the absence of LasR, transcription of *lasI* initiates from an alternate site, which is functional in *P. aeruginosa* and in *E. coli* (10).

In LasR, amino acid residues L36-A127 contribute to the formation of the AHL-binding domain of LasR, with Y56, W60, D73 and S129 being directly involved in the interactions with 3-oxo-C_{12}-HSL (Fig. 1) (4). Binding of 3-oxo-C_{12}-HSL leads to the multimerization of the protein, which then binds the "*las* box" DNA sequence within promoters of the regulated genes (5, 6). The presence of the CT-N_{12}-AG motif appears to be required for DNA binding of LasR (Fig. 2) (6).

Quorum sensing homeostasis is maintained by a negative regulator RsaL, which itself is negatively autoregulated and positively regulated by LasR in the 3-oxo-C_{12}-HSL-dependent manner (Fig. 2) (11). The *rsaL* gene is encoded divergently from *lasI* (Fig. 2) in *P. aeruginosa* and other pseudomonads (11–13). RsaL belongs to a tetrahelical subclass of helix-turn-helix proteins (13).

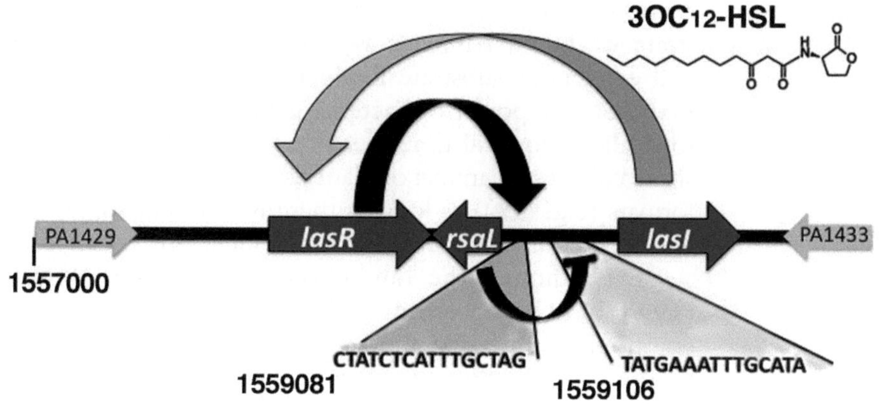

Fig. 2. Relevant features of the Las QS system. At low population densities, *lasI* is expressed from an alternate promoter (10). LasI synthesizes a QS signal, 3-oxo-C_{12}-HSL, which is perceived by LasR (activation of LasR by 3-oxo-C_{12}-HSL is indicated by a *gray arrow*). When complexed with the cognate AHL, LasR binds the *las* box sequence (CTATCTCATTTGCTAG, shown in a *dark-gray triangle*) within the divergent *rsaL-lasI* promoter. Binding of LasR-AHL complex to the *las* box (shown by a *black arrow*) leads to the initiation of transcription from the primary transcriptional start site (10). The *lasI* promoter is negatively regulated by RsaL, which binds to the TATGAAATTTGCATA sequence (shown as a *light gray triangle*) upstream of *lasI*. *Numbers* indicate positions of nucleotides in the published genome of *P. aeruginosa* (accession # NC_002516).

The RsaL dimer binds to the TATGnAAnTTnCATA sequence within promoters of *lasI*, *phzA1*, *phzM* and *hcnA* and reduces expression of the corresponding genes (11, 13). Within the divergent *rsaL-lasI* promoter, LasR and RsaL bind to distinct sites (Fig. 2) (11).

In *P. aeruginosa*, QS regulates dozens of genes and proteins, many of which are known to contribute to the virulence of this pathogen (6, 7, 9). Because control of infections caused by *P. aeruginosa* is complicated by the fact that this pathogen easily evolves resistance to multiple antibiotics, manipulation of QS was suggested as an alternative to antibiotic treatments (14, 15). It is thought that the use of QS inhibitors would not put pathogens under a strong selective pressure, but will reduce expression of virulence genes, thus leading to the control of the infection process (14).

1.3. Reporters for the Identification of QS Signals, Agonists and Antagonists

The search for QS signals, their synthetic and naturally occurring agonists and antagonists has been greatly facilitated by the development of sensitive bioassays which are based on QS systems from *Agrobacterium tumefaciens* or *Chromobacterium violaceum* (1, 16). In addition, Winson et al. (1998) developed semi-synthetic luminescent reporters based on the Rhl and Las QS systems of *P. aeruginosa*, Lux QS system of *Vibrio fischeri* and Ahy QS system of *Aeromonas hydrophila* (17). The reporters consist of the genes encoding AHL receptors and promoters of the nearby genes encoding AHL synthases, which are cloned upstream of a promoterless *luxCDABE* cassette (17). These reporters function in *E. coli*, a heterologous host. There are several advantages of these reporters. The cassette contains *luxCDE* genes that encode a fatty acid reductase complex required for the synthesis of the aldehyde substrate for the LuxAB luciferase (18). Therefore, there is no need to add exogenous substrate for luciferase. The development of the *luxCDABE* reporter of bacterial gene expression (17, 18), therefore, allows for real-time measurement of the reporter activity in shake cultures, in animal or plant tissues and using high throughput formats (e.g. (19)) as long as there is sufficient supply of oxygen, which is required for the function of luciferase.

Even though the luminescent reporters of Winson et al. (1998) have been instrumental for the identification of AHLs and compounds with QS-modulating activities, reliance on these reporters as a sole route for characterizing a compound as an AHL-mimic could be overly simplistic. For example, the inhibition of the Las "indirect bioassay" (a standard assay in which a cognate AHL is added alongside with a presumed QS antagonist) could result from the competition of the two molecules for the AHL-binding pocket of the receptor. However, an inhibition of the indirect bioassay could also result from the inhibition of the multimerization of the LasR protein, disruption of binding of

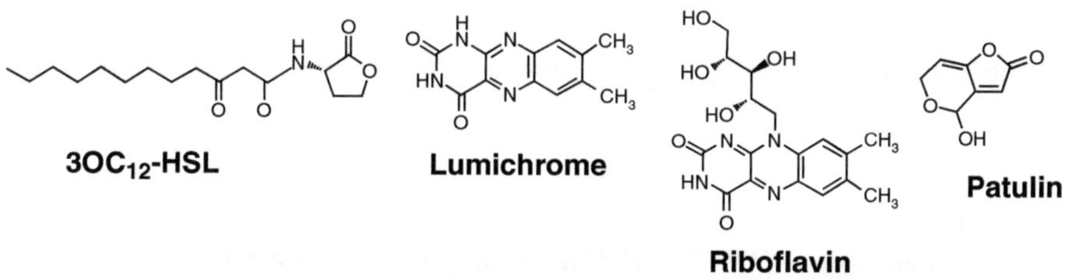

Fig. 3. Signals that affect LasR-mediated QS. 3-oxo-C_{12}-HSL is the AHL produced by LasI which binds LasR and effects LasR-mediated gene expression (4, 5, 8). Lumichrome and riboflavin are AHL-mimics which stimulate LasR-mediated QS, likely by directly interacting with the AHL-binding amino acid residues of LasR (20). Patulin is a eukaryotic compound capable of inhibiting LasR-mediated gene expression (15).

LasR or RsaL within the *lasI* promoter, non-specific binding of the compound to LasR or RsaL, inhibition of *lasR* expression or stability of the *lasR* mRNA, covalent binding of a compound to an AHL leading to its inactivation, inhibition of the LuxAB luciferase, or general toxicity. Biochemical and physiological assays will sort out these alternatives. However, addressing them could be time-consuming and expensive, especially when dozens of putative compounds are identified in a high throughput screen. To help streamline the process of the characterization of the various candidate QS inhibitory compounds, we designed a series of luminescent reporters, which are based on the pSB1075 reporter constructed by Winson et al. (1998). Some of these constructs have already been used to isolate, purify and establish mode of action of lumichrome, the eukaryotic agonist of QS (Fig. 3) (20).

1.4. Construction of the Reporter Constructs

Relevant features of all constructs are shown in Fig. 4. All primers are described in Table 1. Unless otherwise indicated in the text, plasmids were extracted using Qiaprep Spin Miniprep Kit (Qiagen, Valencia, CA). To purify DNA fragments from agarose gels, Qiaquick PCR purification Kit (Qiagen) was used.

To construct pSB1075-Y56F, pSB1075-W60F and pSB1075-D73N, site-directed mutagenesis of *lasR* was carried out using QuickChange XL Site-Directed Mutagenesis Kit (Agilent Technologies, Stratagene, La Jolla, CA) and complimentary primer pairs LasR-Y56F and LasR-Y56F(compli) (to introduce Y56F); LasR-W60F and LasR-W60F(compli) (to introduce W60F); LasR-D73N and LasRD73N(compli) (to introduce D73N) as detailed in the Stratagene manual. Briefly, eighteen cycles of *Pfu* Turbo polymerase-catalyzed PCR were carried with these primer pairs and pSB1075 (purified from a methylation positive *E. coli* DH5α) as a template. The resulting PCR product was cut with *Dpn*I to eliminate methylated parent templates. The *Dpn*I-treated PCR reaction mix was then transformed into chemically-competent *E. coli* XL-10 Gold Ultracompetent cells. Transformants were selected on LB agar with ampicillin (200 μg/ml).

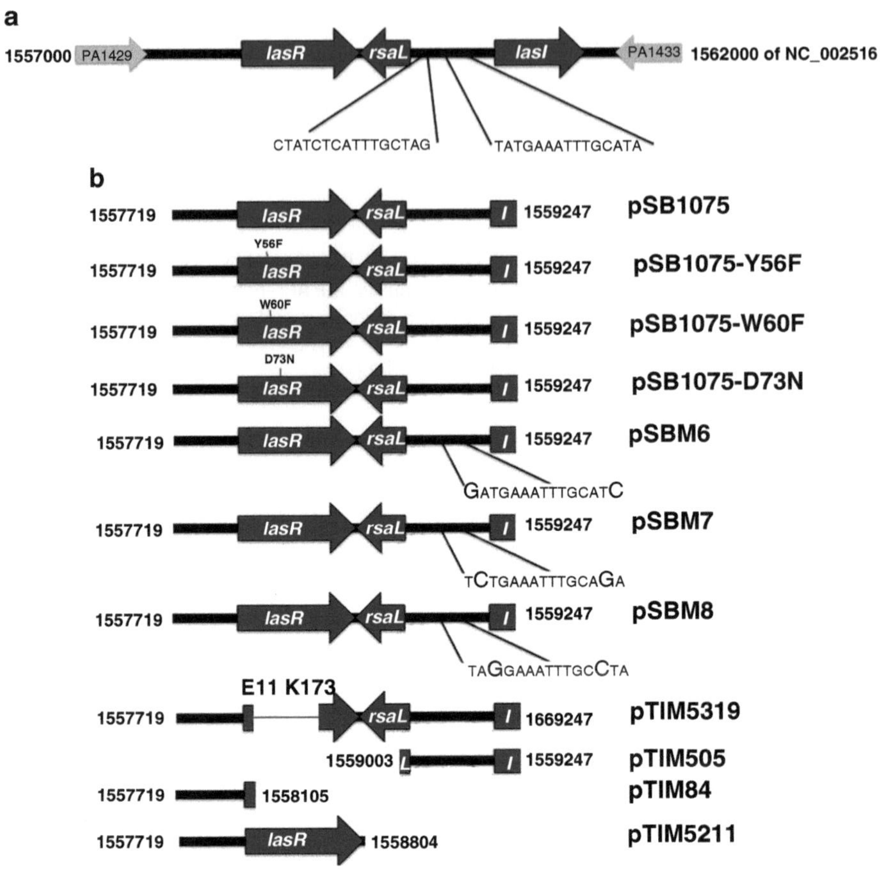

Fig. 4. Features of the luminescent reporters for the characterization of LasR-mediated QS and interference. (**a**) Genomic context for *lasR*, *rsaL* and *lasI*. A fragment spanning nucleotides 1557000–1562000 of the *P. aeruginosa* PA01 genome (NC_002518) is shown. The *left inset* indicates DNA sequence bound by LasR, the *right inset* is the DNA sequence bound by RsaL. (**b**) pSB1075 is the wild type reporter constructed by Winson et al. (1998); it contains *lasR* expressed from the native promoter, *rsaL* and a promoter for *lasI* (17). pSB1075-Y56F, pSB1075-W60F and pSB1075-D73N contain point mutations in *lasR*, the constructs are otherwise isogenic to pSB1075. pSBM6, pSBM7 and pSBM8 contain mutations in the RsaL-binding site upstream of *lasI*. *Capital letters* in the insets indicate mutated nucleotides. The reporters are otherwise isogenic to pSB1075. pTIM5319, pTIM505 and pTIM84 are cloned in pSB377 (18). pTIM5211 carries *lasR* on a low copy number plasmid pWSK129 (*21*). *Diagrams* depict regions of the *P. aeruginosa* PA01 genome contained in the reporters, *numbers to the right and left of each diagram* correspond to the nucleotides of NC_002518.

The mutagenized plasmids were sequenced at the UC Davis Genome Core Center and at the University of Florida ICBR Core. Enzymes and competent cells were purchased from Stratagene, La Jolla, CA.

The reporter constructs, in which mutations were introduced into the RsaL-binding site, were constructed essentially as above using the QuickChange XL Site-Directed Mutagenesis Kit and complimentary pairs of primers 5PlasI-mutVI and 3PlasI-mutVI to constructs pSBM6.

5PlasI-mutVII and 3PlasI-mutVII to construct pSBM7; and 5PlasI-mutVIII and 3PlasI-mutVIII to construct pSBM8. *Pfu* Turbo-catalyzed PCR (18 cycles) reactions were carried out with

Table 1
Primers used to construct reporters

5PlasI-mutVI	GCTAGTTATAAAAT**G**ATGAAATTTGCAT**C**AATTCTTCAGCTTCC	Binds nt 1559091 on the top strand (Accession # NC_002516). Introduced nucleotides are shown in bold
3PlasI-mutVI	GGAAGCTGAAGAATT**G**ATGCAAATTTCAT**C**ATTTTATAACTAGC	Binds nt 1559134 on the bottom strand (NC_002516). Introduced nucleotides are shown in bold
5PlasI-mutVII	GCTAGTTATAAAATT**C**TGAAATTTGCA**G**AAATTCTTCAGCTTCC	Binds nt 1559091 on the top strand (NC_002516). Introduced nucleotides are shown in bold
3PlasI-mutVII	GGAAGCTGAAGAATTT**C**TGCAAATTTCA**G**AATTTTATAACTAGC	Binds nt 1559134 on the bottom strand (NC_002516). Introduced nucleotides are shown in bold
5PlasI-mutVIII	GCTAGTTATAAAATTA**G**GAAATTTGC**C**TAAATTCTTCAGCTTCC	Binds nt 1559091 on the top strand (NC_002516). Introduced nucleotides are shown in bold
3PlasI-mutVIII	GGAAGCTGAAGAATTTAGGCAAATTTCCTAATTTTATAACTAGC	Binds nt 1559134 on the bottom strand (NC_002516). Introduced nucleotides are shown in bold
BA247	GAGTCATTCAATATTGGCAGGTAAACAC	Binds within *luxC* of *Photorhabdus luminescens* nt. 553 (Accesion # M90093)
BA439	CACTAACGTCCCAGCCTTTGCGCTC	Binds nt 1559261 on the bottom strand (NC_002516) (*17*)
BA1403	TTCTCTCGTGTGAAGCCATTGCTCTGAT	Binds nt 1559003 on the top strand (NC_002516)
BA1404	GCGTGGCGATGGGCCGACAGTG	Binds nt 1557719 on the top strand (NC_002516) (*17*)

(continued)

Table 1 (continued)

BA1407	ACCTGAGAGGCAAGATCAGAGAGTAAT	Binds nt 1558804 on the bottom strand (NC_002516)
BA1415	GCGTTCCAGCTCAAGAAAACCGTC	Binds nt 1558105 on the bottom strand (NC_002516)
BA1416	AAACCGGTGGTTCTGACCAGCCGG	Binds nt 1558586 on the top strand (NC_002516)
LasR-D73N	GCTACGCGCGGGTC**AAT**CCGACGGTCAGTCACTGTAC	Binds nt 1558272 on the top strand (NC_002516). Introduced nucleotides are shown in bold
LasR-D73N(compli)	GTACAGTGACTGACCGTCGG**ATT**GACCCGCGCGTAGC	Binds nt 1558308 on the bottom strand (NC_002516). Introduced nucleotides are shown in bold
LasR-W60F	GCAACTACCCGGCAGCC**TTT**CGCGAGCATTACGACCG	Binds nt 15558230 of the top strand (NC_002516). Introduced nucleotides are shown in bold
LasR-W60F(compli)	CGGTCGTAATGCTCGCG**AAA**GGCTGCCGGGTAGTTGC	Binds nt 1558266 on the bottom strand (NC_002516). Introduced nucleotides are shown in bold
LasR-Y56F	GCCTTCATCGTCGGCAAC**TTT**CCGGCAGCCTGGCGC	Binds nt1558217 on the top strand (NC_002516). Introduced nucleotides are shown in bold
LasR-Y56F (compli)	GCGCCAGGCTGCCGG**AAA**GTTGCCGACGATGAAGGC	Binds nt 1558252 on the bottom strand (NC_002516). Introduced nucleotides are shown in bold

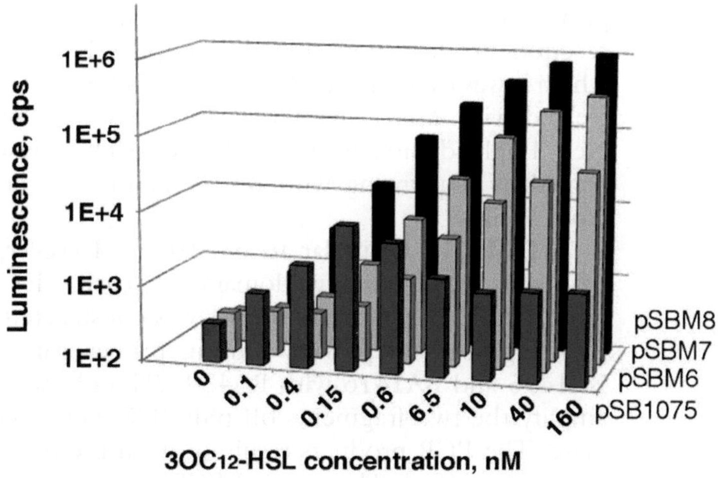

Fig. 5. Responses of reporters with mutations in the RsaL binding site to 3-oxo-C_{12}-HSL. Luminescence of pSB1075 (*dark grey bars*), and its derivatives with mutations in the RsaL binding site, pSBM6 (shown as *medium grey bars*), pSBM7 (*light grey bars*) and pSBM8 (*black bars*), was tested in a direct assay. Luminescence of the reporters after 8 h of incubation is shown.

pSB1075 as a template as described above. The resulting plasmids were sequenced at the UC Davis Genome Core Center and at UF ICBR Core. Responsiveness of the reporters to dilutions of 3-oxo-C_{12}-HSL was tested (Fig. 5)

To construct pTIM505, a fragment spanning the divergent *rsaL-lasI* promoter was amplified with primers BA1403 and BA439 using purified pSB1075 as a template. The PCR product was cloned into pCR2.1, from which it was excised with *Eco*RI and subcloned into pSB377 (18) treated with *Eco*RI and CIAP. The resulting plasmid (pTIM505) was transformed into chemically-competent *E. coli* DH5α. For bioassays, pTIM505 was transformed by electroporation into *E. coli* JM109.

pTIM5211 carries a full-length *lasR* on a low copy number plasmid. The *lasR* gene and its upstream region were PCR amplified using primers BA1404 and BA1407 off purified genomic DNA of *P. aeruginosa* PAO1 as a template. The resulting PCR product (~1.1 kB) was first cloned into pCR2.1, from which it was excised with *Eco*RI subcloned into pWSK129 (21) treated with *Eco*RI. The construct was confirmed by sequencing at the University of Florida Core Facility. The plasmid maintained in *E. coli* DH5α pTIM5211 was also electroporated into *E. coli* JM109 pTIM505, resulting in *E. coli* JM109 pTIM505 pTIM5211. This reporter strain lacks *rsaL*. Because *lasR* is carried on a low-copy number plasmid, potentially artifactual responses associated with the presence of multiple copies of *lasR* could be avoided.

To construct pTIM84, which carries a promoter for *lasR*, it was amplified by *Vent* DNA polymerase-catalyzed PCR with primers BA1404 and BA1415 and pSB1075 as a template. The reaction product was cloned into pCR-BluntII-TOPO, from which it was excised with *Eco*RI and subcloned into pSB377 (18) treated with *Eco*RI and CIAP. Enzymes were purchased from New England Biolabs (Ipswitch, MA). The construct was confirmed by sequencing with BA247 at the University of Florida Core Facility.

pTIM5319 is similar to pSB1075. However, residues that span the AHL-binding domain of LasR (E11-K173) were removed. The truncated product was constructed by amplifying two fragments of pSB1075 using primer pairs BA1404 with BA1415 and BA1416 with BA439. *Pfu* polymerase was used to amplify the two fragments off pSB1075 which served as a template. The PCR products resulting from the two PCR reactions were gel purified. The purified PCR products were then treated for 30 min with T4 Polynucleotide kinase (USB, Affymetrix, Cleveland, OH) in the T4 PNK buffer and ATP, as per manufacturer's instructions. The enzyme was heat-inactivated at 70°C for 15 min, and the reaction was cooled slowly to room temperature. Each reaction was then purified by gel filtration using Illustra Micro-Spin G50 columns (GE Healthcare), and then DNA fragments were concentrated using ethanol precipitation. The two purified phosphorylated PCR products were combined (1:1) and ligated overnight. Dilutions of the ligation reaction were used as templates for Vent *Pfu*-catalyzed PCR reaction with primers BA1404 and BA439. The resulting PCR product was cloned into pCR-TOPOII-Zero Blunt, from which it was excised with *Eco*RI and subcloned into pSB377. The construct was confirmed by sequencing with BA247 and BA1404.

To construct a constitutively luminescent reporter pTIM2442, a fragment containing a promoter for phage λ was cloned into pSB377 treated with *Eco*RI and CIAP. The construct was sequenced with BA247 (Table 1) and was confirmed to be constitutively luminescent regardless of the presence of AHLs (Fig. 6).

2. Materials

1. Glass tubes for growing bacteria.
2. Glass vials for storing solutions of chemicals.
3. Eppendorf tubes.
4. Disposable pipette tips.
5. Syringes.
6. 0.2 micron filters.
7. 96-well black microtiter plates (Corning).

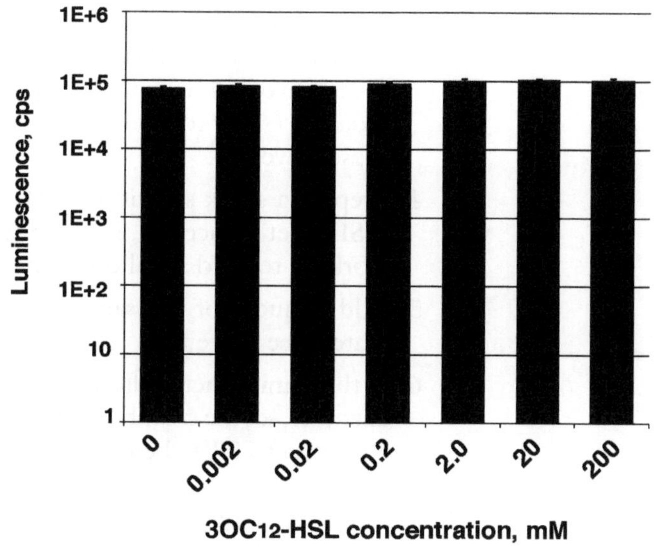

Fig. 6. Luminescence of the constitutive reporter. Exposure of JM109 pTIM2442 to various amounts of 3-oxo-C_{12}-HSL had no effect on luminescence of this constitutive reporter. Luminescence of cultures was measured every 2 h, data corresponding to the 8-h time point are shown. *Error bars* are standard errors of two technical replications.

8. Reporter strains (Fig. 4)- maintain frozen (at –80°C) as glycerol stocks (see Note 1).
9. LB broth (Miller, Fisher Scientific).
10. 3-oxo-C_{12}-HSL (CAS 168982-69-2, Fig. 3) is also known as *Pseudomonas* Autoinducer-2 (PAI-2) or *N*-3-oxo-dodecanoyl-L-HSL (OdDHL). (Cayman Chemical, Ann Arbor, MI or Sigma Aldrich, St. Louis, MO).
11. Patulin (Acros Organic).
12. Multi-mode microtiter plate reader Victor-3 (Perkin Elmer, Fremont, CA), equipped with Wallac1420 Manager Work-station software.

3. Methods

3.1. Sample Preparation

1. If samples are dissolved in water, they can be assayed directly (proceed to step 5).
2. If samples contain ethanol, methanol, or acetonitrile, organic solvents will need to be evaporated. This can be accomplished by directly adding solutions containing compounds of interest to the wells of microtiter plates, and then leaving them on a sterile flow-bench for 1–5 h for the solvents to evaporate (up to 30 μl of a sample dissolved in alcohol or acetonitrile will evaporate within approximately an hour).

3. If samples are in ethyl acetate or chloroform, then the solvents need to be evaporated first, and samples will need to be re-dissolved in an another appropriate solvent prior to the bioassay (so as to avoid corrosion of the microtiter plates by these solvents).
4. Prepare a stock solution (1 M) of the standard, 3-oxo-C12-HSL in ethyl acetate, store in glass vials (see Note 2). Make working tenfold serial dilutions of the AHL in ethanol.
5. Add aliquots of the sample to the wells of black microtiter plates (see Note 3).
6. If the luminometer allows for temperature control within the chamber, set the chamber temperature to 37°C. Let the chamber warm up to 37°C before proceeding with the assays.

3.2. Reporter Preparation

1. Grow reporter strains overnight at 37°C with shaking, in 5–10 ml of LB broth with the appropriate antibiotic(s). Start all cultures directly from glycerol stocks by scraping a loopfull of the frozen stock from the cryovial and inoculating it into LB broth with the antibiotic (see Note 4). Strains carrying pSB1075, pSB1075-Y56F, pSB1075-W60F, pSB1075-D73N, pSBM6, pSBM7, pSBM8, pTIM5319, pTIM505, pTIM84, pTIM2442 are grown with 200 μg/ml ampicillin, strains carrying pTIM5211 are grown with 50 μg/ml kanamycin.
2. Dilute 50 μl of overnight cultures into 5 ml of fresh LB broth with the appropriate antibiotics. LB broth should be at room temperature or pre-warmed to 37°C. Grow bacterial cultures for ~2 h at 37°C with shaking. Cultures should reach optical density (OD_{600}) ~ 0.4 (see Note 5).
3. Dilute 100 μl of the culture into 5 ml of fresh LB broth with appropriate antibiotics. Incubate for 30 min at 37°C with shaking. Cultures should reach OD_{600} ~ 0.05.
4. Dilute 1 ml of cultures into 10 ml of fresh LB broth (LB broth should be at room temperature or at 37°C). This diluted suspension is inoculum. Approximately 10 ml of inoculum will be required to seed a 96-well plate.

3.3. Direct Assays

Direct assays are designed to detect QS agonists, or compounds that indirectly affect QS by activating/reducing expression of *lasR*, *lasI* or luciferase. Direct assays can be carried out with all reporters shown in Fig. 4.

1. Add 100 μl of the inoculum to the wells containing compounds of interest, mix the reporter suspension and the sample by gently pipetting the well content up and down. At least two technical replications for each reporter will need to be carried out (see Note 6).

2. Measure luminescence immediately after adding the reporter. Cover the plate with a lid, place the plate inside a zip-top bag with a moist (but not wet) paper towel. Incubate the plate at 37°C.
3. Continue luminescence measurements every 2 h for 6–8 h (see Note 7).

3.4. Indirect Assays

Indirect assays are designed to detect QS antagonists. Indirect assays are carried out in the presence of the cognate AHL (3-oxo-C_{12}-HSL). Indirect assays will only provide useful information when pSB1075, pSBM6, pSBM7, pSBM8 and the dual-plasmid reporter pTIM5211 pTIM505 are used. Because other reporters will not respond strongly to 3-oxo-C_{12}-HSL, their use in indirect assays is of questionable value.

1. First, follow the protocol for the direct assays and determine the amount of 3-oxo-C_{12}-HSL that is necessary for a half-maximal activation of the reporter (EC_{50}).
2. Add 3-oxo-C_{12}-HSL to the inoculum (see Subheading 3.2, step 4) to reach EC_{50} (see Note 8).
3. Add 100 μl of the inoculum with 3-oxo-C_{12}-HSL to the wells containing compounds of interest, mix the reporter suspension and the sample. At least two technical replications for each reporter will need to be carried out.
4. Measure luminescence immediately after adding the reporter. Cover the plate with a lid, place the plate inside a zip-top bag with a moist (but not wet) paper towel. Incubate the plate at 37°C.
5. Continue luminescence measurements every 2 h for 6–8 h.

3.5. Data Analysis

1. The data obtained in a direct or an indirect bioassay can be presented as either total luminescence, which is a direct output of the luminometer (measured as luminescence counts/second, cps), or as "relative activity". To calculate relative activity (RA), we use the following formula:

$$RA = (L_S - L_{NG}) / L_{NG} \quad \text{(a)}$$

where L_S is luminescence of each sample, L_{NG} is luminescence of negative control (for the direct assays, it is "reporter only" control; for indirect assays, the negative control is "reporter with EC_{50} of 3-oxo-C_{12}-HSL"). Consider examples of data outputs for the direct assay shown in Fig. 7.

2. The formula (a) does not account for cell numbers in each well. This omission is due to the fact that it is difficult to accurately measure OD_{600} in clear-bottom microtiter plates, and even small differences in OD_{600} will affect the accuracy of calculations.

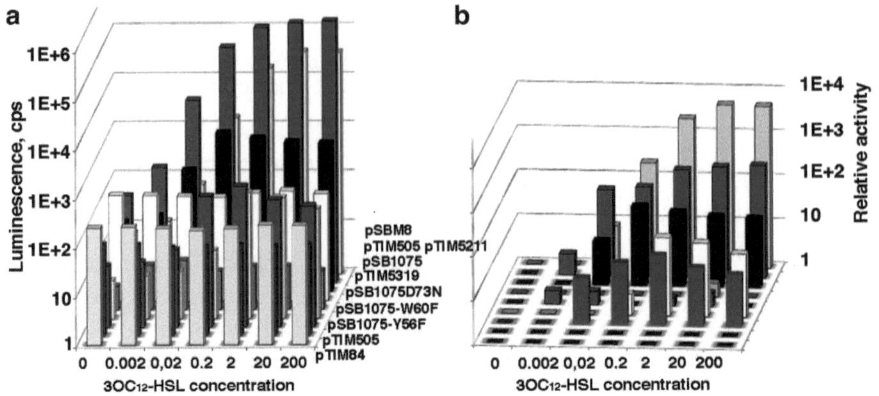

Fig. 7. Responses of the Las QS reporters to 3-oxo-C_{12}-HSL. Reporters were exposed to a dilution series of 3-oxo-C_{12}-HSL. Measurements were taken every 2 h, results obtained at 10 h are shown. Data are shown as luminescence of cultures (**a**) or relative activity (**b**).

In some instances (for example, when studying time course of reporter activation), it is important to present activity adjusted for bacterial growth (or growth inhibition). In these instances we propose to calculate "adjusted relative activity" (ARA):

$$\mathrm{ARA} = \left(L_{\mathrm{S}} / \mathrm{CL}_{\mathrm{S}} - -L_{\mathrm{NG}} / \mathrm{CL}_{\mathrm{NG}}\right) / \left(L_{\mathrm{NG}} / \mathrm{CL}_{\mathrm{NG}}\right) \qquad \text{(b)}$$

where L_S is luminescence of each sample, L_{NG} is luminescence of negative control, CL_{S} is 1/100 luminescence of pTIM2442 (the constitutive control luminescence reporter) exposed to the same amount of the sample, and $\mathrm{CL}_{\mathrm{NG}}$ is 1/100 luminescence of pTIM2442 in the negative control (because luminescence of pTIM2442 is strong, we recommend to divide the luminescence value by at least 100). If a compound affects luminescence in pTIM2442, even without affecting bacterial growth, ARA will score such compounds as not active in the QS assays.

The use of the entire suite of the reporters will aid in formulating a hypothesis to test mode of action of a candidate QS inhibitor (see Note 9).

4. Notes

1. Glycerol stocks are prepared by first centrifuging 1 ml aliquots of overnight cultures of the reporters for 1 min at 10,000 rpm in a bench-top microcentrifuge, then aspirating and discarding the supernatant and resuspending the pellet in LB broth supplemented with 15% glycerol. Four hundred microliters of the bacterial suspension in LB with glycerol are then added to plastic cryovials, which are stored at –80°C.

It is also prudent to maintain frozen preparation of reporter plasmids, which could be transformed into competent cells.

2. We find that AHLs are best maintained as stock solutions in ethyl acetate acidified with glacial acetic acid (1 ml of acetic acid per 1 L of ethyl acetate). We store them at −20°C in glass vials. If borosilicate vials are used for this purpose, prior to adding AHL solutions, glass vials are first treated with 0.2 N HCl, and then rinsed extensively with deionized water. Teflon-lined caps should be used for capping vials containing organic solutions of AHLs. As needed for experiments, serial dilution of 3-oxo-C_{12}-HSL can be prepared in other organic solvents (including ethanol, methanol and acetonitrile), however stability of the compound in these solutions is not known. 3-oxo-C_{12}-HSL is not stable in aqueous solutions where it converts into tetramic acid, an antibiotic (22). At neutral or basic pH, AHLs undergo spontaneous lactonolysis (23).

3. If multiple samples or reporters are added to the same plate, avoid placing them in nearby wells to avoid "light leakage". We use solid black plates (from Corning) for the bioassays to minimize "light contamination" from nearby wells. Even using the same samples, values obtained in wells that are solid black, black with clear bottoms or black with white insets will be different. For example, when aliquots of the same overnight culture of JM109 pTIM2442 were added to the wells of solid black plates, the measured luminescence in the wells with the reporter was $5.5 \times 10^4 \pm 1.6 \times 10^3$ counts per second (cps), in the wells immediately adjacent to those with the reporter, luminescence readings were 8 ± 2 cps, in the second row of wells, luminescent measured at 2 ± 1.6; with the background at 2 ± 5 cps. In black plates with clear bottoms (Corning), the corresponding measurements were $6.7 \times 10^4 \pm 1.1 \times 10^3$ cps; 4 ± 6 cps; 3.3 ± 3.86 cps and background 1.75 ± 1.7 cps. In black plates with white insets, the corresponding values for the same reporter were $5.3 \times 10^5 \pm 1.2 \times 10^4$ cps; 600 ± 77 cps; 320 ± 34 cps; background luminescence in the empty wells away from the wells with the reporter was measured at 200 ± 25 cps. "Light contamination" is especially problematic when sequential fractions (for example, from HPLC) are subject to bioassay on the same 96-well plate. Volatile compounds should be assayed on separate plates.

4. In our experience, data from the reporters, which were started from streaks maintained on LB agar plates at 4°C, have been highly variable. Starting reporters from glycerol stocks each time has helped reduce variability of responses.

5. Diluting reporter cultures, followed by short incubations helps maintain maximum responsiveness and reproducibility of the bioassay. Nevertheless, overnight cultures could be

used directly for bioassays. If overnight cultures are used directly, the background luminescence is 10- to 100-fold higher and the reporters are less responsive to AHLs at lower concentrations. However, relative activity of the reporters in the presence of higher amounts of AHLs will be generally the same regardless of reporter preparation. If overnight cultures are used directly, they will respond to AHLs within 30–120 min; they will remain responsive to AHLs for at least 18 h, however upon extended incubation, relative activity of the reporters is less than 5 units.

6. We recommend to begin the search for QS agonists or antagonists by first using the pSB1075 and pTIM2442 reporters only. Once interesting activities are detected, their mode of action can be characterized using the other reporters. In our experience, the majority of false positive hits also affected the pTIM2442 reporter. Therefore, only activities that affect the pSB1075 reporter and have no effect on the responsiveness of pTIM2442 should be selected for further assays.

7. Typically, responses of the reporters to AHLs can be detected within the first 2 h of incubation. The optimal responsiveness of the reporters is usually 8–10 h after the beginning of the bioassay.

8. There is some variability in potency of 3-oxo-C_{12}-HSL in different batches, especially if the compound was purchased from smaller suppliers or synthesized in house. We suggest to prepare a stock tenfold dilution series of 3-oxo-C_{12}-HSL in absolute methanol or ethanol, establish EC_{50} and use the dilution series for the bioassays. 3-oxo-C_{12}-HSL could be added by directly pipetting stock solution of 3-oxo-C_{12}-HSL into the suspension of the inoculum. However, because most organic solvents are toxic to bacteria, no more than 5 μl of the stock solution could be added to 10 ml of inoculum. Alternatively, a stock solution of 3-oxo-C_{12}-HSL could be first added to an eppendorf tube, the solvent can then be evaporated in a spin-vac or under nitrogen and the residue could be resuspended in a ~200 μl of LB broth and then immediately mixed with the inoculum.

9. In proof-of-principle experiments, we tested responses of the reporters to patulin, a compound with known QS inhibitory activity (15). We tested a range of concentrations of patulin in direct and indirect assays. At 0.3 mM and higher concentrations, the compound inhibited luminescence of JM109 pTIM2442, therefore we focused on the effects of 0.03 mM patulin on the reporters. At 0.03 mM, patulin inhibited responses of pSB1075 to 0.2 mM 3-oxo-C_{12}-HSL with RA = 2; no strong effect on pSB1075 was observed in the absence of the AHL. In the absence of *lasR* and *rsaL*, patulin

had no effect (RA = 0) on the regulation of *lasI* (JM109 pTIM505), suggesting that the effects of patulin are not due to non-specific interference with the transcription from the *lasI* promoter. The reporter with a large deletion in *lasR* (JM109 pTIM5319) was not affected (RA = 0) by the compound at this concentration, indicating that a functional LasR is required for the effects of patulin. Strong inhibitory effects of the compound were observed on the reporters borne on pSBM8, pSB1075-D73N, pSB1075-W60F, pSB1075-Y56F (RA = 16, 4, 10, 7, respectively, in the indirect bioassay; and RA = −8, −1, −4, −2, respectively, in the direct assays). Strong inhibition of the P_{lasR}-*luxCDABE* (JM109 pTIM84) by 0.03 mM patulin was also observed (RA = −10 in the direct assay), suggesting that one of the potential mechanisms by which patulin inhibits Las-mediated QS is through the inhibition of expression of *lasR*.

Acknowledgments

The development of the reporters and experiments presented in this manuscript were supported by USDA NRI grant # 2007-35319-18158 and funding from Protect Our Reefs Foundation.

References

1. Cha, C., Gao, P., Chen, Y. C., Shaw, P. D., and Farrand, S. K. (1998) Production of acyl-homoserine lactone quorum-sensing signals by Gram-negative plant-associated bacteria. *Mol. Plant Microbe Interact.* **11**, 1119–1129.
2. Fuqua, C., and Eberhard, A. (1999) Signal generation in autoinduction systems: synthesis of acylated homoserine lactones by LuxI-type proteins. *in* "Cell-cell signalling in bacteria" (Dunny, G. M., and Winans, S. C., Eds.), pp. 211–42, American Society for Microbiology, Washington, DC.
3. Zhang, R. G., Pappas, T., Brace, J. L., Miller, P. C., Oulmassov, T., Molyneaux, J. M., Anderson, J. C., Bashkin, J. K., Winans, S. C., and Joachimiak, A. (2002) Structure of a bacterial quorum sensing transcription factor complexed with pheromone and DNA. *Nature* **417**, 971–974.
4. Bottomley, M. J., Muraglia, E., Bazzo, R., and Carfi, A. (2007) Molecular insights into quorum sensing in the human pathogen *Pseudomonas aeruginosa* from the structure of the virulence regulator LasR bound to its autoinducer. *J. Biol. Chem.* **282**, 13592–13600.
5. Kiratisin, P., Tucker, K. D., and Passador, L. (2002) LasR, a transcriptional activator of *Pseudomonas aeruginosa* virulence genes, functions as a multimer. *J. Bacteriol.* **184**, 4912–4919.
6. Gilbert, K. B., Kim, T. H., Gupta, R., Greenberg, E. P., and Schuster, M. (2009) Global position analysis of the *Pseudomonas aeruginosa* quorum-sensing transcription factor LasR. *Mol. Microbiol.* **73**, 1072–1085.
7. Cao, H., Baldini, R. L., and Rahme, L. G. (2001) Common mechanisms for pathogens of plants and animals. *Annu. Rev. Phytopathol.* **39**, 259–284.
8. Williams, P., and Camara, M. (2009) Quorum sensing and environmental adaptation in *Pseudomonas aeruginosa*: a tale of regulatory networks and multifunctional signal molecules. *Curr. Opin. Microbiol.* **12**, 182–191.
9. Juhas, M., Wiehlmann, L., Huber, B., Jordan, D., Lauber, J., Salunkhe, P., Limpert, A. S., von Gotz, F., Steinmetz, I., Eberl, L., and Tummler, B. (2004) Global regulation of quorum sensing and virulence by VqsR in *Pseudomonas aeruginosa*. *Microbiology* **150**, 831–841.

10. Seed, P. C., Passador, L., and Iglewski, B. H. (1995) Activation of the *Pseudomonas aeruginosa lasI* gene by LasR and the *Pseudomonas* Autoinducer PAI – an autoinduction regulatory hierarchy. *J. Bacteriol.* **177**, 654–659.
11. Rampioni, G., Schuster, M., Greenberg, E. P., Bertani, I., Grasso, M., Venturi, V., Zennaro, E., and Leoni, L. (2007) RsaL provides quorum sensing homeostasis and functions as a global regulator of gene expression in *Pseudomonas aeruginosa. Mol. Microbiol.* **66**, 1557–1565.
12. Dubern, J. F., Lugtenberg, B. J. J., and Bloemberg, G. V. (2006) Quorum sensing by 2-alkyl-4-quinolones in *Pseudomonas aeruginosa* and other bacterial species. *J. Bacteriol.* **188**, 2898–2906.
13. Rampioni, G., Polticelli, F., Bertani, I., Righetti, K., Venturi, V., Zennaro, E., and Leoni, L. (2007) The *Pseudomonas* quorum-sensing regulator RsaL belongs to the tetrahelical superclass of H-T-H proteins. *J. Bacteriol.* **189**, 1922–1930.
14. Rasmussen, T. B., and Givskov, M. (2006) Quorum sensing inhibitors: a bargain of effects. *Microbiology* **152**, 895–904.
15. Rasmussen, T. B., Skindersoe, M. E., Bjarnsholt, T., Phipps, R. K., Christensen, K. B., Jensen, P. O., Andersen, J. B., Koch, B., Larsen, T. O., Hentzer, M., Eberl, L., Hoiby, N., and Givskov, M. (2005) Identity and effects of quorum-sensing inhibitors produced by *Penicillium* species. *Microbiology* **151**, 1325–1340.
16. McClean, K. H., Winson, M. K., Fish, L., Taylor, A., Chhabra, S. R., Camara, M., Daykin, M., Lamb, J. H., Swift, S., Bycroft, B. W., Stewart, G. S., and Williams, P. (1997) Quorum sensing and *Chromobacterium violaceum*: exploitation of violacein production and inhibition for the detection of *N*-acylhomoserine lactones. *Microbiology* **143**, 3703–3711.
17. Winson, M. K., Swift, S., Fish, L., Throup, J. P., Jorgensen, F., Chhabra, S. R., Bycroft, B. W., Williams, P., and Stewart, G. S. (1998) Construction and analysis of *luxCDABE*-based plasmid sensors for investigating *N*-acyl homoserine lactone-mediated quorum sensing. *FEMS Microbiol. Lett.* **163**, 185–192.
18. Winson, M. K., Swift, S., Hill, P. J., Sims, C. M., Griesmayr, G., Bycroft, B. W., Williams, P., and Stewart, G. S. (1998) Engineering the *luxCDABE* genes from *Photorhabdus luminescens* to provide a bioluminescent reporter for constitutive and promoter probe plasmids and mini-Tn*5* constructs. *FEMS Microbiol. Lett.* **163**, 193–202.
19. Teplitski, M., Robinson, J. B., and Bauer, W. D. (2000) Plants secrete substances that mimic bacterial N-acyl homoserine lactone signal activities and affect population density-dependent behaviors in associated bacteria. *Mol. Plant Microbe Interact.* **13**, 637–648.
20. Rajamani, S., Bauer, W. D., Robinson, J. B., Farrow, J. M., Pesci, E. C., Teplitski, M., Gao, M., Sayre, R. T., and Phillips, D. A. (2008) The vitamin riboflavin and its derivative lumichrome activate the LasR bacterial Quorum-Sensing receptor. *Mol. Plant Microbe Interact.* **21**, 1184–1192.
21. Wang, R. F., and Kushner, S. R. (1991) Construction of versatile low-copy-number vectors for cloning, sequencing and gene expression in *Escherichia coli. Gene* 100, 195–199.
22. Kaufmann, G. F., Sartorio, R., Lee, S. H., Rogers, C. J., Meijler, M. M., Moss, J. A., Clapham, B., Brogan, A. P., Dickerson, T. J., and Janda, K. D. (2005) Revisiting quorum sensing: discovery of additional chemical and biological functions for 3-oxo-*N*-acylhomoserine lactones. *Proc. Natl. Acad. Sci. U. S. A.* **102**, 309–314.
23. Yates, E. A., Philipp, B., Buckley, C., Atkinson, S., Chhabra, S. R., Sockett, R. E., Goldner, M., Dessaux, Y., Camara, M., Smith, H., and Williams, P. (2002) *N*-acyl homoserine lactones undergo lactonolysis in a pH-, temperature-, and acyl chain length-dependent manner during growth of *Yersinia pseudotuberculosis* and *Pseudomonas aeruginosa. Infect. Immun.* **70**, 5635–5646.

Part II

Determining the Function of Autoinducers In vivo

Chapter 10

Modulation of Mammalian Cell Processes by Bacterial Quorum Sensing Molecules

Vladimir V. Kravchenko, Richard J. Ulevitch, and Gunnar F. Kaufmann

Abstract

Microbial pathogens use a wide repertoire of pathogen-associated molecular patterns (PAMPs) that affect host cell responses through activation of intracellular signaling events in a PAMP-specific manner. Here we describe a set of western blot-based methodologies for the evaluation of biochemical effects specifically induced by *N*-(3-oxo-acyl) homoserine lactones (3-oxo-AHLs) small molecules secreted by a number of Gram-negative bacteria, including the opportunistic human pathogen *Pseudomonas aeruginosa*. First, we will highlight the AHL-mediated effects on proapoptotic and stress pathways. Secondly, we will demonstrate that AHLs possess the ability to alter stimulus-induced NF-κB signaling, a key biochemical marker of inflammation and innate immune responses.

Key words: Acyl homoserine lactones, Macrophages, NF-kappa B, *Pseudomonas aeruginosa*, MAP kinase, Apoptosis

1. Introduction

The innate immune system is the first line of defense against infectious diseases. It is activated in response to pathogen-associated molecular patterns (PAMPs) through a complex set of signaling networks, including induction of the nuclear factor kappa B (NF-κB) pathway (1–3). For example, macrophages, critical cellular components of the innate immune system, use the Toll-like receptor 4 (TLR4) to recognize lipopolysaccharide (LPS) as a generic signal for an infection caused by Gram-negative bacteria. The TLR4-mediated recognition of LPS results in rapid activation of the transcription factor NF-κB and the expression of genes encoding proinflammatory cytokines, including tumor necrosis factor-α (TNF). In turn, TNF signaling initiates the next round of NF-κB activation through the engagement of TNF receptors

Kendra P. Rumbaugh (ed.), *Quorum Sensing: Methods and Protocols*, Methods in Molecular Biology, vol. 692,
DOI 10.1007/978-1-60761-971-0_10, © Springer Science+Business Media, LLC 2011

that is abundantly expressed on all cell types. These NF-κB-dependent multilayered processes, in concert with other signaling cascades, such as the p38 protein kinase pathway, result in coordinated physiological responses that are critical for pathogen elimination. However, although the adherence of *Pseudomonas aeruginosa* activates NF-κB signaling in normal and cystic fibrosis (CF) cells (4), this pathogen is able to establish persistent infections that occur in several pathological conditions, such as chronic colonization of the airways of CF patients or biofilm formation in patients with catheters and on respirators.

The key breakthrough that opened the way for our modern understanding of *P. aeruginosa*'s pathogenesis was the identification of *N*-(3-oxododecanoyl)-*S*-homoserine lactone (3-oxo-C_{12}-HSL) as a bacterial molecule essential for the control of virulence gene expression through a process called "quorum sensing" (QS) (5–7). Extensive studies have demonstrated that in addition to its function as a bacterial QS regulator, 3-oxo-C_{12}-HSL (and several other AHLs) exerts strong effects on mammalian cells (8–18), although the mechanism of action is poorly understood (19).

Several important characteristics with respect to AHL-mediated effects on mammalian cells might be deciphered by elucidating the molecular mechanisms involved. First, consistent with the chemical nature of AHLs, 3-oxo-C_{12}-HSL freely enters mammalian cells and displays intracellular activity (20, 21). However, hydrolysis at physiological pH (17, 22, 23) and by mammalian enzymes (24) will rapidly reduce levels of the active compound, as the structural integrity of the homoserine lactone ring motif is absolutely required for 3-oxo-C_{12}-HSL-mediated effects on mammalian cells (13, 17, 25). Therefore, many of the 3-oxo-C_{12}-HSL effects on cultured mammalian cells require micromolar concentrations (10–50 μM) that are higher than those that have been detected in planktonic *P. aeruginosa* cultures (1–5 μM), although concentrations of up to 600 μM in biofilms have been reported (26). Secondly, it has been observed in a wide variety of mammalian cell types that 3-oxo-C_{12}-HSL, as well as other members of AHL family, induces apoptosis and rapid activation of at least two stress signaling markers: the phosphorylation of eukaryotic translation initiation factor 2α (eIF2α) and mitogen-activated protein kinase p38 (12, 15–17). Thus, monitoring these cellular responses can be a useful assay for validation of the biological activity of AHLs under the applied experimental conditions.

Given the central relevance of the NF-κB pathway for inflammation and innate immunity, it is strongly recommended to investigate and compare the effects of bacterial products (3-oxo-C_{12}-HSL versus LPS for instance (17, 25)) on NF-κB signaling, e.g., by monitoring the coordinated phosphorylation, degradation, and resynthesis of alpha subunit of the NF-κB inhibitor protein (IκBα), as this is a key functional feature of the NF-κB pathway in response to proinflammatory stimuli (2, 3).

2. Materials

2.1. Reagents, Standard Buffers, and Antibodies

1. *N*-(3-oxododecanoyl)-*S*- and -*R*-homoserine lactones (C12 and C12R, respectively), *N*-butyryl-*S*-homoserine lactone (BHL), 3-oxo-dodecanoic acid (2-oxo-pyrrolidin-3-yl)-amide (lactam, C12-LM), *N*-(3-oxo decanoyl)-*S*-homoserine lactone (C10), or *N*-(3-oxotetradecanoyl)-*S*-homoserine lactone (C14) is dissolved in dimethyl sulfoxide (DMSO; ATCC, Cat. No. 4-X-5) at 200× the desired concentration, stored in aliquots at −20°C, and then added to tissue culture dishes as required (unless stated otherwise, final concentration of DMSO in all tissue culture experiment is 0.5%; see also Note 1). In addition, a quantitative QCL-1,000 chromogenic *Limulus* amoebocyte lysate assay (BioWhittaker, Walkersville, MD) should be conducted to demonstrate that the preparations of C10, C12, C12R, C12-LM, C12-TA, C14, and BHL are endotoxin free. Chemical structure of the AHLs and analogs are shown in Fig. 1 (see also Note 2).
2. Lipopolysaccharide (LPS) from *Salmonella minnesota* Re595 or *E. coli* 0111:B4 (List Biological Laboratory, Campbell, CA) is dissolved at 5 mg/mL in sterile water (Baxter Healthcare Corporation, Deerfield, IL) containing 5 mM ethylenediamine tetraacetic acid (EDTA; LifeTechnologies, USA) and stored in aliquots at −20°C. Working solutions are prepared by dilution in sterile water at 1,000× the desired

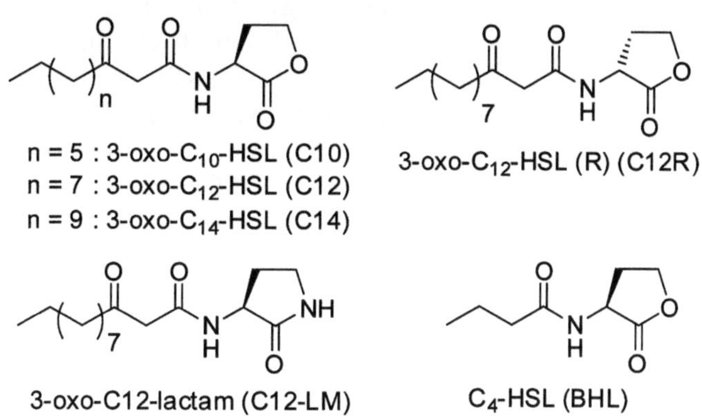

Fig. 1. Chemical structures of naturally occurring *N*-(3-oxo-acyl)-*S*-homoserine lactones (C10, C12, and C14), *N*-butyryl-*S*-homoserine (BHL) and synthetic analogs of C12 [*N*-(3-oxododecnoyl)-*R*-homoserine lactones (C12R) and 3-oxo-dodecanoic acid (2-oxo-pyrrolidin-3-yl)-amide (C12-LM)] are shown.

concentration and stored at −20°C. The recommended concentration of LPS in the experiments on AHLs is 100 ng/mL (for stimulation of macrophages) or 1 μg/mL (for the stimulation of primary embryonic fibroblasts). Before addition to the cells, the aliquot of appropriate LPS stock is defrosted by ultrasonic treatment in a water bath (Branson ultrasonic cleaner, model Bransonic 12, for example) for 10–30 min at room temperature and kept on ice during all experimental time, but not longer than 6 h. The unused amount of LPS solution can be stored at −20°C.

3. Tumor necrosis factor-alpha (TNF) from any commercial sources can be used instead of LPS as a proinflammatory stimulus for all types of cells. The recommended concentration of TNF in the experiments is 40 ng/mL. The required aliquot of TNF must be prepared immediately before the experiment [for dilution use cold 10 mM sodium phosphate buffer, pH 7.5, containing 0.5 mM ethylenediamine tetraacetic acid (EDTA)] and kept on ice for not longer than 10 h. Never use an aliquot of TNF that is older than 10 h.
4. Tris-buffered saline (TBS) with Tween-20: 25 mL of 1 M Tris–HCl, pH 7.5, and 1 mL of Tween-20 (Fisher) into 950 mL of 0.9% sodium saline solution. This standard western blotting solution is referred to as TBST in this protocol.
5. Primary antibodies (Abs) for the detection of inhibitor of NF-κB alpha (IκBα) (Santa Cruz Biotechnology, Inc., Santa Cruz, CA; dilution 1:2,000); phospho-IκBα (p-IκBα) (Cell Signaling Technology, Danvers, MA; dilution 1:1,000); eukaryotic translation initiation factor 2 alpha (eIF2α) (Cell Signaling Technology; dilution 1:1,000); phospho-eIF2α (p-eIF2α) (Cell Signaling Technology; dilution 1:1,000); mitogen-activated protein kinase p38 (p38) (Cell Signaling Technology; dilution 1:1,000); phospho-p38 (p-p38) (Cell Signaling Technology; dilution 1:1,500); poly(ADP-ribose) polymerase (PARP) (Cell Signaling Technology; dilution 1:1,500); protein kinase C delta (PKCδ) (Santa Cruz Biotechnology, Inc.; dilution 1:2,000); and actin (Sigma; dilution 1:2,500). All primary antibodies are diluted in TBST buffer supplemented with 1% (w/v) nonfat dry milk and 0.025% (w/v) sodium azide (see also Note 3).
6. Secondary Abs: anti-Rabbit and anti-Mouse IgG conjugated to horse radish peroxidase (Bio-Rad, Hercules, CA) are diluted (1:7,500 and 1:5,000, respectively) in TBST supplemented with 1% (w/v) nonfat dry milk. The dilutions of secondary Abs are prepared immediately before their application and discarded after use.
7. Amersham-enhanced chemiluminescent (ECL) plus Western Blotting Detection System (GE Healthcare, UK) and BioMax

XAR film (Kodak, Rochester, NY) are used according to manufacturer's protocols.

8. PVDF membranes (Amersham, GE Healthcare, UK) and 3MM Chr chromatography paper (Whatman, Maidstone, UK) are used for the assembly of the protein transfer system in Bio-Rad western's blot apparatus.
9. Cell lysis buffer LB1: 25 mM HEPES, pH 7.5, 137 mM NaCl, 3 mM EDTA, 3 mM β-glycerolphosphate, 1% (v/v) Triton X-100, 0.1 mM sodium dodecavanadate, 1 μg/mL aprotenin, 2 μM MG132 (Calbiochem, La Jolla, CA), 1 mM phenylmethylsulfonyl fluoride (PMSF). Store in aliquots at −20°C. After defrosting, keep the aliquots on ice.
10. Cell lysis buffer LB2: 10 mM Tris–HCl, pH 6.8, 250 mM dithiotreitol, 4% (w/v) sodium dodecyl sulfate (SDS), 15% glycerol (v/v), 0.1 mM sodium dodecavanadate, 2 μM MG132, 1 mM PMSF, 0.05% (w/v) bromophenol blue. Store in aliquots at −20°C. After defrosting, keep the aliquots on ice.

2.2. Cell Culture

1. Fetal bovine serum (FBS) (HyClone) is heat treated for 30 min at 55°C and stored in 50 mL aliquots at −20°C.
2. Dulbecco's Modified Eagle's Medium (DMEM with 4.5 g/L glucose and puruvate, Gibco BRL, Invitrogen Corporation, USA) supplemented with L-glutamine, penicillin/streptomycin, nonessential amino acids (the 100× stocks from Invitrogen), 10 mM HEPES, pH 7.4 (1 M stock solution from Invitrogen), and 10% FBS is prepared by filtration through a 0.22 μm filter system (Corning) and stored at 4°C (not longer than 2 weeks). This standard tissue culture medium is referred to as "growth medium" (GM) in this protocol. For some experiments, GM containing only 1% FBS (referred to as GM with 1% FBS) will be required.
3. L922 conditioned medium is collected from confluent monolayer of L929/NCTC clone 929 (ATCC) mouse fibroblasts growing in GM. Pooled (usually from 20 T150 flasks) solutions of this conditioned medium are passed through a 0.22 μm filter system and stored in aliquots at −20°C.
4. Bone marrow-derived macrophages (BMDM) and murine embryonic fibroblasts (MEFs) are prepared from C57BL/6 mice (Jackson Laboratory) and used for experiments on AHLs and analogs. MEFs are maintained in GM [pass (1:3) every 3 days]; BMDM are cultured in 70% GM- and 30% L929-conditioned medium (medium has to be changed every 2 days). BMDM are split (1:2) when the cellular monolayer becomes confluent (usually after 4–5 days). BMDM can be used for up to passage five; usually, MEFs can be used up to

passage seven as primary embryonic fibroblasts, although these cells reproducibly show the same biochemical responsiveness to LPS and AHLs after immortalization (passages as late as 20–50).

5. Alternative mammalian cell lines that can be used: normal human bronchial epithelial cells (NHBE) from Cambrex (East Rutherford, NJ); THP-1 (monocyte, human), RAW 264.7 (monocyte/macrophage, mouse), and WI-38 (SV40 transformed lung fibroblasts, human; CCL-75.1) cell lines from ATCC.
6. Tissue culture dishes, flasks, and 12- or 24-wells plates from Corning. Cell scrapers from Sarstedt (Newton, NC).
7. Cell dissociation buffers: 0.05% trypsin containing 1 mM EDTA or Enzyme-free Hank's-based cell dissociation buffer (EFCDB) (Invitrogen). Alternatively, solution of 2 mM EDTA in the 1× Dulbecco's phosphate-buffered saline (Invitrogen) can be used as EFCDB.

2.3. Reagents, Buffers, and Manual Instructions for SDS–Polyacrylamide Gel Electrophoresis and Western Blotting

1. Laemmli's SDS–Polyacrylamide Gel Electrophoresis (SDS–PAGE) buffer system (*27*) is routinely used in our experiments for analyzing cellular extracts by gel electrophoresis of proteins.
2. Separating buffer (4×): 1.5 M Tris–HCl, pH 8.7, 0.4% SDS. Store at room temperature.
3. Stacking buffer (4×): 0.5 M Tris–HCl, pH 6.8, 0.4% SDS. Store at room temperature.
4. Running buffer: A 10× stock can be made by dissolving 30.25 g of Tris (base), 144.1 g of glycine, and 10 g of SDS in 900 mL of deionized water. The volume is adjusted to 1,000 mL with deionized water. Store at room temperature.
5. 30% Acrylamide/bis solution (ratio 29:1) (this is a neurotoxin when unpolymerized and so care should be taken to prevent exposure).
6. *N,N,N,N'*-Tetramethyl-ethylenediamine (TEMED).
7. Ammonium persulfate: Prepare 30% solution in deionized water and immediately freeze in single use (100–200 μL) aliquots at –20°C.
8. Prestained molecular weight marker (Bio-Rad).
9. SDS–polyacrylamide gel – commercially available or can be prepared according to standard protocols (28).
10. Transfer buffer: 4 L (this volume assumes the use of Bio-Rad's TRAN-BLOT-CELL apparatus) of solution for transfer proteins from SDS–polyacrylamide gel to the PVDF membrane is made by dissolving 12.1 g of Tris (base), 57.65 g of glycine in 3,000 mL of distilled water; finally, add 800 mL

of ethanol and 1 mL of 20% of SDS, adjust the volume to 4,000 mL with deionized water and store at 4°C.

11. Western blot striping reagent (SignaGen Laboratories).

3. Methods

3.1. Cell Stimulation and Preparation of Samples for Western Blotting Analysis

1. Cell passage and/or setup: Monolayer of BMDM (about 95% confluent) in 100 mm dish is rinsed with EFCDB (7 mL), then fresh 7 mL of EFCDB are added, and cells are harvested using a cell scaper and transferred into a 50 mL centrifuge tube (Corning, Lowell, MA) containing 30 mL of culture medium (CM = 70% GM- and 30% L929-conditioned medium). Cells are collected by centrifugation (800 × *g*, 5 min at 20°C) and resuspended in 5 mL of fresh CM. Aliquots of the cell suspension are used to setup 100 mm dish (2.5 mL per dish containing 9 mL of CM) or multiwell plates for the experiments (e.g., 0.1 mL per well in 24-wells plate containing 1 mL of CM in each well); monolayer of MEFs (about 95% confluent) in T75 flask is rinse with EFCDB (7 mL), then fresh 3.5 mL of EFCDB plus 3.5 mL of trypsin/EDTA solution are added, and cells are gently scraped off and transferred into a 50 mL centrifuge tube (Corning) containing 30 mL of GM. Finally, cells are collected by centrifugation (800 × *g*, 5 min at 20°C) and resuspended in 6 mL of fresh GM. Aliquots of the cell suspension are used to setup T75 flasks (2 mL preflask containing 20 mL of GM) or multiwell plates for experiment (e.g., 0.2 mL per well in 12-wells plate containing 1 mL of GM in each well). Immediately mix (gently using pipette) the cellular suspension after the addition of each aliquot of cells into fresh dish, flask, or well. Incubate the new cultures in Forma (or similar) Incubator (37°C, 5.5–6.0% of CO_2, 95% humidity).
2. The cultures of BMDM or MEFs usually reach the required density after 48 h. At this point, remove old medium by aspiration and gently add (use pipette with a fixed volume) 1 mL of fresh GM for BMDM or 2 mL of GM with 1% FBS for MEFs. BMDM are incubated for an additional 2 h, while MEFs should be incubated for an additional 36–48 h.
3. Before the start of the experiment, make sure that all required reagents and materials are thawed and ready: the appropriate aliquots of DMSO (as a vehicle control), chemical compounds, and LPS or TNF; a labeled microcentrifuge tube (1.5 mL Eppendorf) for each sample; cell lysis buffers; a vacuum aspirator; a dry thermoblock at 100°C; and a tank with liquid nitrogen.

4. This example of cell stimulation assumes that the use of BMDM incubated in 24-wells plate containing 1 mL of GM (see above). Add DMSO or an appropriate C12 stock (5 μL of each) into the medium (do not touch the plastic inside the well), and gently mix the contents immediately. Right after that a proinflammatory stimulus, such as LPS, other TLR ligands or TNF, can be added into an appropriate sample followed by mixing the content in the same manner as described above. Incubate the sample(s) according to the desired protocol (dose, concentration titration, or kinetics). To stop cell stimulation, place the plate on ice (following procedures must all be carried out on ice) and remove the medium by aspiration. The cells are lysed by adding 40 μL of lysis buffer LB1 (add first) with following addition of 40 μL of LB2. Incubate the plate on ice for 5 min, collect the contents of each well using a 1 mL fixed-volume pipette, and transfer it into appropriately labeled Eppendorf tube.
5. Briefly spin the sample tubes in a microcentrifuge and transfer them into a dry thermoblock (to prevent loss of the samples, put a heavy plastic or metal plate on top of the tubes during heat exposure). Incubate the samples at 100°C for 10 min. Finally, the samples are cooled on ice for 2–3 min, briefly spun down, frozen in liquid nitrogen bath, and stored at –20°C. Before loading onto the gel, thaw the sample at room temperature (about 10 min) and then heat to 100°C in a dry thermoblock for 2–3 min.
6. Thirty minutes before loading the sample, assemble the gel unit in the electrophoresis apparatus and add running buffer to the bottom and top reservoirs.
7. Prepare 9% SDS–polyacrylamide gel (make sure to reserve a well for the prestained molecular weight marker), load marker and samples (10 μL per lane), run the gel, and transfer the SDS–PAGE separated proteins into PVDF membrane. Example of SDS–PAGE run: Load protein samples (10 μL of each sample in a well) on a 9% gel (150-mm height, 140-mm length, and 1.5-mm thick; the teeth of the comb 10×3 mm). The dye front can be run at 100 V through the stacking gel (about 30–45 min) and then at 115 V through the separating gel until the dye marker runs off the gel (about 5 h). Alternatively, an appropriate premade SDS gels and SDS–PAGE systems can be used. For detail instructions, how to perform the protein transfer from SDS–polyacrylamide gels we recommend standard protocols. (28) Alternatively, the protocols and instructions provided with other SDS–PAGE and protein-transfer systems can be used. Example: proteins separated on a 9% gel can be transferred to PVDF membrane in a Bio-Rad TRAN-BLOT-CELL unit (a volume of the transfer-tank about 2,000 mL) at 35 V overnight (but not longer than 16 h) at room temperature.

8. Use whole membrane or cut the membrane into two or more parts (for example between 75 and 50 kDa marker), immerse the membrane in at least 30 mL TBST and place on a slowly rotating platform for a minimum of 15 min at room temperature. Discard old buffer, rinse the membrane with 30 mL of fresh TBST, and add 30 mL of 4% (w/v) nonfat dry milk (milk) in TBST to block the membrane. Rotate in the blocking solution for 20 min, then rinse at least five times in TBST. After this point, the membrane is ready for western blotting assay.

3.2. Activation of Proapoptotic Signaling by C12 in Macrophage: Western Blotting Analysis for Cleavage of PARP and PKCδ

1. Immerse the membrane in a solution containing anti-PARP or anti-PKCδ Abs and incubate overnight on a slow rotating platform setting at 4°C. On the next day, all manipulations of the membrane are conducted at room temperature.
2. Remove primary Abs solution (see Note 2) and wash the membrane at least seven times for 5–15 min each with 30 mL TBST using a slowly rotating platform.
3. Immerse the washed membrane in a solution containing secondary Abs, and slowly rotate for 20 min, discard the solution of Abs, add 30 mL TBST, and place on a slow rotating platform for 20 min. Finally, wash the membrane at least seven times for 5–15 min each with 30 mL TBST. After this point, the membrane is ready for development with ECL reagent.
4. Put ECL-developed membrane between the leaves of an acetate sheet protector, mark the orientation with a luminescent tape, place the membrane in an X-ray film cassette with film, and expose the film for 15 s, 30 s, 1 min, or until a satisfactory image has been obtained.
5. Although the membrane can be stripped by using appropriate protocols (see below), usually after extensive washing (seven times and leave membrane overnight in TBST) these membranes can be probed for actin (a loading control) on the next day.
6. Prototypic examples of western blotting analysis for PARP and PKCδ cleavage are shown in Fig. 2.

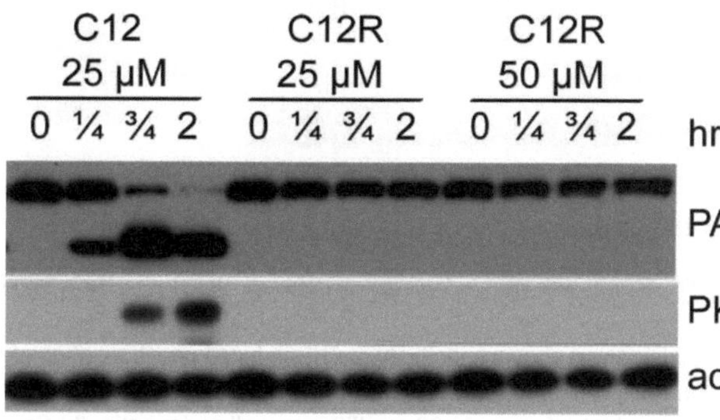

Fig. 2. Western blot analysis demonstrates stereospecificity of C12-induced cleavage of PARP and generation of the truncated form of PKCδ in bone marrow-derived macrophages.

3.3. Activation of Stress Pathways by C12 in Macrophages

1. For probing the membrane for phosphorylated forms of eEF2α (p-eIF2α) and MAPK p38 (p-p38), first, use the protocol described above and carry out steps 1–4 using solution of primary Abs for the detection of p-eIF2α.
2. The bands of p-eIF2 and p-p38 should be perfectly separated on a 9% gel (150-mm height, 140-mm length, and 1.5-mm thick), therefore stripping of the membrane is not required.
3. Wash extensively the p-eIF2α-stained membrane in TBST and subsequently probe for p-p38.
4. Strip the membrane according to manufacturer's instructions.
5. Block the stripped membrane again for 10 min in TBST/2% milk, rinse five times (2–5 min each) in TBST.
6. The membrane is then ready to be probed for total eIF2α, p-p38, or actin with washes, secondary antibodies, and ECL detection as above.
7. The prototypic example of western blotting analysis of C12-induced eIF2α and p38 phosphorylation in macrophages is shown in Fig. 3.

3.4. Evaluation of Anti-inflammatory Activity of C12 in LPS-Activated Macrophages

1. This experiment illustrated the western blot analysis of protein extracts from macrophages treated with LPS in the presence of C12 (see step 3 in Subheading 3.1). Similar analysis can be conducted using cellular extracts from macrophages or other cell types stimulated by a combination of a TLR ligand or TNF and C12, C14, or analogs.
2. First, perform the western blot assay using mouse monoclonal Abs specific for the phosphorylated form of IκBα as above (see steps 1–4 in Subheading 3.2). After developing, rinse the membrane in TBST, briefly block in 2% milk/TBST for 5–10 min, and then rinse again five times in TBST.

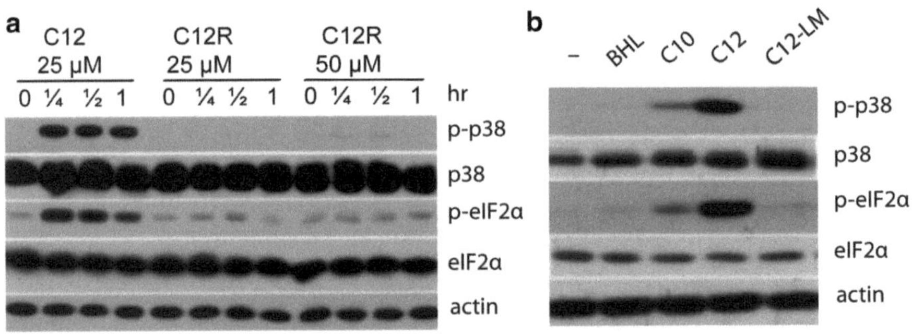

Fig. 3. Western blot analysis shows specificity of AHL-induced phoshorylation of p38 and eIF2α in bone marrow-derived macrophages. (**a**) Cells were treated with C12 or C12R as indicated, and cellular extracts were prepared and analyzed for phosphorylated and total forms of p38 or eIF2α as well as for actin. (**b**) Cells were treated with 50 μM of BHL, C10, C14, or C12-LM for 30 min, and cellular extracts were analyzed as above.

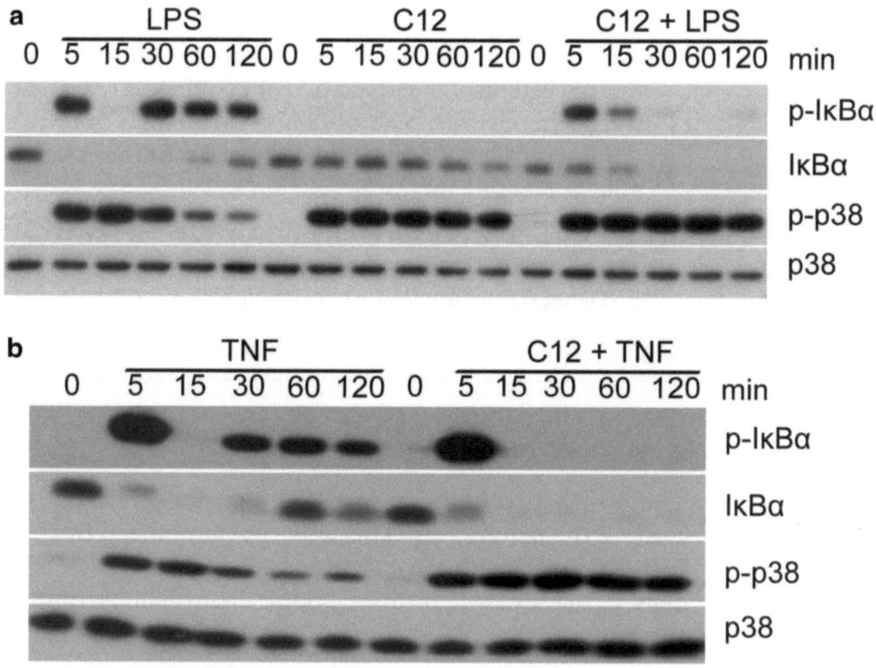

Fig. 4. Western blot analysis shows C12-mediated effect on the coordinated phosphorylation, degradation, and synthesis of the IκBα protein in bone marrow-derived macrophages (**a**) or normal human bronchial epithelial cells (**b**) stimulated with LPS or TNF and its combination with C12 as indicated.

3. Second, probe this membrane with rabbit polyclonal Abs specific for total IκBα protein (stripping is not require due to species differences in secondary Abs in these two assays).
4. Optional, rinse, briefly block, and rinse again the IκBα-probed membrane (the striping is not required) above. Use this membrane for western blot with Abs specific for p-38.
5. Finally, after reblocking by 2% milk/TBST and extensive washing (seven times and leave the membrane overnight in TBST), the membrane can be probed for total p38 or actin (as loading control) on the next day.
6. The prototypic example of western blotting analysis of the coordinated phosphorylation, degradation, and synthesis of IκBα protein in cells treated with LPS or TNF in the presence of C12 is shown Fig. 4.

4. Notes

1. Before addition to the cultured cells, the aliquot of the appropriate stock is thawed using a dry hot-block at 37°C (approximately 10 min) and kept at room temperature during all experimental time, but not longer than 6 h. The unused amount of a compound can be stored at –20°C.

2. The AHLs as well as the control compounds should be synthesized according to published methodologies (13, 23). Special attention needs to be paid to the correct (i.e., natural) stereochemistry of the obtained molecules. A few AHLs are commercially available; however, many of them are only available as racemic mixtures.
3. The solution of primary antibody can be saved (after use), stored at 4°C for 1 or 2 months and repeatedly used for up to 5–10 western blotting experiments.

Acknowledgments

The authors would like to thank Colin A. Lowery for critical reading the manuscript. This work was supported by the National Institutes of Health (AI079436 to VVK and AI080715 to GFK). This is manuscript number 20550 from The Scripps Research Institute.

References

1. Janeway, C. A., Jr., and Medzhitov, R. (2002) Innate immune recognition, *Annu Rev Immunol* **20**, 197–216.
2. Hoffmann, A., and Baltimore, D. (2006) Circuitry of nuclear factor kappaB signaling, *Immunol Rev* **210**, 171–186.
3. Schmitz, M. L., and de la Vega, L. (2008) A bacterial small molecule undermining immune response signaling, *Chembiochem* **9**, 2575–2577.
4. DiMango, E., Ratner, A. J., Bryan, R., Tabibi, S., and Prince, A. (1998) Activation of NF-kappaB by adherent *Pseudomonas aeruginosa* in normal and cystic fibrosis respiratory epithelial cells, *J Clin Invest* **101**, 2598–2605.
5. Pearson, J. P., Gray, K. M., Passador, L., Tucker, K. D., Eberhard, A., Iglewski, B. H., and Greenberg, E. P. (1994) Structure of the autoinducer required for expression of *Pseudomonas aeruginosa* virulence genes, *Proc Natl Acad Sci U S A* **91**, 197–201.
6. Smith, R. S., and Iglewski, B. H. (2003) *Pseudomonas aeruginosa* quorum sensing as a potential antimicrobial target, *J Clin Invest* **112**, 1460–1465.
7. Shiner, E. K., Rumbaugh, K. P., and Williams, S. C. (2005) Inter-kingdom signaling: deciphering the language of acyl homoserine lactones, *FEMS Microbiol Rev* **29**, 935–947.
8. DiMango, E., Zar, H. J., Bryan, R., and Prince, A. (1995) Diverse *Pseudomonas aeruginosa* gene products stimulate respiratory epithelial cells to produce interleukin-8, *J Clin Invest* **96**, 2204–2210.
9. Palfreyman, R. W., Watson, M. L., Eden, C., and Smith, A. W. (1997) Induction of biologically active interleukin-8 from lung epithelial cells by Burkholderia (Pseudomonas) cepacia products, *Infect Immun* **65**, 617–622.
10. Telford, G., Wheeler, D., Williams, P., Tomkins, P. T., Appleby, P., Sewell, H., Stewart, G. S., Bycroft, B. W., and Pritchard, D. I. (1998) The *Pseudomonas aeruginosa* quorum-sensing signal molecule *N*-(3-oxododecanoyl)-L-homoserine lactone has immunomodulatory activity, *Infect Immun* **66**, 36–42.
11. Smith, R. S., Fedyk, E. R., Springer, T. A., Mukaida, N., Iglewski, B. H., and Phipps, R. P. (2001) IL-8 production in human lung fibroblasts and epithelial cells activated by the Pseudomonasautoinducer *N*-3-oxododecanoyl homoserine lactone is transcriptionally regulated by NF-kappa B and activator protein-2, *J Immunol* **167**, 366–374.
12. Tateda, K., Ishii, Y., Horikawa, M., Matsumoto, T., Miyairi, S., Pechere, J. C., Standiford, T. J., Ishiguro, M., and Yamaguchi, K. (2003)

The *Pseudomonas aeruginosa* autoinducer *N*-3-oxododecanoyl homoserine lactone accelerates apoptosis in macrophages and neutrophils, *Infect Immun* **71**, 5785–5793.

13. Chhabra, S. R., Harty, C., Hooi, D. S., Daykin, M., Williams, P., Telford, G., Pritchard, D. I., and Bycroft, B. W. (2003) Synthetic analogues of the bacterial signal (quorum sensing) molecule *N*-(3-oxododecanoyl)-L-homoserine lactone as immune modulators, *J Med Chem* **46**, 97–104.
14. Ritchie, A. J., Yam, A. O., Tanabe, K. M., Rice, S. A., and Cooley, M. A. (2003) Modification of in vivo and in vitro T- and B-cell-mediated immune responses by the *Pseudomonas aeruginosa* quorum-sensing molecule *N*-(3-oxododecanoyl)-L-homoserine lactone, *Infect Immun* **71**, 4421–4431.
15. Li, L., Hooi, D., Chhabra, S. R., Pritchard, D., and Shaw, P. E. (2004) Bacterial *N*-acylhomoserine lactone-induced apoptosis in breast carcinoma cells correlated with down-modulation of STAT3, *Oncogene* **23**, 4894–4902.
16. Shiner, E. K., Terentyev, D., Bryan, A., Sennoune, S., Martinez-Zaguilan, R., Li, G., Gyorke, S., Williams, S. C., and Rumbaugh, K. P. (2006) *Pseudomonas aeruginosa* autoinducer modulates host cell responses through calcium signalling, *Cell Microbiol* **8**, 1601–1610.
17. Kravchenko, V. V., Kaufmann, G. F., Mathison, J. C., Scott, D. A., Katz, A. Z., Wood, M. R., Brogan, A. P., Lehmann, M., Mee, J. M., Iwata, K., Pan, Q., Fearns, C., Knaus, U. G., Meijler, M. M., Janda, K. D., and Ulevitch, R. J. (2006) *N*-(3-oxo-acyl) homoserine lactones signal cell activation through a mechanism distinct from the canonical pathogen-associated molecular pattern recognition receptor pathways, *J Biol Chem* **281**, 28822–28830.
18. Jahoor, A., Patel, R., Bryan, A., Do, C., Krier, J., Watters, C., Wahli, W., Li, G., Williams, S. C., and Rumbaugh, K. P. (2008) Peroxisome proliferator-activated receptors mediate host cell proinflammatory responses to *Pseudomonas aeruginosa* autoinducer, *J Bacteriol* **190**, 4408–4415.
19. Cooley, M., Chhabra, S. R., and Williams, P. (2008) *N*-Acylhomoserine lactone-mediated quorum sensing: a twist in the tail and a blow for host immunity, *Chem Biol* **15**, 1141–1147.
20. Williams, S. C., Patterson, E. K., Carty, N. L., Griswold, J. A., Hamood, A. N., and Rumbaugh, K. P. (2004) *Pseudomonas aeruginosa* autoinducer enters and functions in mammalian cells, *J Bacteriol* **186**, 2281–2287.
21. Ritchie, A. J., Whittall, C., Lazenby, J. J., Chhabra, S. R., Pritchard, D. I., and Cooley, M. A. (2007) The immunomodulatory *Pseudomonas aeruginosa* signalling molecule *N*-(3-oxododecanoyl)-L-homoserine lactone enters mammalian cells in an unregulated fashion, *Immunol Cell Biol* **85**, 596–602.
22. Yates, E. A., Philipp, B., Buckley, C., Atkinson, S., Chhabra, S. R., Sockett, R. E., Goldner, M., Dessaux, Y., Camara, M., Smith, H., and Williams, P. (2002) *N*-acylhomoserine lactones undergo lactonolysis in a pH-, temperature-, and acyl chain length-dependent manner during growth of *Yersinia pseudotuberculosis* and *Pseudomonas aeruginosa*, *Infect Immun* **70**, 5635–5646.
23. Kaufmann, G. F., Sartorio, R., Lee, S. H., Rogers, C. J., Meijler, M. M., Moss, J. A., Clapham, B., Brogan, A. P., Dickerson, T. J., and Janda, K. D. (2005) Revisiting quorum sensing: discovery of additional chemical and biological functions for 3-oxo-*N*-acylhomoserine lactones, *Proc Natl Acad Sci U S A* **102**, 309–314.
24. Ozer, E. A., Pezzulo, A., Shih, D. M., Chun, C., Furlong, C., Lusis, A. J., Greenberg, E. P., and Zabner, J. (2005) Human and murine paraoxonase 1 are host modulators of *Pseudomonas aeruginosa* quorum-sensing, *FEMS Microbiol Lett* **253**, 29–37.
25. Kravchenko, V. V., Kaufmann, G. F., Mathison, J. C., Scott, D. A., Katz, A. Z., Grauer, D. C., Lehmann, M., Meijler, M. M., Janda, K. D., and Ulevitch, R. J. (2008) Modulation of gene expression via disruption of NF-kappaB signaling by a bacterial small molecule, *Science* **321**, 259–263.
26. Charlton, T. S., de Nys, R., Netting, A., Kumar, N., Hentzer, M., Givskov, M., and Kjelleberg, S. (2000) A novel and sensitive method for the quantification of *N*-3-oxoacyl homoserine lactones using gas chromatography–mass spectrometry: application to a model bacterial biofilm, *Environ Microbiol* **2**, 530–541.
27. Laemmli, U. K. (1970) Cleavage of structural proteins during the assembly of the head of bacteriophage T4, *Nature* **227**, 680–685.
28. Sambrook, J., Fritsch, E. F., and Maniatis, T. (1989) *Molecular Cloning: A Laboratory Manual*, Vol. 3, 2nd ed., Cold Spring Harbor Laboratory Press, Cold Spring Harbor, NY.

Chapter 11

Imaging *N*-Acyl Homoserine Lactone Quorum Sensing In Vivo

Louise Dahl Christensen, Maria van Gennip, Tim Holm Jakobsen, Michael Givskov, and Thomas Bjarnsholt

Abstract

In order to study *N*-acyl homoserine lactone (AHL)-based quorum sensing in vivo, we present a protocol using an *Escherichia coli* strain equipped with a *luxR*-based monitor system, which in the presence of exogenous AHL molecules expresses a green fluorescent protein (GFP). Lungs from mice challenged intratracheally with alginate beads containing both a *P. aeruginosa* strain together with the *E. coli* monitor strain can be investigated at different time points postinfection. Epifluorescent or confocal scanning laser microscopy (CSLM) is used to detect the GFP-expressing *E. coli* monitor strain in the lung tissues, indicating production and excretion of AHLs in vivo by the infecting *P. aeruginosa*.

Key words: Quorum sensing, *Pseudomonas aeruginosa*, Pulmonary infection model, Epifluorescence microscopy, CSLM, In vivo

1. Introduction

The biofilm mode of growth is known to protect *Pseudomonas aeruginosa* against the host defense and antibiotics (1–6). Due to the high density of cells in biofilms, *P. aeruginosa* is able to make use of its cell–cell communication [quorum sensing (QS)] systems. *P. aeruginosa* employs two *N*-acyl-L-homoserine lactone (AHL) signal-molecule-based QS sensors, encoded by the *las* and *rhl* systems. Both systems are organized with a LuxR-type transcriptional regulator, LasR and RhlR, and a cognate LuxI-type synthetase, LasI and RhlI, which mainly synthesize the signal molecules *N*-(3-oxododecanoyl)-L-homoserine lactone (3-oxo-C_{12}-HSL) and *N*-butanoyl-L-homoserine lactone (C_4-HSL), respectively (7, 8). In addition to the AHL signal molecules, a

Kendra P. Rumbaugh (ed.), *Quorum Sensing: Methods and Protocols*, Methods in Molecular Biology, vol. 692,
DOI 10.1007/978-1-60761-971-0_11, © Springer Science+Business Media, LLC 2011

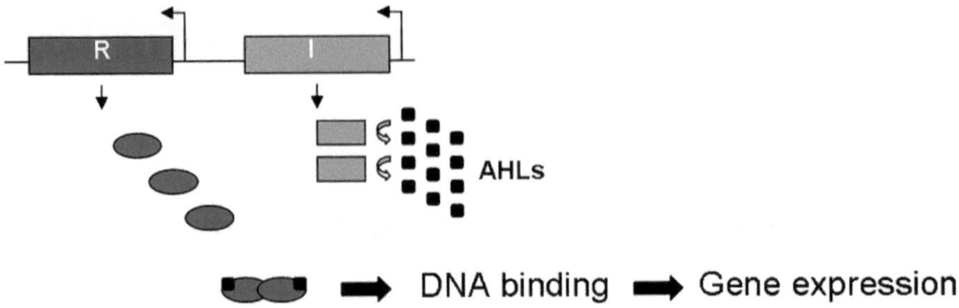

Fig. 1. The basics of AHL regulation. The AHL regulatory components are encoded by an *R* gene and an *I* gene. The gene products of these two genes are the regulator (R protein) and the signal synthetase (I protein), respectively.

third signaling compound, denoted the *Pseudomonas* quinolone signal (PQS), has been identified (9).

In simple terms, AHL regulation is comprised of an *I* gene encoding the AHL synthetase and an *R* gene encoding the receptor (see Fig. 1). During growth of the bacteria, the AHLs, produced by the I synthetase, accumulate in the medium until a critical threshold concentration is reached. At this concentration, the R protein forms a multimeric complex with the AHLs. This complex exhibits DNA binding properties and is believed to interact with specific recognition sequences positioned in the promoter regions of QS-regulated target genes. Binding can either (dependent on the genes) upregulate or downregulate transcription initiation (see review (10)).

The ability to monitor the production of AHLs by pathogenic Gram-negative bacteria during infection can contribute to the understanding of the host–pathogen interaction. In principle, the presence of exogenous AHLs can be detected by a reporter gene fused to any QS target gene maintained in a bacterial background devoid of the signal generator (*I*). We have previously designed a sensitive green fluorescent protein (GFP) reporter, which enables visualization of cell–cell communication at the single-cell level, and it has been used to detect in vitro and in vivo production of AHLs by *P. aeruginosa* (11–16).

2. Materials

2.1. Bacterial Strains

1. Any AHL producing *P. aeruginosa* strain can be used; we use *P. aeruginosa* (PAO1), originally obtained from Professor Barbara Iglewski (University of Rochester Medical Center, NY, USA). This strain is well tolerated by the mice, and it is QS-proficient; however, it features a reduced production of C_4-HSL (17), resulting in lowered rhamnolipid production which we have shown to be extremely toxic for the mice (18).

After choosing a bacterial strain, a dose–response experiment must be conducted to optimize the inoculum since the virulence of bacterial strains significantly affects the load of bacteria that can get instilled in the lungs without leading to mortality. It must be emphasized that our end point of the mouse model is not death but visualization of QS or gradual eradication of the bacteria. A dose–response experiment should also be conducted for the *E. coli* AHL monitor strain.

2. AHL monitor strain: The unstable GFP derivative GFP(ASV), containing alanine (A), serine (S), and valine (V) at the C terminus, is used as a reporter (15). The GFP-based monitor plasmid pJBA132, which carries the *Vibrio fischeri* region encoding *luxR* and a fusion between *PluxI* and *gfp* [*PluxI-gfp*(ASV)], has been cloned into the stable broad-host-range vector pME6031 (pVS1 replicon, accession no. AF118811). The *E. coli* monitor strain MT102 is a restriction-minus derivative of MC1000 (*araD139* (Δ*ara-leu*)7697 Δ*ac*, *thi*, *hsdR*) (for further details, see references. 12, 14). The JB525-*gfp*(ASV) is *E. coli* MT102 containing the plasmid pJBA132.

2.2. Alginate Beads

1. Protanal: Seaweed alginate with approximately 60% guluronic acid content (10/60) (Protanal, FMC BioPolymer) is dissolved in 0.9% NaCl to a concentration of 11 mg/mL and autoclaved. Store at 4°C until use. It can be stored for several months.
2. Tris-buffer: Mix 3.03 g (0.1 M) of Trizma® base with 200 mL of Milli-Q water. When dissolved, adjust the pH to 7.0 by adding 25% HCl. Adjust total volume to 250 mL by adding more Milli-Q water and autoclave. Store at 4°C until use. It can be stored for several months.
3. Tris-buffer with 0.1 M $CaCl_2$: Mix 100 mL of Tris-buffer with 1.47 g (0.1 M) $CaCl_2$ on a magnetic stirrer.
4. Bead making: The viscous suspension of bacteria and protanal is placed in a cylindrical reservoir and forced through an 18-G cannular by compressed air (0.4 bar), while another jet of air is forced coaxially along the needle to blow off the alginate droplets (1.0 bar) (19) (for a figure, see reference 20). Clean the equipment in 70% ethanol (vol/vol).

2.3. Mice

1. Female BALB/c mice at 9–11 weeks of age: BALB/c mice have been shown to have a Th_2 immune response as seen in chronically infected cystic fibrosis patients and are susceptible to infection with *P. aeruginosa* (21). The mice should be maintained on standard mouse chow and water ad libitum for a minimum 1 week before challenge. All animal studies should be performed in accordance with relevant guidelines and regulations.

2.4. Analgesia, Anesthesia, and Euthanasia

1. Anaesthetic: Mix one part Hypnorm® (0.315 mg/mL fentanyl citrate and 10 mg/mL fluanisone, Roche) with two parts of sterile water. Pour them together gently. Then, add one part 5 mg/mL midazolam (Roche). Pour together gently! A fresh solution should be made every time. Store at room temperature until use.
2. Analgesia: 5 mg/mL Bupivacaine. Store at room temperature.
3. Euthanasia: 200 mg/mL Pentobarbital with 20 mg/mL lidocaine hydrochloride. Store at room temperature.

2.5. Infection Procedure

1. A bead tipped needle is used when installing the alginate beads in the mouse lung.
2. Sutur: Ethicon Vicryl, 3-0, FS-2 (Johnson-Johnson Intl.).
3. Surgical equipment can be chosen as desired. We use Nasal scissors (B. Braun Aesculap), Durogrip needle holder (B. Braun Aesculap), and tweezers (B. Braun Aesculap).
4. Heating plates: LP reptiles vivarium heating mat 1,200 × 280 mm (LP Racks).

2.6. Quantitative Bacteriology

1. Blue agar plates, which are selective for Gram-negative bacilli (22) and therefore makes the counting of *P. aeruginosa* easy. Luria–Bertani (LB) plates can also be used.
2. SilentCrusher M, tool 12F (Heidolph, Germany).
3. 15- and 50-mL Centrifuge tubes.
4. Tubes for making serial dilutions.
5. Ice.

2.7. Freeze Microtomy

1. Tissue-Tek Cryomold standard (25 × 20 × 5 mm). Disposable vinyl specimen molds for placement of lungs.
2. Tissue-Tek O.C.T.™ gel for embedment of lungs before freezing.
3. Tissue-Tek Cryo 2000 for sectioning of the lungs.
4. Hexane.
5. Dry ice.

2.8. Epifluorescence Microscopy

1. We use the Olympus BX 40 with the universal DP71 camera to visualize GFP of the AHL monitor strains. Images can be obtained using objectives in the range 40–100×.

2.9. Confocal Scanning Laser Microscopy

1. Image acquisitions of bacteria in the lungs can be performed with confocal scanning laser microscopy (CSLM). We currently use a Leica SP5 (TCS4D; Leica Lasertechnik, GmbH, Heidelberg, Germany) equipped with a detector and a filter

set for monitoring GFP. In addition, a reflection detector for acquiring bright-field images is installed. Images can be obtained using objectives in the range 40–100×; we prefer oil-based objectives. Image scanning should be carried out with the 488-nm laser line from an Ar/Kr laser. Imaris software package (Bitplane AG) can used to generate pictures of the bacteria.

3. Methods

The GFP from the jellyfish *Aequorea victoria* was chosen as the reporter since it only requires a trace amount of oxygen to mature and thereby no external compounds needs to be added to detect green fluorescence (23). However, since the GFP is extremely stable (derived from *gfpmut3* (24)), different *gfp* genes with varying half-lives ranging from 40 min to a few hours was constructed (15). One of these variants is the unstable version *gfp*(ASV). The construct of a live bacterial AHL sensor, using the *luxR* gene from *V. fischeri* as the quorum sensor, acts in the presence of exogenous AHLs by LuxR and positively affects the expression of the *luxI* promoter (*luxR*-P_{luxI}), which in turn induces the expression of the *gfp*-reporter (14). The sensitivity of this *gfp*-reporter enables the visualization of AHLs produced by *P. aeruginosa* in vivo. In practice, this means that no signal should be observed in mice infected with alginate beads containing either the *E. coli* monitor strain or a *P. aeruginosa* strain alone. However, when a mixture of the two strains are used to inoculate the lungs of mice, one should be able to detect a GFP signal produced by the *E. coli* monitor strain in response to AHLs produced by *P. aeruginosa*.

Other AHL monitor strains have been constructed to study the ability of antipathogenic drugs to block QS. One example is the *P. aeruginosa* strain constitutively expressing a red fluorescent protein, which is advantageous for easy identification of the bacteria in the tissue samples. Moreover, the same strain contains a *lasR–PlasBgfp*(ASV) reporter fusion to detect an activated QS system. In practice, this means that the strain always fluoresces red, and when QS is activated, it will produce a GFP signal in addition. When QS is blocked by the presence of exogenous QSI compounds, the GFP signal will disappear. For further details, see reference 13.

The different monitor strains can also be used in in vitro continuous-culture flow cell experiments (see details in Chapter 19) where both the AHL production from different strains and QSIs can be investigated. In the present method section, the protocol for in vivo detection of AHLs produced by *P. aeruginosa* is described. When isolating lungs for ex vivo inspection, lungs should also be isolated, at the same time, for quantitative bacteriology, to verify that bacteria are present and that the monitor strain is functional.

3.1. Preparation of Bacteria in Alginate Beads

1. Plate bacteria from a freezer stock onto a blue agar plate and incubate at 37°C overnight. Use your choice of *P. aeruginosa* strain and the *E. coli* monitor strain JB525-*gfp*(ASV).
2. Use one bacterial colony from the plate to inoculate an overnight culture in 100 mL of LB medium in a 250-mL conical flask. Allow the culture to grow for 18–20 h at 37°C on an orbital shaker at 180 rpm.
3. Transfer the bacterial overnight culture to two 50-mL centrifuge tubes and centrifuge for 10 min at 5,000 × *g*. Discard the supernatant and resuspend the bacterial pellet in 4.5 mL of LB medium.
4. Mix 0.5 mL of resuspended bacterial pellet with 4.5 mL of sterile protanal (see Subheading 2.2, item 1) for production of beads.
5. When making the beads (see Subheading 2.2, item 4), the droplets are directed into a cross-linking solution of 0.1 M $CaCl_2$ in Tris–HCL buffer (0.1 M, pH 7.0). Allow the beads to cure for 1 h in the calcium bath with continuous stirring at minimum rotation.
6. Transfer the beads to two 50-mL centrifuge tubes and centrifuge for 10 min at 600 × *g*.
7. Discard the supernatant and wash the beads in 0.9% NaCl. Centrifuge for 10 min at 600 × *g*. Do this twice.
8. Resuspend the beads in 7.5 mL of 0.9% NaCl using a 5-mL pipette. Homogenize the beads gently using a 5-mL pipette, then a 1-mL pipette, and last a 200-μL pipette to avoid clumps. *Do not use a vortexer for mixing*. Make two tenfold serial dilutions using two different pipette tips.
9. Plate appropriate dilutions using sterile glass spreaders on blue plates. Plate each dilution double. Store the beads at 4°C until next day and incubate the plates at 37°C.

3.2. Infection Procedure

1. Count the plates and calculate the colony forming units (CFU)/mL.
2. Right before challenge, adjust the beads containing *P. aeruginosa* and the beads containing the *E. coli* monitor strain to the appropriate concentration in 0.9% NaCl (see Note 1).
3. The bead solution containing both bacteria is mixed at ratio 1:2 (*P. aeruginosa* and *E. coli*).
4. Use epifluorescence or CSLM to verify that the beads are not fluorescing green and thus insure that the monitor is not turned on before infection.

5. There should be three groups of mice in an experiment using a *P. aeruginosa* strain and one AHL monitor strain. One group of mice is infected with *P. aeruginosa* alone, one group with the AHL monitor strain alone, and one group with the mixture of *P. aeruginosa* and the AHL monitor strain (see Note 2).
6. Sedate the mice using 10 mL/kg body weight Hypnorm®/midazolam (see Subheading 2.4, item 1) injected subcutaneously (s.c.) in the groin area (see Note 3).
7. Fixate each mouse on its back. Use a rubber band fastened between two needles to hold the head of the mouse fixated. We use a styrofoam tray (29 × 39 cm) as the operation area.
8. Sterilize the throat of the mouse with 70% (vol/vol) ethanol.
9. Make an incision in the skin above the trachea with a pair of scissors and expand the incision using the pair of scissors by making it larger in the cutting area.
10. Move the muscles to the sides to locate the trachea using the pair of scissors. Again do this by making the cutting area larger. Do not cut the muscles.
11. Use the scissors to make an incision in the membrane surrounding the trachea.
12. Make a way into the trachea using an 18-G needle.
13. Use a bead tipped needle to install 0.04 mL of the bead suspension in the left lung, 11 mm from the tracheal penetration site.
14. Suture the incision with silk.
15. Sterilize the surgical equipment between mice using 70% (vol/vol) ethanol.
16. Euthanize four mice right after challenge and isolate the lungs (see Subheading 3.3). The lungs are used to estimate the infection inoculum.

    Alternative: Make two tenfold serial dilutions of the bead suspension and plate the appropriate dilutions on blue plates.
17. Use an 18-G needle to apply one drop of bupivacaine on the incision site for postoperative pain.
18. Inject the mice s.c. with 0.75 mL of 0.9% NaCl in the neck-skin area to avoid dehydration.
19. Place the mice on a row on top of the bedding in the cage and cover them with a paper napkin for recovering after surgery. Place the part of the cage where the mice are placed on top of the heating pad. There should also be an area of the cage that is free from heating. Heating should be no more than 28°C.

    Check the mice after 4–6 h and thereafter minimum once a day (see Note 4).

3.3. Quantitative Bacteriology

1. Inject pentobarbital (10.0 mL/kg body weight) i.p. to euthanize the mice. Ensure that local animal-care regulations are followed.
2. Fixate each dead mouse on its back and apply 70% (vol/vol) ethanol on the chest and stomach to avoid contamination.
3. To isolate the lungs, first remove the skin with a pair of scissors. Then, open the thoracic cavity without damaging the lungs. Take hold of the arteries below the lungs using tweezers and cut the arteries with a pair of scissors. Lift the lungs upright and cut the arteries next to the heart. The lungs should now be isolated with the tweezers.
4. Place the isolated lungs in 5 mL of 0.9% NaCl in a 15-mL centrifuge tube and keep it on ice until homogenization.
5. Sterilize the surgical equipment between mice using 70% (vol/vol) ethanol.
6. Homogenize the lungs using a SilentCrusher M tool 12F (Heidolph, Germany). The crusher is washed in 35 mL of 70% (vol/vol) ethanol for 5 s, for 20 s in 35 mL of 70% (vol/vol) ethanol, for 5 s in 35 mL of 0.9% NaCl, and for 5 s in 30 mL of 0.9% NaCl. Use 50-mL centrifuge tubes for the washing. All the tubes are changed after five to eight samples. Between groups, the tool 12F must be removed, taken apart, cleaned, and sterilized. Please note that the homogenizer must only be used when placed in fluid.
7. Make tenfold serial dilutions of each lung homogenate and plate all the dilutions onto blue agar plates. Incubate the plates at 37°C overnight.
8. Count the plates and calculate the CFU/lung (see Notes 5 and 6).

3.4. Freeze Microtomy

1. After a chosen period of time (see Note 7), the mice are euthanized using pentobarbital (10.0 mL/kg body weight) injected intraperitoneally (i.p.). Ensure that local animal-care regulations are followed.
2. Fixate each dead mouse on its back and apply 70% (vol/vol) ethanol on the chest and stomach to avoid contamination.
3. Isolate the lungs by first removing the skin with a pair of scissors and tweezers. Then, open the thoracic cavity without damaging the lungs. Take hold of the arteries below the lungs using a tweezers and cut the arteries with the scissors. Lift the lungs upright and cut the arteries next to the heart. The lungs should now be isolated with the tweezers.
4. Whole lungs are placed in the Tissue-Tek mold and covered with Tissue-Tek O.C.T.™ gel. Hexane is cooled down using dry ice (see Note 8), and the embedded lungs are placed in the cooled hexane for 5–10 min. Keep it on dry ice afterwards. This should happen as soon as possible after isolation of the lungs.

5. Cool the Tissue-Tek Cryo 2000 to operating temperature and make different sections of the lung tissue. Frozen sections should be 10–50 μM thick.
6. Place the sections on cover glasses for microscopic inspection.
7. Study the lung sections with epifluorescence or CSLM (see Note 9).

4. Notes

1. With respect to the Iglewski PAO1, we adjust the beads to a concentration of 1.5×10^8 CFU/mL.
2. A power analysis should be made to estimate the group sizes.
3. This is the concentration we use, but is has to be adjusted depending on the mouse strain. Please consult other references regarding this subject.
4. Usually the mice appear sick within the first 24 h after infection. We believe that this is due to the fact that they develop pneumonia and have been sedated.
5. In case of the Iglewski PAO1, homogenates from most of the mice will display CFUs when diluted to 10^{-3}. If no CFU can be counted, the mice are treated with too little beads.
6. Bacteria from lung tissues containing mixed inoculates of *P. aeruginosa* and *E. coli* can be discriminated based on the colony color on the blue agar plates combined with the ability of *E. coli* colonies to express GFP in the presence of exogenous AHLs. Therefore, the plates can be supplemented with 10 nM 3-oxo-C_6-HSL.
7. The experiments should not run longer than 3 days since it is not possible to detect the *E. coli* monitor strain.
8. Make sure that the hexane is cold enough.
9. Ensure that what are observed through the microscope are in fact bacteria. This can be verified by the morphology and by looking at the bacteria under normal light.

Acknowledgment

We acknowledge the contributions made by Drs. Morten Hentzer, Hong Wu, Thomas Bovbjerg Rasmussen, Jens Bo Andersen, and Allan Bech Christensen with regard to the initial developments of the model systems.

References

1. Costerton, J. W., Stewart, P. S. and Greenberg, E. P. (1999) Bacterial biofilms: a common cause of persistent infections, *Science.* **284**, 1318–1322.
2. Høiby, N. (1974) *Pseudomonas aeruginosa* infection in cystic fibrosis. Relationship between mucoid strains of *Pseudomonas aeruginosa* and the humoral immune response, *Acta Pathol Microbiol Scand B Microbiol Immunol.* **82**, 551–558.
3. Singh, P. K., Parsek, M. R., Greenberg, E. P. and Welsh, M. J. (2002) A component of innate immunity prevents bacterial biofilm development, *Nature.* **417**, 552–555.
4. Bjarnsholt, T., Jensen, P. O., Burmølle, M., Hentzer, M., Haagensen, J. A., Hougen, H. P., et al. (2005) *Pseudomonas aeruginosa* tolerance to tobramycin, hydrogen peroxide and polymorphonuclear leukocytes is quorum-sensing dependent, *Microbiology.* **151**, 373–383.
5. Høiby, N., Krogh Johansen, H., Moser, C., Song, Z., Ciofu, O. and Kharazmi, A. (2001) *Pseudomonas aeruginosa* and the in vitro and in vivo biofilm mode of growth, *Microbes Infect.* **3**, 23–35.
6. Pamp, S. J., Gjermansen, M., Johansen, H. K. and Tolker-Nielsen, T. (2008) Tolerance to the antimicrobial peptide colistin in *Pseudomonas aeruginosa* biofilms is linked to metabolically active cells, and depends on the pmr and mexAB-oprM genes, *Mol Microbiol.* **68**, 223–240.
7. Passador, L., Cook, J. M., Gambello, M. J., Rust, L. and Iglewski, B. H. (1993) Expression of *Pseudomonas aeruginosa* virulence genes requires cell-to-cell communication, *Science.* **260**, 1127–1130.
8. Pearson, J. P., Passador, L., Iglewski, B. H. and Greenberg, E. P. (1995) A second *N*-acylhomoserine lactone signal produced by *Pseudomonas aeruginosa*, *Proc Natl Acad Sci U S A.* **92**, 1490–1494.
9. Diggle, S. P., Winzer, K., Chhabra, S. R., Worrall, K. E., Cámara, M. and Williams, P. (2003) The *Pseudomonas aeruginosa* quinolone signal molecule overcomes the cell density-dependency of the quorum sensing hierarchy, regulates *rhl*-dependent genes at the onset of stationary phase and can be produced in the absence of LasR, *Mol Microbiol.* **50**, 29–43.
10. Hentzer, M. and Givskov, M. (2003) Pharmacological inhibition of quorum sensing for the treatment of chronic bacterial infections, *J Clin Invest.* **112**, 1300–1307.
11. Hentzer, M., Riedel, K., Rasmussen, T. B., Heydorn, A., Andersen, J. B., Parsek, M. R., et al. (2002) Inhibition of quorum sensing in *Pseudomonas aeruginosa* biofilm bacteria by a halogenated furanone compound, *Microbiology.* **148**, 87–102.
12. Wu, H., Song, Z., Hentzer, M., Andersen, J. B., Heydorn, A., Mathee, K., et al. (2000) Detection of *N*-acylhomoserine lactones in lung tissues of mice infected with *Pseudomonas aeruginosa*, *Microbiology.* **146 (Pt 10)**, 2481–2493.
13. Wu, H., Song, Z., Hentzer, M., Andersen, J. B., Molin, S., Givskov, M., et al. (2004) Synthetic furanones inhibit quorum-sensing and enhance bacterial clearance in *Pseudomonas aeruginosa* lung infection in mice, *J Antimicrob Chemother.* **53**, 1054–1061.
14. Andersen, J. B., Heydorn, A., Hentzer, M., Eberl, L., Geisenberger, O., Christensen, B. B., et al. (2001) gfp-based *N*-acyl homoserine-lactone sensor systems for detection of bacterial communication, *Appl Environ Microbiol.* **67**, 575–585.
15. Andersen, J. B., Sternberg, C., Poulsen, L. K., Bjorn, S. P., Givskov, M. and Molin, S. (1998) New unstable variants of green fluorescent protein for studies of transient gene expression in bacteria, *Appl Environ Microbiol.* **64**, 2240–2246.
16. Hentzer, M., Wu, H., Andersen, J. B., Riedel, K., Rasmussen, T. B., Bagge, N., et al. (2003) Attenuation of *Pseudomonas aeruginosa* virulence by quorum sensing inhibitors, *EMBO J.* **22**, 3803–3815.
17. Kohler, T., van Delden, C., Curty, L. K., Hamzehpour, M. M. and Pechere, J. C. (2001) Overexpression of the MexEF-OprN multidrug efflux system affects cell-to-cell signaling in *Pseudomonas aeruginosa*, *J Bacteriol.* **183**, 5213–5222.
18. van Gennip, M., Christensen, L. D., Alhede, M., Phipps, R., Jensen, P. Ø., Christophersen, L., et al. (2009) Inactivation of the *rhlA* gene in *Pseudomonas aeruginosa* prevents rhamnolipid production, disabling the protection against polymorphonuclear leukocytes, *APMIS.* **117**, 537–546.
19. Pedersen, S. S., Shand, G. H., Hansen, B. L. and Hansen, G. N. (1990) Induction of experimental chronic *Pseudomonas aeruginosa* lung infection with *P. aeruginosa* entrapped in alginate microspheres, *APMIS.* **98**, 203–211.
20. Bjarnsholt, T., van Gennip, M., Jakobsen, T. H., Christensen, L. D., Jensen, P. Ø. and Givskov, M. (2010) *In vitro* screens for quorum

sensing inhibitors and *in vivo* confirmation of their effect, *Nat Protoc.* **5**, 282–293.

21. Moser, C., Johansen, H. K., Song, Z., Hougen, H. P., Rygaard, J. and Høiby, N. (1997) Chronic *Pseudomonas aeruginosa* lung infection is more severe in Th2 responding BALB/c mice compared to Th1 responding C3H/HeN mice, *APMIS.* **105**, 838–842.
22. Høiby, N. (1974) Epidemiological investigations of the respiratory tract bacteriology in patients with cystic fibrosis, *Acta Pathol Microbiol Scand B Microbiol Immunol.* **82**, 541–550.
23. Chalfie, M., Tu, Y., Euskirchen, G., Ward, W. W. and Prasher, D. C. (1994) Green fluorescent protein as a marker for gene expression, *Science.* **263**, 802–805.
24. Cormack, B. P., Valdivia, R. H. and Falkow, S. (1996) FACS-optimized mutants of the green fluorescent protein (GFP), *Gene.* **173**, 33–38.

Chapter 12

Defining the Structure and Function of Acyl-Homoserine Lactone Autoinducers

Mair E.A. Churchill, Hiruy M. Sibhatu, and Charis L. Uhlson

Abstract

Quorum sensing plays a central role in regulating many community-derived symbiotic and pathogenic relationships of bacteria, and as such has attracted much attention in recent years. Acyl-homoserine lactones (AHLs) are important signaling molecules in the quorum sensing gene-regulatory processes found in numerous gram-negative species of bacteria that interact with eukaryotic organisms. AHLs are produced by acyl-homoserine lactone synthases. Bacteria can have multiple genes for AHL synthase enzymes, and such species are likely to produce several different types of AHLs. Determination of the types and the relative amounts of AHLs produced by AHL synthases in bacteria under varied conditions provides important insights into the mechanism of AHL synthase function and the regulation of transcriptional cascades initiated by quorum sensing signaling. This chapter describes a mass spectrometry method for determining the types and relative amounts of AHLs present in a sample.

Key words: Acyl-homoserine lactone, AHL, Quorum sensing, Mass spectrometry, AHL synthase

1. Introduction

Understanding the molecular mechanism of AHL-dependent signaling in bacterial quorum sensing requires knowledge of the recognition of AHLs by receptors as well as the AHLs that are present in the environment. The acyl-homoserine lactones, AHLs, differ in their length of acyl chain as well as the substituents at the 3-position of the acyl chain (Fig. 1a) (1). Some species of bacteria have a single AHL synthase enzyme and produce predominantly one type of AHL, whereas others have multiple AHL synthases and produce many types of AHLs (2, 3). Furthermore, the amount and types of AHLs that are produced in bacteria can differ depending on the particular AHL synthases that are expressed and the availability of the substrates needed for AHL synthesis.

Kendra P. Rumbaugh (ed.), *Quorum Sensing: Methods and Protocols*, Methods in Molecular Biology, vol. 692,
DOI 10.1007/978-1-60761-971-0_12, © Springer Science+Business Media, LLC 2011

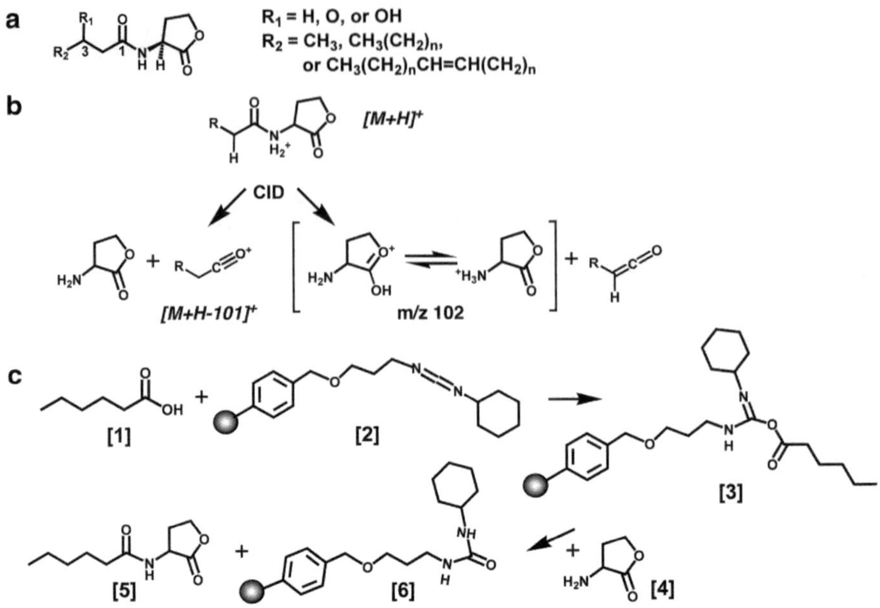

Fig. 1. Scheme of AHL structure, mass spectrometry, and chemical synthesis. (**a**) Structure of AHLs. AHLs vary by substitution at the C3 position (R1) and the length and unsaturation of the acyl chain (R2). HSL refers to D/L-homoserine lactone; AHL refers to *N*-acyl-D/L-homoserine lactone, with any acyl chain length (*indicated by the number of carbons CN*) or degree of substitution, and the extent of deuterium labeling indicated by D_n. (**b**) Mass spectrometry scheme proposed for collisionally induced decomposition of AHLs into the two major fragment ions observed for the 3-oxo-HSLs (lactone and acyl). (**c**) Carbodiimide-based chemical synthesis of D_3-C6-HSL. The fatty acid (1) reacts with the resin-bound carbodiimide reagent (2) to give an acylating intermediate (3). Next, nucleophilic attack by the HSL-hydrobromide amine (4) on the acylating intermediate releases the amide product AHL (5). A urea-like byproduct remains coupled to the resin-bound carbodiimide (6). Any excess or unreacted fatty acid was removed during filtration because it is left in the form of the acylating intermediate bound to the resin (6). The final AHL product (5) was obtained by filtration of the reaction mixture and solid-phase extraction using a normal-phase SPE.

The ability to identify populations of AHLs and to analyze changes in AHLs produced with mutant forms of AHL synthase genes or under different metabolic conditions is important for understanding quorum sensing signaling. However, it is difficult to identify all of the AHLs produced and to determine how much of each AHL species is present using conventional bioassay procedures (4, 5).

This chapter describes a relatively unbiased method for determining the types and relative amounts of AHLs present in a sample. Mass spectrometry provides a complementary approach to traditional bioassays. It permits the semi-quantitative analysis of the changes to the distribution of AHLs by a system either grown under various conditions, or one in which the AHL synthase activity had been changed. The distribution of AHLs from experimental samples can be determined by liquid chromatography-electrospray ionization-tandem mass spectrometry (LC/MS/MS) using a triple–quadrupole mass spectrometer (5). It is necessary first to extract the AHLs from the system of interest with an added internal standard, and this is described for laboratory bacterial cultures in Subheadings 3.2 and 3.3. It is important to have appropriate reference standards and internal standards and to determine the HPLC retention times for each AHL. Therefore, the approach for synthesizing several types of AHLs and stable-isotope-labeled internal standards is described in Subheading 3.1, since several AHLs, including internal standards for mass spectrometry, are not commercially available. Second, the AHLs require purification prior to this mass spectrometry analysis (Subheading 3.4). Finally, in Subheading 3.5, the methods used to characterize the AHLs using liquid chromatography/mass spectrometry/mass spectrometry (LC/MS/MS) are described. This approach is based on monitoring multiple major fragment ions that are produced during the collision-induced decomposition process in the mass spectrometer (Fig. 1b) (5).

A common approach for dissecting the AHL-mediated quorum sensing pathways is to examine the AHLs that exist in a wild-type bacterium and in bacteria that have been engineered, harboring individual components of the signaling system to determine their specificity (5–8). The simple example used to illustrate the method here is the analysis of the AHLs produced by the enzyme *Las*I expressed recombinantly in *E. coli*. The extracted and purified samples (Fig. 4) contain 3-oxo-C12-HSL and other 3-oxo-HSLs that are made under the conditions of expression in the two independent experiments (5). These approaches can also be adapted to examine biological samples from the environment and patients.

2. Materials

2.1. Synthesis of AHLs Using Carbodiimide Chemistry

1. *N*-Cyclohexylcarbodiimide-*N*′-propyloxymethyl polystyrene resin (PS-carbodiimide resin; Argonaut Technologies).
2. Homoserine lactone starting material: α-amino-γ-butyrolactone hydrobromide (HSL-HBr) (Aldrich).
3. Acyl-chain starting materials for deuterium-labeled mass spectrometry standards include D_3-hexanoic acid or those with other lengths of acyl chain (Cambridge Isotope).
4. Acyl-chain starting materials for production of reference standards including unsubstituted and 3-hydroxy-AHLs: Fatty acids were purchased from Sigma and the 3-hydroxy-C6 fatty acid was purchased from Larodan Lipids (Sweden) and was supplied as 25 mg in a liquid.
5. *Solvents*: methylene chloride (CH_2Cl_2), *N*,*N*-dimethylformamide (DMF), triethylamine (TEA), and methanol (Sigma) (see Note 1).

6. Teflon fritted filters and glass reaction vials of appropriate scale. MUST USE ONLY ORGANIC-SAFE GLASSWARE (see Note 2).

2.2. Production of AHLs from Bacterial Culture

1. Any media may be used; the selection of the media depends on the growth of the desired bacterial strain.
2. Amberlite XAD 16 (Sigma; (XAD16)) resins must be added to media after autoclaving, at 5 g/L.
3. Methanol (Fisher, optima 0.2 micron filtered) is used to initially resuspend the resin and cells.
4. Syringe filters (non sterile filter, 0.2 μm, Nalgene) are used to filter the methanol-resuspended supernatant.

2.3. Solid Phase Extraction Purification of AHLs

1. Isooctane (Fisher; HPLC grade, submicron filtered) and diethyl ether anhydrous (Fisher) are mixed at a ratio of 1:1 and are used in the purification procedure. MUST USE ONLY GLASSWARE (see Notes 1 and 2).
2. Ethyl acetate (Fisher HPLC Grade) is used as an elution buffer. MUST USE ONLY GLASSWARE.
3. Sep-Pak plus, silica cartridges (Waters) used to purify AHLs.
4. Supelco Visiprep vacuum purification system.
5. Fisher brand disposable culture tubes (Borosilicate Glass) to collect AHLs from Sep-Pak plus cartridges.
6. Glass pipettes (Fisher 10 mL and 9 in. disposable Pasteur pipettes).

2.4. Liquid Chromatography/Mass Spectrometry/Mass Spectrometry

1. Reagents for standard curve 10×32 culture tubes (Fisher Borosilicate Glass), 3-oxo-C6-HSL (Aldrich), C6-HSL (Fluka), and C12-HSL (Sigma) and 3-oxo-C12-HSL (Sigma) and deuterium-labeled HSL (see Subheading 2.1).
2. High-Pressure Liquid Chromatography (HPLC) Shimdzu SCL-10AVP controller and LC10ASVP system.
3. 1.0×150 mm 5 μm C18 Columbus reversed-phase column, Phenomenex at a flow rate of 0.050 μL/min.
4. PAL autosampler (LEAP Technologies); vials used for autosampler: Supelco 10× 32 0.9 mL crimp vials.
5. HPLC solvent A: 8.3 mM acetic acid, buffered to pH 5.7 with ammonium hydroxide solvent B: Methanol
6. Mass spectrometer; Applied Biosystems MDS Sciex Q Trap LC/MS/MS system (Applied Biosystems, Foster City, CA).
7. Areas of peaks are integrated using a quantitation software package Analyst 1.5 (Applied Biosystems).

3. Methods

3.1. Synthesis of 3-Hydroxy, Unsubstituted, and Deuterium-Labeled AHLs Using Carbodiimide Chemistry (Fig. 1c)

1. 4.2 mg (23 μmol) of HSL-HBr was weighed in a small glass reaction vial and was completely dissolved (5–10 min) by vortexing in 200 μL of methylene chloride (CH_2Cl_2) with 10% *N*,*N*-dimethyl-formamide (DMF) and 4.6 μmol of triethylamine (TEA) (6.4 μL).
2. For the deuterium-labeled or other unsubstituted fatty acids that are supplied as neat liquids (25 mg; density ~0.93 g/mL), 4.4 μL (~35 μmol) was added to 400 μL of CH_2Cl_2 with 10% DMF in a separate glass reaction vial. To this reaction vial, 40 mg of PS-carbodiimide was added and stirred for 5 min at room temperature (Go to step 4).
3. For the unlabeled unsubstituted or 3-hydroxy fatty acids, which are supplied as liquid or wax, 25 mg was weighed out and resuspended in 100 μL of methanol, transferred to a clean glass vial, dried down by vacuum evaporation and then resuspended in 25 μL of methanol, and treated as 1 mg/μL. The 25 μL was added to 400 μL of CH_2Cl_2 with 10% DMF in a separate glass reaction vial. To this reaction vial, 40 mg of PS-carbodiimide was added and stirred for 5 min at room temperature (Go to step 4).
4. After the fatty acid had been allowed to equilibrate with the resin, the 200 μL solution containing the HSL-HBr was added and stirred for 30 h at room temperature.
5. The reaction mixture was then diluted into 6 mL of isooctane:ethyl ether (1:1) and passed through the teflon frit of the glass tube directly onto the activated normal phase silica SPE cartridge for purification as described in Subheading 3.4.

3.2. Preparative Scale Production of AHLs from Bacterial Culture

1. Prepare desired media and autoclave. Add 5 g of Amberlite XAD 16 resin per liter of media after the media has cooled down to room temperature (Fig. 2).
2. Grow desired strain as if growing bacteria for recombinant protein expression. Each strain grows according to specific medium, antibiotic, temperature, and IPTG induction concentration, as well as induction OD.
3. Harvest resin by centrifugation at 4066 ×*g* for 10 min.
4. Store at −80°C until ready to extract AHLs.
5. Resuspend in a 10-fold weight to volume of methanol, and vortex and shake vigorously until homogenized. Then, shake and vortex vigorously at 20 min intervals, for 2 h. For example, 10 g of harvested cell culture should be resuspended in 100 mL of methanol.

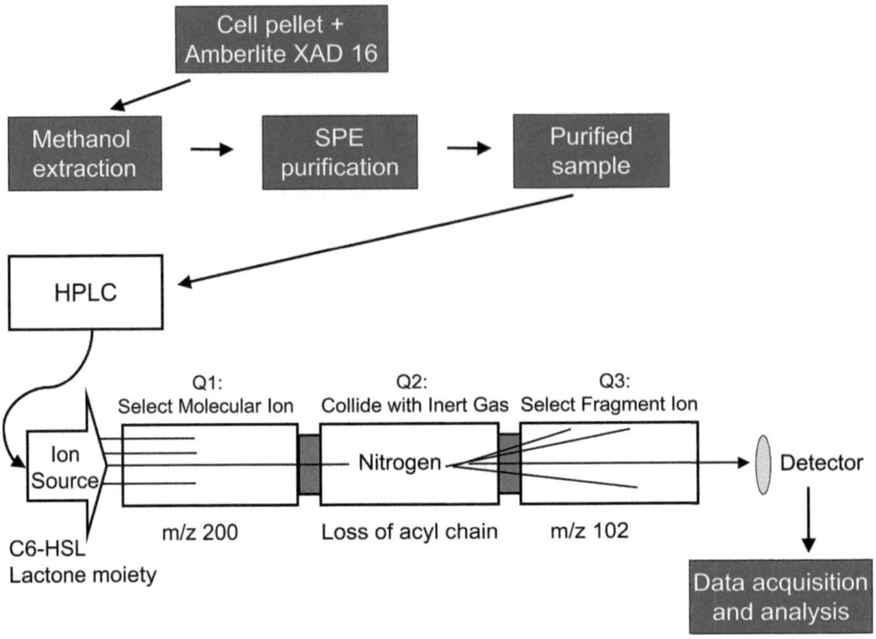

Fig. 2. Flowchart of the methods described in this chapter. The cell cultures were grown in the presence of Amberlite XAD 16, which adsorbs AHLs. In order to extract the AHLs from the resin and cell debris, the harvested pellet from the bacterial culture and Amberlite XAD 16 was resuspended in methanol and left overnight. The methanol solution was filtered and centrifuged, and then subjected to solid phase extraction to remove any remaining impurities and to concentrate the AHLs. The purified sample was then injected onto the reversed phase HPLC column to separate the AHLs based on their lipophilicity and introduce them into the mass spectrometer. In the multiple reaction monitoring (MRM) experiment, the sample was passed through the ion source into Quadrupole1 (Q1), then the selected ions proceeded into Quadrupole 2 (Q2) where they collide with inert gas (nitrogen). After fragmentation, selected ions move into Quadrupole 3 (Q3) and are visualized by the detector. Data were processed by applying standard isotope dilution curves to the analytical data (as in Figs. 3 and 4).

6. Leave overnight at room temperature with a final vigorous shake.
7. After leaving overnight shake and vortex vigorously then centrifuge at 4500 ×*g* for 10 min.
8. Syringe filter supernatant into test tubes and dry down each tube to a final volume of approximately 2 mL.
9. Collect all 2 mL samples together then centrifuge at 4500 ×*g* for 10 min.
10. Collect supernatant into new glass culture tubes.
11. Store supernatant at −20°C for a maximum of 16 h until SPE purification.

3.3. Analytical Scale Identification of AHL Products (Fig. 2)

1. Grow desired strain as if growing bacteria for recombinant protein expression. Each strain grows according to specific medium, antibiotic, temperature, and IPTG induction concentration, as well as induction OD. The only difference is that 5 g of Amberlite XAD 16 resin per liter of growth medium is added immediately before inoculation.

2. Harvest the cells and resin by centrifugation at 4066 ×*g* for 10 min.
3. Store at –80°C until ready to extract AHLs.
4. Resuspend the samples in a 10-fold weigth to volume of methanol. Vortex and shake vigorously until homogenized and shake and vortex vigorously at 20 min intervals, for 2 h.
5. Add 2 nmol of the internal standards (D_3-C6-HSL and D_3-C12-HSL).
6. Leave overnight at room temperature with a final vigorous shake.
7. Shake and vortex vigorously and centrifuge at 4500 ×*g* for 10 min.
8. Filter the supernatant using a syringe into test tubes and dry down each tube to a final volume of approximately 2 mL.
9. Collect all 2 mL samples and centrifuge at 4500 ×*g* for 10 min.
10. Store at –20°C for a maximum of 16 h until SPE purification.

3.4. Solid Phase Extraction Purification of AHLs (Fig. 2)

1. Wash all the equipment with 1:1 isooctane/ethyl ether solution.
2. Equilibrate Sep-Pak Cartridges (WAT020520) plus Silica, with 8 mL of:
   (a) 1:1 isooctane:ethyl ether
   (b) Ethyl acetate
   (c) 1:1 isooctane:ethyl ether.
3. Take 0.2 mL of sample and bring up to 8 mL using 1:1 isooctane:ethyl ether then load on Sep-Pak Cartridges (WAT020520) plus Silica.
4. Wash twice with 8 mL of 1:1 isooctane:ethyl ether.
5. Elute with 8 mL ethyl acetate.
6. Dry down using Speed Vac concentrator under vacuum, until completely dry.

3.5. Liquid Chromatography/Mass Spectrometry/Mass Spectrometry

1. The standard curve is prepared by making a 1 mg/mL stock solution of each of the following reference standards (RS) AHLs in methanol: 3-oxo-C6-HSL, C6-HSL, C12-HSL, and 3-oxo-C12-HSL (Fig. 3a).
2. The stock standard solutions were serially diluted to yield concentrations of 10, 2, 0.4, 0.08, 0.016, and 0 nmol (Table 1).
3. For each standard curve point (SC1-5) and sample, 2 nmol of D_3-C6-HSL internal standard in methanol is added to each vial (Figs. 3 and 4).

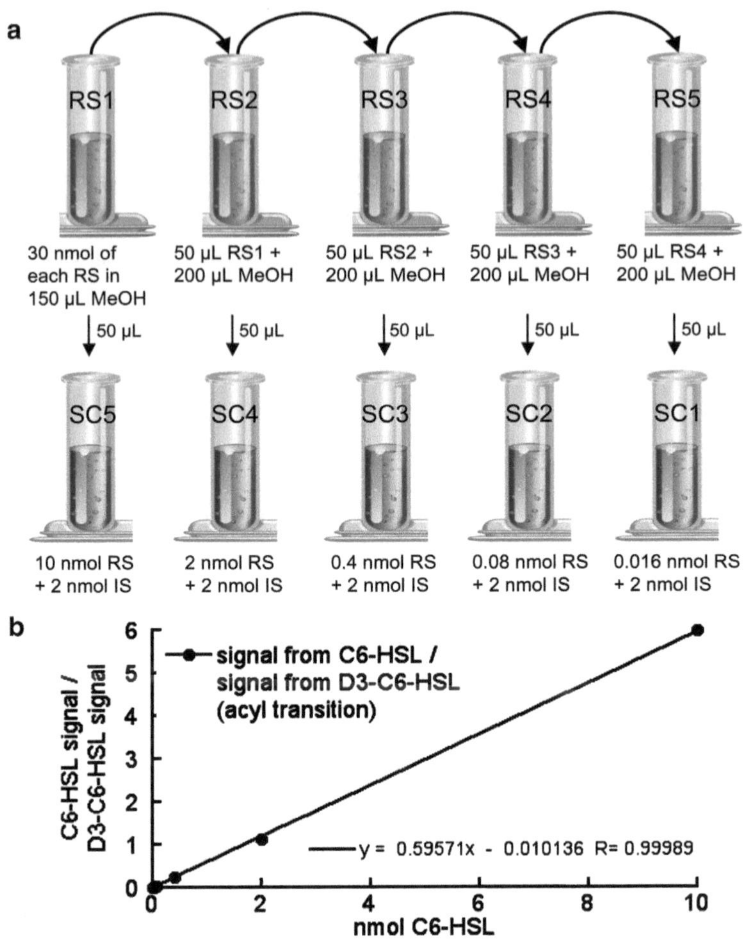

Fig. 3. Preparation of a standard curve for mass spectrometry of AHLs. (**a**) Step-by-step preparation of internal standard and reference standard mixtures for mass spectrometry. First a series of serial dilutions (1:5) are prepared from a stock solution (vial 1a) containing 30 nmol each of reference standard (RS) in 150 μL methanol (Table 1). Here, the four reference standards being used are 3-oxo-C6-HSL, 3-oxo-C12-HSL, C6-HSL, and C12-HSL. The samples in vials 2a–5a are then made by serial dilution of 50 μL from the preceding vial into 200 μL of methanol. To produce the standard curve samples (SC1-5), 50 μL of each samples 1a–5a is transferred to a new vial and a fixed amount of the stable isotope-labeled internal standard (IS) is added to each vial, as well as an empty control vial (SC0, not shown in the figure). Here the IS was D_3-C6-HSL and 2 nmol was added to each of the SC vials. The samples are then dried down before use. (**b**) Standard curve of C6-HSL. The ratio of the integrated ion transition peak area of a particular transition for the RS (signal of C6-HSL $[M+H]^+ \rightarrow [M+H-101]^+$) to the integrated ion transition peak area of the same transition of the IS (signal of D_3-C6-HSL $[M+H]^+ \rightarrow [M+H-101]^+$) was plotted and fitted with a linear equation (*black line*). This standard curve was then applied to analytical data.

4. Standards and samples are dried down under nitrogen and resuspended in crimp vials with 60 μL of HPLC solvent A and 10 μL of HPLC solvent B.
5. 30 μL of the sample is injected onto a 1.0 × 150 mm 5 μm C18 reversed-phase column.

Table 1
Sample standard curve for C6-HSL

nmol of C6-HSL	Area of C6-HSL/area of D_3-C6-HSL
0	0
0.016	0.00999
0.08	0.0463
0.04	0.257
2	1.11
10	5.96

Table 2
Chromatography profile used in LC/MS experiments

Step	Time (min)	Flow rate (μL/min)	A%	B%
0	0	50	95	5
1	5	50	95	5
2	35	50	5	95
3	55	50	5	95
4	56	50	95	5
5	60	50	95	5

6. The high-pressure liquid chromatography (HPLC) flow rate is 50 μL/min with initial conditions at time 0 of 5% B. A typical HPLC program is shown in Table 2, but briefly the gradient is isocratic for 5 min, ramps up to 95% B in 30 min, is isocratic at 95% B for 20 min, and at 56 min, the concentration of B goes to 5% and the system re-equilibrates for 4 min with the effluent flowing directly into the mass spectrometer.
7. Once the sample has passed into the mass spectrometer, a Multiple Reaction Monitoring (MRM) experiment is used to collect the data; the parameters are ion spray voltage of 4200 V, focusing potential of 200 V, and collision energy of 25 V. The collision gas thickness in the collision cell is 1.6×1015 molecules/cm^2; nitrogen is used as the collision gas.
8. The ions monitored in Q1 include the parent AHL $[M+H]^+$. In Q3, both the acyl moiety $[M+H-101]^+$ and the lactone moiety at m/z 102 are monitored and appear at characteristic retention times (Fig. 4a). The retention times and masses

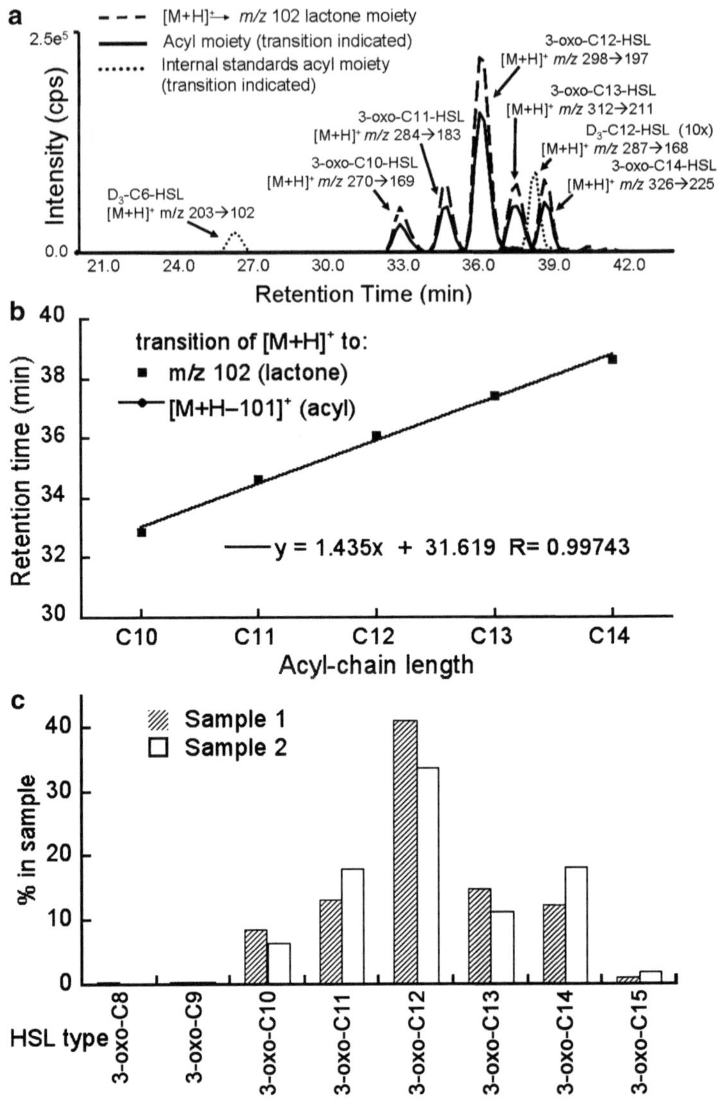

Fig. 4. Example of AHL analysis of AHLs produced by *LasI* expressed in *E. coli*. (**a**) Total ion current observed in the MRM analysis of reversed phase chromatographic separation of AHLs. AHLs were examined using the MRM mode that was monitoring the transitions for precursor $[M+H]^+ \rightarrow m/z$ 102 (shown as a *dashed line*) and the transition from $[M+H]^+ \rightarrow [M+H-101]^+$ (shown as *solid lines*), and the IS peaks (*dotted lines*). The identified AHLs are labeled. In this experiment, it is clear that the integrated peak areas for the ion transitions of 3-oxo-AHLs are of nearly equal size (approximately 40:60) for the two transitions (to the acyl and lactone moieties, respectively). (**b**) Retention times of the transitions observed for the lactone and acyl moieties plotted versus the number of carbon atoms in the acyl chain of each AHL. The plot shows that all the AHLs with the same substitution lie on a straight line, which coincides with a line that was generated by using the reference standards. Deviation from the line of the reference standards indicates that the compound, which has been named in the data analysis, is likely to be incorrect. (**c**) *Bar graph* representation of the AHLs purified from two independent experiments of *LasI* expressed in *E. coli*, with two slightly different growth and extraction conditions. Sample 1 and sample 2 differ in the number of hours that *LasI* was grown as

monitored in Q1 and Q3 are illustrated for a selection of AHLs in Table 3 that are used in the example presented in this chapter (Fig. 4b and Note 3).

9. Quantitation is accomplished by applying the standard curves to the analytical data. The amount of the unknown in each sample is determined by direct application of the slope and intercept of the standard curve to the ratios determined for the signals (from both transitions) of unknowns to internal standard, according to the following equation:

   Amount of unknown = ((Area counts of unknown/Area counts of internal standard) x (slope of standard curve) + intercept of standard curve)

10. Typically the ratios are calculated directly from the quantitative analysis routines included with mass spectrometer software, such as Analyst Software. Standard curves can be applied to the data using spreadsheet software, such as Microsoft Excel. For example, the standard curve for C6-HSL was generated by plotting the ratio of the integrated signals for the reference standard for C6-HSL to that of the labeled internal standard versus the amount of the reference standard for the transition $[M+H]^+ \rightarrow [M+H-101]^+$, as shown in Fig. 3b and Table 3. A similar curve was also generated and used for the $[M+H]^+ \rightarrow m/z$ 102 transition (lactone moiety; not shown).

11. Finally the results can be presented in a graphical format. Fig. 4c compares the 3-oxo-AHLs produced by *Las*I expressed recombinantly in *E. coli* from two independent experiments. It plots the type of AHL identified, based on the fragmentation pattern (Fig. 4a), retention time (Fig. 4b), and comparison to the reference standards (Table 3 and Note 4). The amount of AHL plotted is the sum of the amount determined for both of the monitored transitions $[M+H]^+ \rightarrow m/z$ 102 (lactone moiety) and $[M+H]^+ \rightarrow [M+H - 101]^+$ (acyl moiety) as a percentage of the total AHL in the sample. This graph illustrates the utility of the method and gives an indication of the reproducibility achieved in detection of AHLs from similar samples.

Fig. 4. (continued) well as the addition of Amberlite XAD 16. In sample 1, the Amberlite XAD 16 resin was added immediately after induction and the culture was grown for only 8 h, whereas for sample 2 the Amberlite XAD 16 resin was added right before inoculation and the culture was grown for 24 h. The height of the *bars* in the *graph* represents the percentage of each AHL and not the total amounts of AHL in the sample.

Table 3
Subset of transitions monitored in MRM experiments used in Fig. 4

	Transition to lactone moiety			Transition to acyl moiety		
	Q1 (*m/z*)	Q3 (*m/z*)	Retention time (min)	Q1 (*m/z*)	Q3 (*m/z*)	Retention time (min)
Standards						
D_3-C6-HSL	203.3	102.1	26.2	203.3	102.2	26.2
D_3-C12-HSL	287.3	102.1	38.17	287.3	186.2	38.17
C6-HSL	200.3	102.1	26.14	200.3	99.2	26.15
C12-HSL	284.4	102.1	37.75	284.3	183.2	37.75
3-oxo-C6-HSL	214.3	102.1	19.24	214.3	113.2	19.24
3-oxo-C12-HSL	298.3	102.1	35.66	298.3	197.2	35.67
Sample						
3-oxo-C6-HSL	214.3	102.1	19.27	214.3	113.2	19.30
3-oxo-C8-HSL	242.3	102.1	27.88	242.3	141.2	27.87
3-oxo-C9-HSL	256.3	102.1	30.73	256.3	155.2	30.69
3-oxo-C10-HSL	270.3	102.1	32.87	270.3	169.2	32.86
3-oxo-C11-HSL	284.3	102.1	34.64	284.3	183.2	34.64
3-oxo-C12-HSL	298.3	102.1	36.09	298.3	197.2	36.07
3-oxo-C13-HSL	312.3	102.1	37.44	312.3	211.2	37.42
3-oxo-C14-HSL	326.3	102.1	38.64	326.3	225.2	38.64
3-oxo-C15-HSL	340.3	102.1	39.69	340.3	239.2	39.68

4. Notes

1. All organic solvents should be tested for the presence of AHLs. AHL contaminants have been found in commercially purchased solvents.
2. Use glass when working with organic solvents.
3. It is important to use a sufficient number of reference standards to define the retention times for a given series of AHLs.
4. True quantitative analysis can only be obtained with an identical isotope-labeled internal standard. The values obtained from this analysis are semi-quantitative.

Acknowledgments

The authors would like to thank Drs. Ty Gould and Robert C. Murphy for their contributions in initially developing this method, Dr. Jake Herman for help in refining the method, and Dr. Joseph Hankin, Chris Johnson, and Wesley Martin for their assistance with the mass spectrometry. We also thank Dr. Robert C. Murphy for helpful suggestions and for use of the mass spectrometry equipment, which is supported by the Lipid Maps Large Scale Collaborative Grant (NIH GM069338 to R.C.M.). This work was supported by Grants from the National Science Foundation #0703467 and #0821220.

References

1. Fuqua, C. and Eberhard, A. (1999) Signal generation in autoinduction systems: synthesis of acylated homoserine lactones by LuxI-type proteins, In G. Dunny and S.C. Winans (eds.), Cell–cell signaling in bacteria. ASM Press, Washington, DC. 211–230.
2. Pearson, J.P., Passador, L., Iglewski, B.H., and Greenberg, E.P. (1995). A second *N*-acylhomoserine lactone signal produced by *Pseudomonas aeruginosa*. *Proc Natl Acad Sci USA* **92**, 1490–1494.
3. Churchill, M.E.A. and Herman, J.P. (2008) Acyl-homoserinelactone biosynthesis: structure and mechanism. Chapter 17, In Stephen Winans and Bonnie Bassler (eds.), Chemical communication among bacteria. ASM Press, Washington, DC.
4. Shaw, P.D., Ping, G., Daly, S.L., Cha, C., Cronan, J.E., Rinehart, K.L., and Farrand, S.K. (1997) Detecting and characterizing *N*-acyl-homoserine lactone signal molecules by thin-layer chromatography. *Proc Natl Acad Sci USA* **94**, 6036–6041.
5. Gould, T.A., Herman, J.P., Krank, J., Murphy, R.C., and Churchill, M.E.A. (2006) Specificity of acyl-homoserine lactone syntheses examined by mass spectrometry. *J Bacteriol* **188**, 773–783.
6. Watson, W.T., Minogue, T.D., Val, D.L., Beck von Bodman, S., and Churchill, M.E.A. (2002) Structural basis and specificity of acyl-homoserine lactone signal production in bacterial quorum sensing. *Mol Cell* **9**, 685–694.
7. Khan, S.R., Herman, J.P., Krank, K., Serkova, N.J., Churchill, M.E.A., Suga, H., and Farrand, S.K. (2007) N-(3-hydroxy-hexanoyl)-L-homoserine lactone is the biologically relevant quormone that regulates the phz operon of *Pseudomonas chlororaphis* (*aureofaciens*) Strain 30-84. *Appl Environ Microbiol* **73**, 7443–55.
8. Duerkop, B.A., Herman, J.P., Ulrich, R.L., Churchill, M.E.A., and Greenberg, E.P. (2008) The *Burkholderia mallei* BmaR3-BmaI3 quorum-sensing system produces and responds to *N*-3-hydroxy-octanoyl homoserine lactone. *J Bacteriol* **190**, 5137–5141.

Chapter 13

Global Expression Analysis of Quorum-Sensing Controlled Genes

Martin Schuster

Abstract

Transcript profiling has improved our understanding of quorum-sensing gene regulation on a global scale. Here, we describe methods for high-quality sample preparation for use in DNA microarray and real-time PCR experiments.

Key words: Transcriptome, Microarray, RNA isolation, cDNA synthesis, Real-time PCR, Quorum sensing, *Pseudomonas aeruginosa*

1. Introduction

Global transcript profiling using microarrays has become a standard procedure in molecular microbiology. It also had a significant impact on the field of quorum sensing (1, 2). Quorum sensing is a mechanism by which bacteria coordinate gene expression in response to cell density and environmental viscosity (3–5). The opportunistic pathogen *Pseudomonas aeruginosa* serves as a model to study quorum-sensing gene expression. Shortly after the *P. aeruginosa* genome had been fully sequenced (6) and a high-quality microarray became available (Affymetrix GeneChip), several groups set out to identify quorum-controlled genes in *P. aeruginosa* (7–9). From these studies, we know that quorum sensing in *P. aeruginosa* is a global regulatory circuit, affecting the expression of at least 6% of the genome. We also gained further insight into the timing of gene expression and signal specificity.

Here we provide detailed protocols on sample preparation for bacterial transcript profiling, with a particular emphasis on quorum-sensing gene expression. These include RNA isolation

Kendra P. Rumbaugh (ed.), *Quorum Sensing: Methods and Protocols*, Methods in Molecular Biology, vol. 692, DOI 10.1007/978-1-60761-971-0_13, © Springer Science+Business Media, LLC 2011

and cDNA synthesis for microarray and real-time PCR analysis. We also provide a protocol for measuring transcript abundance by real-time PCR, which can be used for expression analysis of a concise set of genes or for confirmation of microarray data. These protocols have been developed and used for *P. aeruginosa*, but will certainly be applicable to other culturable Gram-negative bacteria as well. We have used these protocols for the identification of quorum-sensing controlled genes (8) and of genes regulated by the stationary-phase sigma factor RpoS (10). Our protocols have also found application in several other transcriptome studies (11–16). It is beyond the scope of this chapter to provide detailed information about microarray processing and data analysis. Array platforms differ by type and the particular organism in question. In addition, tasks such as microarray hybridization, washing, and scanning are typically performed by core facilities at academic institutions and do not require knowledge by the individual researcher. They are typically performed according to standard instructions in manufacturer's manuals. Analysis options for microarray data abound and need to be customized depending on the specific requirements of each individual project.

While global transcript profiling by microarray technology is still widespread and timely, new technologies are emerging that supplement, and may even some day replace, microarray profiling. Transcriptome analysis by high-throughput parallel sequencing, "RNASeq," is one such technology (17, 18). The advantages are potentially higher sensitivity, recognition of novel transcripts (e.g., noncoding RNAs), and independence from strain-specific templates. The disadvantages are that extremely large amounts of data are generated, that data analysis is computationally intensive, and that no streamlined analysis tools are available yet. Especially in terms of continuity, ease of use, established technology, and secondary analysis, microarrays are currently a reasonable choice for transcript profiling.

2. Materials

2.1. RNA Isolation

1. RNase Away (Fisher) or equivalent.
2. 1.5 ml Microcentrifuge tubes.
3. 0.5 ml Thin-walled microcentrifuge tubes.
4. Aerosol-resistant filter tips, 20, 200, and 1,000 μl.
5. 15 and 50 ml Conical tubes.
6. Microtip sonicator (e.g., Branson Sonifier 250).
7. RNA Protect Bacteria Reagent (Qiagen).

8. RNeasy Mini Kit (Qiagen) (see Note 1).
9. RNase-free DNase set (Qiagen).
10. RNase-free water.
11. Lysozyme.
12. 100% Ethanol.
13. RQ DNase (Promega).
14. 25 mM Nucleotide mix for PCR.
15. Expand Long Template PCR Kit (Roche).
16. Northern Max-Gly Gel Prep/Running Buffer (Ambion).
17. 0.24–9.5 kb RNA Ladder (LifeTechnologies).
18. Glyoxal Sample Loading Dye (Ambion).
19. RNase-free agarose, e.g., Agarose LE (Roche).
20. Optional: Bioanalyzer (Agilent).

2.2. cDNA Synthesis

1. 100 mM dATP, dCTP, dGTP, and dTTP (GE Healthcare). Prepare a mix of all four dNTPs at final concentration of 10 mM each.
2. Superscript III Reverse Transcriptase (Invitrogen Life Technologies).
3. RNase inhibitor.
4. $(NS)_5$ Oligonucleotide. $(NS)_5$ is a semirandom decamer (N = A, T, C, or G; S=C or G) that has on average a 75% G+C content to compensate for the high G+C content of the *P. aeruginosa* genome. The oligonucleotide is ordered as a custom oligo (5′-NSNSNSNSNS), resuspended in RNase-free TE buffer to 2 μg/μl, and diluted to a working concentration of 250 ng/μl in RNase-free dH_2O. Organisms with very different G+C content will require different primers.
5. Spike transcripts (optional). See protocol from the microarray manufacturer.
6. QIAquick PCR Purification Kit (Qiagen).
7. 10× One-Phor-All Buffer (GE Healthcare).
8. DNase I, RNase-free (Roche).
9. 1 N NaOH solution.
10. 1 N HCl solution.
11. Ready-load 100 bp DNA ladder (Invitrogen Life Technologies).
12. GelStar Nucleic acid gel stain (FMC).

2.3. Array Processing and Data Analysis

1. Respective DNA microarray format (e.g., *P. aeruginosa* GeneChip, Affymetrix).

2. Reagents for DNA labeling (e.g., GeneChip DNA Labeling Kit, Affymetrix).
3. Equipment and reagents for array hybridization and washing (e.g., Affymetrix Fluidics Station; see individual protocols for further details).
4. Primary analysis software (e.g., Expression Console, Affymetrix).
5. Secondary analysis software from various sources depending on specific needs (e.g., Cyber-T, freely available at http://cybert.microarray.ics.uci.edu/; GeneSpring, Agilent).

2.4. Real-Time PCR

1. Power SYBR Green PCR Master Mix (Applied Biosystems), contains SYBR Green I Dye, AmpliTaq Gold DNA Polymerase, dNTPs.
2. ABI Prism Sequence Detection System (Applied Biosystems).
3. Sequence Detection System (SDS) software (Applied Biosystems).
4. Forward and reverse primers designed with Primer Express Software. The software uses the following design parameters:
   (a) Length of amplicons between 50 and 150 bp.
   (b) G/C content between 20 and 80%.
   (c) No runs of identical nucleotides.
   (d) T_M between 58 and 60°C.
   (e) The five nucleotides at the 3′ end should have no more than two G and/or C bases (this is not automatically screened for by the software).
5. Regular primer synthesis, e.g., 25 nM scale, desalted. Resuspend lyophilized primers to 100 μM in TE, pH 7. Make a working stock of 5 μM in water.
6. NuSieve 3:1 high resolution agarose (FMC).
7. Optical 96-well reaction plates and optical caps (Applied Biosystems).
8. Genomic Tips DNA Purification Kit (Qiagen).

3. Methods

3.1. Bacterial Culture

1. Both planktonic and biofilm bacteria can be used (see Note 2). For planktonic culture, grow *P. aeruginosa* strains overnight. Typically 3-5 ml of LB in 18 mm glass test tubes. Cells from *in vitro P. aeruginosa* biofilms are suitable for the RNA isolation procedure described below. For example, RNA has been purified successfully from biofilms grown in silicone tubes for up to 3 days (11).

2. Start new culture by diluting the overnight culture in appropriate medium.
3. At mid-logarithmic phase (OD_{600} between 0.4 and 0.8), start another culture by diluting to an OD_{600} of 0.01 in prewarmed medium.
4. Grow cultures to appropriate density as previously determined by growth curves. Cells harvested at mid-logarithmic phase tend to yield the most consistent, reproducible transcript profiles (but see Note 2).

3.2. RNA Isolation

The following steps should be done consistently without interruption to minimize variation between RNA preps. Harvesting and lysis of cells is based on the enzymatic lysis and mechanical disruption protocol described in the RNA Protect Bacteria Reagent Handbook. Subsequent purification of RNA is based on the RNeasy Mini Kit. Perform all steps at room temperature unless stated otherwise. The protocol is designed for three Mini spin columns per sample, resulting in ≥100 μg of purified RNA altogether. If less RNA is required, the protocol can be scaled down.

1. Remove a volume of culture corresponding to 2×10^9 cells/ml, and add to two volumes of RNA Protect Bacteria Reagent in 15 ml conical tubes (50 ml tubes can be used for larger culture volumes at low cell densities) (see Note 3). Mix immediately by vortexing for 5 s or inverting the tube multiple times. Cell numbers are optimized for *P. aeruginosa*. They may need to be adjusted for other bacterial species. It is important to not overload the column as this may result in incomplete lysis and RNA degradation.
2. Incubate for 5 min at room temperature. Centrifuge for 10 min at 5,000 × *g*.
3. Decant the supernatant carefully. Remove remaining supernatant by gently dabbing the inverted tube onto a paper towel. The pellet looks atypically white. At this point, pellets can be stored at −70°C for up to a month.
4. Add 570 μl (3 × 190 μl) of TE buffer (10 mM Tris–HCl, 1 mM EDTA, pH 8.0) and 30 μl (3 × 10 μl) of lysozyme from a 20 mg/ml stock for a final concentration of 1 mg/ml.
5. Mix by vortexing for 10 s. Incubate at room temperature for at least 10 min with intermittent vortexing. The pellet will look flaky and will not resuspend completely. Incubation time can be extended without adverse effects.
6. Add 2.1 ml (3 × 700 μl) of buffer RLT (10 μl of β-mercaptoethanol per 1 ml buffer RLT must be added before use) to the sample and vortex vigorously. The samples may appear cloudy.

7. Sonicate samples twice for 5–10 s using a microtip sonicator (duty cycle constant, output 0.4). The samples will warm up slightly but should not become hot.
8. Add 1.4 ml (3 × 470 μl) 100% ethanol and mix immediately by pipetting, or inverting the tube.
9. Load 700 μl aliquots of the lysate onto three RNeasy Mini columns. Each column must be loaded and centrifuged twice to accommodate the entire sample volume of approximately 4.2 ml. Work diligently to avoid RNA degradation from this point on (see Note 4).
10. Follow the RNeasy Mini Protocol for the isolation of total RNA from bacteria as described in the RNeasy Mini Handbook. Include the optional on-column DNase digestion using the RNase-free DNase Set. Using 700 μl instead of 350 μl buffer RW1 may increase purity as determined by the 260/230 ratio.
11. Elute twice with 30 μl of RNase-free water. For the second elution, use the eluate from the first elution. The total volume from the three columns combined should be 90 μl.
12. Determine total RNA concentration and purity by spectrophotometric analysis at 230, 260, and 280 nm. Do a 1:50 dilution in 100 μl H_2O. To estimate concentration, 1 U of A_{260} equals 40 μg/ml RNA. Expect about 2 μg/μl of RNA for cultures harvested in mid-logarithmic phase. Stationary phase cultures yield significantly less RNA (approximately 1 μg/μl or less). For purity, the 260/280 and 260/230 ratios should be above 1.8. Note that these ratios are more accurate when determined in buffer such as Qiagen's EB.

3.3. DNase Digestion of Isolated RNA

1. Fifty micrograms of total RNA are digested with RQ DNase (see Note 5). Use 0.1 U/1 μg of RNA and incubate for 1 h at 37°C. A typical 50 μl reaction may include 40 μl of RNA, 5 μl of DNase, and 5 μl of 10× buffer.
2. Remove enzyme and buffer components according to the RNA Clean-Up Protocol in the RNeasy Mini Handbook. Use one spin-column per sample. Elute twice in 30 μl H_2O as described above.
3. Spec RNA samples again as described above.

3.4. DNA Contamination PCR

To detect small amounts of DNA contamination that would not be visible on an agarose gel, a fragment of the ribosomal protein gene, *rplU*, is amplified by PCR using the Expand Long Template system, or an equivalent PCR system that is suitable for the amplification of high GC-content DNA.

1. Prepare the following reaction mix for PCR

   Primers:

   rplU forward, 5′ CGCAGTGATTGTTACCGGTG 3′

   rplU reverse, 5′ AGGCCTGAATGCCGGTGATC 3′

Components	Volume/ amount (μl)
10× Buffer no. 3	5
RNA sample (no template as negative control and purified chromosomal DNA as positive control)	1
Oligonucleotides (25 mM each)	1
rplU Forward primer (15 pmol/μl)	1
rplU Reverse primer (15 pmol/μl)	1
Expand long template polymerase	0.75
dH_2O	Up to 50

2. Perform a PCR amplification in a thermocycler using the following conditions

   5 min at 95°C

   30 cycles of:

   30 s at 94°C

   40 s at 58°C

   60 s at 68°C

   10 min at 68°C.

3. Analyze 10 μl of PCR product on a 1.2% TAE or TBE-agarose gel. The expected product size is 300 bp.

3.5. RNA Quality Control

For quality control, RNA samples are glyoxylated and analyzed by agarose gel electrophoresis using the NorthernMax-Gly system. Alternatively, a Bioanalyzer can be used to estimate RNA quality. This method is less susceptible to RNase contamination. The instrument is typically available at core facilities in academic institutions. In the following, the steps for running an RNA gel are described.

1. Cast a 1% agarose gel using RNase-free agarose and 1× NorthernMax-Gly Gel Prep/Running buffer. The same 1× buffer is used for running gels.
2. Glyoxylate 3 μg of RNA by diluting 3 μg RNA in a total of 5 μl of H_2O. Add 5 μl glyoxal sample loading dye to each sample and heat at 50°C for 30 min.
3. Glyoxylate a suitable RNA ladder to run for size comparison.
4. Run gel at 5 V/cm as measured between electrodes. Gels do not need to be stained as the loading buffer contains ethidium bromide.

5. The 23s and 16s rRNA should give clear bands with the 23s band showing close to twice the intensity. The tRNA and 5s rRNA bands should barely be visible as the RNeasy columns do not retain small RNA species below 200 bases.

3.6. cDNA Synthesis

1. The following protocol starts with 12 μg of size-fractionated RNA per sample. The incubation is performed in a thermocycler with RNase-free PCR tubes. Prepare the following mixture for primer annealing: Control transcripts of defined concentrations can be spiked in according to detailed instructions from the microarray manufacturer.

Components	Volume/amount	Final concentration
Fractionated RNA	12 μg	0.33 μg/μl
$(NS)_5$ Primer (250 ng/μl)	3.6 μl	25 ng/μl
dH_2O	Up to 36 μl	–

2. Place the mix at 70°C for 10 min, 25°C for 10 min and hold at 4°C.
3. Prepare the following reaction mix for cDNA synthesis:

Components	Volume/amount (μl)	Final concentration
5× First strand buffer	12	1×
100 mM DTT	3	5 mM
10 mM dNTPs	3	0.5 mM
RNaseIN (40 U/μl)	2.0	1.33 U/μl
SuperScript III (200 U/μl)	4	13.3 U/μl

4. Incubate the reaction at 25°C for 10 min and 46°C for 3 h; then inactivate the enzyme at 70°C for 10 min and hold at 4°C.
5. Remove the RNA template by alkaline hydrolysis. Transfer reactions to 1.7 ml microfuge tubes, add 20 μl of 1 N NaOH and incubate at 65°C for 30 min. Add 20 μl of 1 N HCl for the titration and add H_2O to final volume of 100 μl.

3.7. Purification and Quantitation of cDNA Synthesis Products

1. Use the QIAquick PCR Purification Kit to clean-up the cDNA synthesis product (for details, see protocol provided by the supplier).
2. Removal of all residual ethanol from the column is critical, especially if subsequent DNase I digestion is required (see below). To ensure ethanol removal, spin the column for at least 10 min at high speed in a clean microfuge tube following

the wash step. It may also be necessary to blot any remaining ethanol from the inner beveled edge of the column with a wipe.

3. Elute cDNA from the QIAquick column with 30 μl of EB Buffer, two times for a final elution volume of 60 μl.
4. Quantify the purified cDNA product by 260 nm absorbance (1.0 A_{260}U = 33 μg/ml of single strand cDNA). Use a 1:25 dilution in dH_2O. Typical yields of cDNA are 6–8 μg per 60 μl of eluted product. If less product is required or less input RNA is available, the entire cDNA synthesis reaction can be performed using 30 μl instead of 60 μl total volume. The concentrations of all components remain the same. The amount of input RNA is 6 μg instead of 12 μg.
5. Save 3.5 μl for analysis by agarose gel electrophoresis in step 3.9.

3.8. cDNA Fragmentation

This step is required for *P. aeruginosa* GeneChip arrays.

1. Prepare the following reaction mix on ice.

Components	Volume/amount	Concentration
10× One-Phor-All Buffer	5 μl	1×
cDNA	35 μl (3–6 μg)	–
DNase I (Roche)	X μl	0.06 U/μg of cDNA
dH_2O	Up to 50 μl	–

Use 1× One-Phor-All buffer for the dilution of DNase I from stock solution prior to the fragmentation reaction. The amount of DNase I required can vary from supplier to supplier and a titration experiment should be performed with each new batch of enzyme. As a starting point, dilute DNase I to 0.06 U/μl and add 1.0 μl per μg of cDNA in the fragmentation reaction.

2. Add diluted DNase I last, mix, and immediately begin incubation.
3. Incubate the reaction at 37°C for 10 min, then inactivate the enzyme at 98–100°C for 10 min.

3.9. Quality Control/Agarose Gel Electrophoresis of cDNA

For quality control, cDNA samples are analyzed by agarose gel electrophoresis. If cDNA is to be fragmented, evaluation of size distribution is crucial for successful array hybridization. To obtain good array hybridization, the majority of the DNA fragments should be within 50–200 bases.

1. Cast a 2% agarose gel containing GelStar Nucleic Acid Gel Stain at a dilution of 1:10,000 using either TAE or TBE buffer.

2. Analyze 5 μl (~500 ng) of fragmented cDNA and 3.5 μl of unfragmented cDNA (~500 ng) as a control, in a suitable gel loading buffer. Use 2.5 μl (250 ng) of Ready-load 100 bp DNA Ladder as marker for size determination.

3.10. Further Sample Processing and Microarray Data Analysis

In the case of *Pseudomonas* GeneChips, cDNA samples are terminally labeled using biotin-ddUTP. The labeled DNA samples are then hybridized to the array, washed, and scanned. These latter steps are typically conducted by a DNA core facility at an academic institution. The respective protocols in the Affymetrix Expression Analysis Technical Manual are followed. Finally, data are collected and processed with basic software (e.g., Affymetrix Expression Console). Basic quality controls are performed, including probe array image inspection, control oligo performance, background, and noise levels. A high signal in many intergenic regions may indicate chromosomal DNA contamination. Once basic data normalization has been performed, gene expression results from individual arrays are typically compared to reveal significant differences (see Note 6). For this purpose, we have used the web-based program Cyber-T. For more advanced analysis and visualization of gene expression patterns (e.g., the kinetics of gene induction), GeneSpring (Agilent) is very useful.

3.11. Real-Time PCR

The protocol described here utilizes SYBR Green as a fluorescent dye. Amplification of PCR products is monitored by measuring the increase in fluorescence caused by the binding of SYBR Green to double-stranded DNA. Quantitation of transcript abundance is based on the relative standard curve method. The relationship between the relative amount of input cDNA and the threshold PCR cycle (C_T) is determined using a standard curve with defined dilutions of purified DNA. Experiments were performed with the ABI Prism 7500 Sequence Detection System.

Normalization of the transcript data from different samples can be accomplished two different ways. Normalization of transcript levels to those of a calibrator gene is commonly done in one-step reverse transcription-PCR to account for differences in reverse transcription efficiencies. Finding an appropriate reference gene that display similar, robust expression levels across all samples is essential here. Examples include GAPDH and 16s rDNA. rRNA may not be the best choice, however, because of the extremely high abundance compared to the mRNA species in question. Data can also be normalized by simply calibrating the amount of input cDNA prior to PCR. This necessitates a two-step reaction in which cDNA synthesis is separated from real-time PCR. After cDNA synthesis, the amount of product is quantified, and a defined amount is used as template for subsequent real-time PCR.

3.11.1. PCR to Detect Nonspecific Amplification

Nonspecific amplification, including primer–dimer formation, is an issue when using SYBR Green instead of a specific Taq-Man probe. The absence of nonspecific amplification should therefore be confirmed by analyzing the amplification products of a regular (end-point) PCR by agarose gel electrophoresis. In addition, nonspecific amplification can also be detected by running a dissociation curve at the end of real-time PCR run (see below).

1. Set up a PCR reaction mix as follows:

Components	Volume/amount (μl)	Final concentration
Power SYBR Green PCR Master Mix (2×)	12.5	1×
Forward primer (5 μM)	1.5	300 nM
Reverse primer (5 μM)	1.5	300 nM
cDNA template (0.2 ng/μl)	5	1 μg (amount)
dH_2O	to 25	–

2. Perform a PCR amplification in a thermocycler using the following conditions (identical to those in the real-time PCR):

   10 min at 95°C – initial denaturation

   40 cycles of:

   15 s at 95°C – denaturation

   60 s at 60°C – annealing and extension

3. Analyze 10 μl of the PCR reaction on a 4% NuSieve 3:1 agarose gel. In the case of nonspecific amplification (multiple bands), perform a primer concentration optimization. Lowering the primer concentrations (as low as 50 nM) has proven successful in previous applications.

3.11.2. Real-Time PCR SetUp

1. Make up a dilution series from chromosomal DNA for use as standard (purified with Genomic Tips Kit). Make 5–7 dilutions that are likely to cover the estimated range of the experimental samples. We routinely make six tenfold dilutions ranging from 0.0001 to 10 ng (final amount). If possible, use the same dilution series for all standard curves (freeze and reuse DNA dilutions).
2. Set up a 25 μl reaction as follows (the suggested volume in the SYBR Green protocol is 50 μl, but 25 μl works just as well and saves reagent). Make a Master Mix of the SYBR Green PCR Master Mix, forward and reverse primers, and water. Add the cDNA template (from step 3.7, nonfragmented) to each well individually. Include a no template control. Mix well. Do every reaction in duplicate. In our hands, triplicates did not yield significantly more accurate numbers.

Components	Volume/ amount (μl)	Final concentration
Power SYBR Green PCR Master Mix (2×)	12.5	1×
Forward primer (5 μM)	1.5	300 nM (or as determined by primer optimization)
Reverse primer (5 μM)	1.5	300 nM (or as determined by primer optimization)
cDNA template (0.2 ng/μl)	5	1 ng (amount)
dH_2O	to 25	–

3. Perform a PCR amplification in a real-time PCR system using the following conditions:

   10 min at 95°C – initial denaturation

   40 cycles of:

   15 s at 95°C – denaturation

   60 s at 60°C – annealing and extension

   Also program the cycler to run a dissociation step at the end to evaluate nonspecific amplification. A dissociation curve with more than one peak is indicative of multiple amplification products.

3.11.3. Quantitation

1. To generate the standard curves, plot the C_T vs. log relative cDNA concentration. Current analysis software, e.g., Applied Biosystems SDS, does this automatically. Do a linear regression to determine the relative concentration of the unknown sample.
2. If using reference genes, normalize the concentrations determined for the unknown samples by dividing the unknown sample concentrations by the standard (reference) concentrations. Then calculate the ratio of the normalized unknown "experiment" and "baseline" samples, e.g., induced vs. uninduced or mutant vs. wild type.

4. Notes

1. For bacterial RNA isolation from complex mixtures of host-bacterial samples, kits that remove mammalian RNA are available (e.g., MICROBEnrich, Ambion). In addition, bacterial mRNA can be further enriched by selective removal of rRNA (e.g., MICROBExpress, Ambion).

2. Careful experimental design, including the choice of strains and growth conditions, is essential for meaningful microarray results. In the case of quorum sensing, two types of quorum-sensing deficient mutants may be chosen – signal synthesis and signal receptor mutants. Signal synthesis mutants offer the advantage that quorum-sensing responsive genes can simply be identified by the addition or omission of purified quorum signal(s), if available. When using signal receptor mutants, it is important that the respective strains are isogenic to the wild type. Another consideration is what time during culture growth to select for transcript profiling, as timing is key in quorum-sensing gene induction (8). Preliminary experiments, for example by real-time PCR, would be appropriate to establish the expression kinetics of individual quorum-sensing indicator genes.
3. RNA quality, in general, is crucial for successful array experiments. Immediate stabilization (e.g., by RNA Protect Reagent) of the biological sample is very important to preserve the native abundance of mRNA species. Due to the short half-life of most mRNAs, even the brief handling of samples (e.g., centrifugation to pellet cells) prior to sample denaturation can alter RNA content. As outlined above, RNA integrity should be assessed by either Bioanalyzer or agarose gel electrophoresis.
4. For RNA work, the usual precautions should be taken to avoid RNase contamination – wearing gloves, treating surfaces with RNase away, using RNase-free water, and, if possible, designating a separate work space or biosafety cabinet.
5. It is essential to remove contaminating chromosomal DNA from the RNA sample. In our hands, standard purification with RNeasy columns, even including the optional on-column RNase treatment, does not result in sufficient elimination of chromosomal DNA. We therefore propose an additional DNase treatment in solution.
6. How many and which replicates should be run? More replicates will lower the false-discovery rate and will identify a larger percentage of the differentially expressed genes (19). Biological rather than technical replicates should be run. Pooling of RNA samples is often used out of necessity, to minimize cost or to reduce sample variation. An apparent disadvantage of pooling is that transcripts identified as significantly changed in individual samples are often not identified in the pooled samples, because information about small, but statistically significant, changes in expression are lost in the pooled samples. However, when designed properly, and depending on the particular question asked, pooling can be beneficial (20, 21).

Acknowledgements

Parts of the cDNA synthesis procedure are based on a protocol by Matt Wolfgang, University of North Carolina. Thanks to Joon-Hee Lee, University of Pusan, South Korea, for his suggestions on real-time PCR.

References

1. Goodman, A. L., and Lory, S. (2004) Analysis of regulatory networks in *Pseudomonas aeruginosa* by genomewide transcriptional profiling, *Curr. Opin. Microbiol. 7*, 39–44.
2. Schuster, M., and Greenberg, E. P. (2006) A network of networks: quorum-sensing gene regulation in *Pseudomonas aeruginosa*, *Int. J. Med. Microbiol. 296*, 73–81.
3. Williams, P., Winzer, K., Chan, W. C., and Camara, M. (2007) Look who's talking: communication and quorum sensing in the bacterial world, *Philos. Trans. R. Soc. London. 362*, 1119–1134.
4. Waters, C. M., and Bassler, B. L. (2005) Quorum sensing: cell-to-cell communication in bacteria, *Annu. Rev. Cell Dev. Biol. 21*, 319–346.
5. Hense, B. A., Kuttler, C., Muller, J., Rothballer, M., Hartmann, A., and Kreft, J. U. (2007) Does efficiency sensing unify diffusion and quorum sensing?, *Nat. Rev. 5*, 230–239.
6. Stover, C. K., Pham, X. Q., Erwin, A. L., Mizoguchi, S. D., Warrener, P., Hickey, M. J., Brinkman, F. S., Hufnagle, W. O., Kowalik, D. J., Lagrou, M., Garber, R. L., Goltry, L., Tolentino, E., Westbrock-Wadman, S., Yuan, Y., Brody, L. L., Coulter, S. N., Folger, K. R., Kas, A., Larbig, K., Lim, R., Smith, K., Spencer, D., Wong, G. K., Wu, Z., and Paulsen, I. T. (2000) Complete genome sequence of *Pseudomonas aeruginosa* PAO1, an opportunistic pathogen, *Nature. 406*, 947–948.
7. Hentzer, M., Wu, H., Andersen, J. B., Riedel, K., Rasmussen, T. B., Bagge, N., Kumar, N., Schembri, M. A., Song, Z., Kristoffersen, P., Manefield, M., Costerton, J. W., Molin, S., Eberl, L., Steinberg, P., Kjelleberg, S., Hoiby, N., and Givskov, M. (2003) Attenuation of *Pseudomonas aeruginosa* virulence by quorum sensing inhibitors, *EMBO J. 22*, 3803–3815.
8. Schuster, M., Lohstroh, C. P., Ogi, T., and Greenberg, E. P. (2003) Identification, timing and signal specificity of *Pseudomonas aeruginosa* quorum-controlled genes: a transcriptome analysis, *J. Bacteriol. 185*, 2066–2079.
9. Wagner, V. E., Bushnell, D., Passador, L., Brooks, A. I., and Iglewski, B. H. (2003) Microarray analysis of *Pseudomonas aeruginosa* quorum-sensing regulons: effects of growth phase and environment, *J. Bacteriol. 185*, 2080–2095.
10. Schuster, M., Hawkins, A. C., Harwood, C. S., and Greenberg, E. P. (2004) The *Pseudomonas aeruginosa* RpoS regulon and its relationship to quorum sensing, *Mol. Microbiol. 51*, 973–985.
11. Bagge, N., Schuster, M., Hentzer, M., Ciofu, O., Givskov, M., Greenberg, E. P., and Hoiby, N. (2004) *Pseudomonas aeruginosa* biofilms exposed to imipenem exhibit changes in global gene expression and beta-lactamase and alginate production, *Antimicrob. Agents. Chemother. 48*, 1175–1187.
12. Sonnleitner, E., Schuster, M., Sorger-Domenigg, T., Greenberg, E. P., and Blasi, U. (2006) Hfq-dependent alterations of the transcriptome profile and effects on quorum sensing in *Pseudomonas aeruginosa*, *Mol. Microbiol. 59*, 1542–1558.
13. Rampioni, G., Schuster, M., Greenberg, E. P., Bertani, I., Grasso, M., Venturi, V., Zennaro, E., and Leoni, L. (2007) RsaL provides quorum sensing homeostasis and functions as a global regulator of gene expression in *Pseudomonas aeruginosa*, *Mol. Microbiol. 66*, 1557–1565.
14. Palmer, K. L., Mashburn, L. M., Singh, P. K., and Whiteley, M. (2005) Cystic fibrosis sputum supports growth and cues key aspects of *Pseudomonas aeruginosa* physiology, *J. Bacteriol. 187*, 5267–5277.
15. Banin, E., Vasil, M. L., and Greenberg, E. P. (2005) Iron and *Pseudomonas aeruginosa* biofilm formation, *Proc. Natl. Acad. Sci. U.S.A. 102*, 11076–11081.
16. Yarwood, J. M., Volper, E. M., and Greenberg, E. P. (2005) Delays in *Pseudomonas aeruginosa* quorum-controlled gene expression are

conditional, *Proc. Natl. Acad. Sci. U.S.A. 102*, 9008–9013.

17. Mardis, E. R. (2008) Next-generation DNA sequencing methods, *Annu. Rev. Genomics Hum. Genet. 9*, 387–402.
18. Pettersson, E., Lundeberg, J., and Ahmadian, A. (2009) Generations of sequencing technologies, *Genomics. 93*, 105–111.
19. Wolfinger, R. D., Gibson, G., Wolfinger, E. D., Bennett, L., Hamadeh, H., Bushel, P., Afshari, C., and Paules, R. S. (2001) Assessing gene significance from cDNA microarray expression data via mixed models, *J. Comput. Biol. 8*, 625–637.
20. Kendziorski, C., Irizarry, R. A., Chen, K. S., Haag, J. D., and Gould, M. N. (2005) On the utility of pooling biological samples in microarray experiments, *Proc. Natl. Acad. Sci. U.S.A. 102*, 4252–4257.
21. Peng, X., Wood, C. L., Blalock, E. M., Chen, K. C., Landfield, P. W., and Stromberg, A. J. (2003) Statistical implications of pooling RNA samples for microarray experiments, *BMC Bioinformatics. 4*, 26.

Chapter 14

Small RNA Target Genes and Regulatory Connections in the *Vibrio cholerae* Quorum Sensing System

Brian K. Hammer and Sine Lo Svenningsen

Abstract

The two-component quorum sensing (QS) system, first described in the marine bacterium *Vibrio harveyi* and evolutionarily conserved among members of the genus *Vibrio*, has been best studied in the human pathogen *Vibrio cholerae* (1, 2). In the *V. cholerae* QS system, the response to the accumulation of extracellular autoinducers triggers a signaling cascade resulting in the transcription of four small regulatory RNAs (sRNAs). Our results support the model that the QS sRNAs bind to the 5′ untranslated region of multiple mRNAs and alter the fate of one in a positive manner and several others in a negative manner. This mechanism ensures the proper timing of the QS response, which includes the expression of traits critical for virulence and for the formation of biofilms (2–6).

Key words: *Vibrio cholerae*, Small RNA, mRNA, Quorum sensing, Virulence, SOE, GFP, Northern blot, Site-directed mutagenesis

1. Introduction

Vibrio cholerae produces two autoinducers, CAI-1 (*c*holerae *a*uto*i*nducer 1) and AI-2 (*a*uto*i*nducer 2), which are synthesized by the CqsA and LuxS synthases, respectively. At low cell density (low AI levels), the cognate sensor kinase receptors, CqsS and LuxP/Q, respectively, act as kinases that transfer phosphate to the σ^{54}-dependent response regulator, LuxO, via the phosphotransfer protein LuxU. LuxO-P activates transcription of four small regulatory RNAs (sRNAs) named Qrr1–4 (*Q*uorum *r*egulatory *R*NAs) that do not encode for proteins. Rather, Qrr sRNAs base pair with and repress the polycistronic mRNA encoding LuxO and LuxU, as well as the mRNA-encoding

Kendra P. Rumbaugh (ed.), *Quorum Sensing: Methods and Protocols*, Methods in Molecular Biology, vol. 692, DOI 10.1007/978-1-60761-971-0_14, © Springer Science+Business Media, LLC 2011

HapR, a transcriptional regulator of numerous genes, including many that participate in biofilm formation and virulence (1–8). Base pairing of the sRNAs also activates translation of the mRNA of a biofilm-promoting diguanylate cyclase, Vca0939, by preventing the formation of a secondary structure that inhibits translation (9). The RNA chaperone Hfq is required for Qrr function (2). At high cell density (high AI levels), binding of AIs switches the receptors to act as phosphatases, which, via LuxU, remove phosphate from LuxO, inactivating it and halting sRNA transcription. In the absence of Qrr sRNAs, Vca0939 is not made, while LuxO and HapR are no longer repressed. QS control of the four Qrr sRNAs provides a regulatory mechanism for controlling the timing and pattern of expression of target genes that are important for *V. cholerae* development and pathogenesis.

The effects of sRNAs on mRNA levels are determined in living cells by altering sRNA levels and then monitoring the expression of translational fusions of the target mRNA to the gene-encoding green fluorescent protein (GFP). Fluorescence levels of bacterial batch cultures are determined with a microtiter plate reader capable of measuring absorbance at 600 nm and GFP-mediated fluorescence with a standard FITC filter set (ex: 485 nm/em: 525 nm). A flow cytometer or cell sorter can also be used to measure mean fluorescence per individual cell. Northern blot analysis of cell extracts is often used when more dynamic and/or sensitive analyses are desired. New computational algorithms with straightforward web interfaces are available for predicting sRNA/mRNA base pairing that can be tested experimentally by site-directed mutagenesis.

2. Materials

2.1. Construction of GFP-Translational Fusion to sRNA Target Gene

1. High-fidelity polymerase PCR reagents (Phusion, New England Biolabs, Ipswich, MA).
2. Restriction enzymes, DNA ligase, and buffers (Promega, Madison, WI).
3. Kanamycin or other appropriate antibiotics.
4. 2,6-Diaminopimelic acid (Acros Organics, Geel, Belgium).
5. Agarose gel extraction kit (Zymo Research, Orange, CA).
6. DNA clean-up kit (Zymo Research, Orange, CA).
7. Plasmid preparation kit (Qiagen, Valencia, CA).
8. Conjugative cloning vector for gene–*gfp* fusions (see Note 3).

9. *gfp* gene source (pEVS141 (10)).
10. *Escherichia coli* Dap⁻ auxotroph donor strain (11).

2.2. Manipulating sRNA Levels

1. High-fidelity polymerase PCR reagents (Phusion, New England Biolabs, Ipswich, MA).
2. PCR purification kit (Qiagen, Valencia, CA).
3. Restriction enzymes, DNA ligase, and buffers (Promega, Madison, WI).
4. Black microtiter plates (Corning-CoStar, Corning, NY).
5. Conjugative cloning vector for sRNA expression (see Note 3).
6. 2,6-Diaminopimelic acid (Acros Organics, Geel, Belgium).
7. *Escherichia coli* Dap⁻ auxotroph donor strain (11).

Access to:

1. Microplate reader (Synergy 4, BioTek, Winooski, VT).
2. Flow Cytometer/Cell Sorter (FACSVantage SE DIVA, BD, Franklin Lakes, NJ).

2.3. Northern Blot of mRNA Half-Life

1. Template DNA PCR product.
2. DEPC-treated water (Qiagen, Valencia, CA).
3. RNase AWAY (Invitrogen, Carlsbad, CA).
4. Trizol.
5. Formamide.
6. Formaldehyde.
7. Chloroform.
8. Isopropanol.
9. 96% ethanol.
10. 10% SDS.
11. Rifampicin (Sigma-Aldrich, St. Louis, MO).
12. PCR purification kit (Qiagen, Valencia, CA).
13. α-^{32}P-dATP (10 mCi/mL) (Perkin-Elmer, Waltham, MA).
14. Positively charged nylon membrane (Applied Biosystems/Ambion, Austin, TX).
15. RNA ladder (Applied Biosystems/Ambion, Austin, TX).
16. 50× Denhardts solution (Applied Biosystems/Ambion, Austin, TX).
17. Absorbent paper (Whatman).
18. 20× SSC: 3 M NaCl, 0.3 M Na-citrate.
19. 10× MOPS buffer (Millipore, Billerica, MA).
20. Autoradiographic film (Biomax MS , Kodak, Rochester, NY).

Access to:

1. Phosphoimager (Molecular Dynamics STORM®, General Electric, Niskayuna NY).

2.4. Computational Prediction of sRNA/mRNA Pairing Sites

1. http://snowwhite.wellesley.edu/*targetRNA*/index.html.

2.5. Experimental Validation of sRNA/mRNA Pairing Sites

1. QuikChange site-directed mutagenesis kit (Stratagene, La Jolla, CA).
2. Plasmid preparation kit (Qiagen, Valencia, CA).
3. Kanamycin or other appropriate antibiotics.
4. 2,6-Diaminopimelic acid (Acros Organics, Geel, Belgium).
5. *Escherichia coli* Dap^- auxotroph donor strain (11).

3. Methods

After a candidate sRNA-regulated target gene has been identified by methods described elsewhere (such as genetic screens or microarrays) (12), both genetic and biochemical methods are necessary to validate experimentally that the particular sRNA alters expression of the target mRNA by base pairing and to characterize the sRNA-mediated regulation. Duplex formation by sRNA/mRNA pairing typically occurs in the 5′ untranslated region (UTR) directly upstream of the ATG start codon, and for repressed targets, like the *hapR* mRNA, often overlaps the ribosome binding site (2). Translational fusions of putative target genes to a visual reporter, such as the gene for the green fluorescent protein (GFP), are constructed and monitored in bacteria in which expression of the sRNAs can be experimentally manipulated. The stability of an mRNA is often altered by duplex formation with an sRNA (13), so Northern blotting can also be used as a sensitive method to document changes in RNA levels.

Bacterial sRNA/mRNA duplexes are typically an imperfect pairing of a region of the sRNA with an mRNA sequence that has partial complementarity. Pairing imperfection has limited the success of predictive computational algorithms (12). Nonetheless, some of these computational methods, such as *TargetRNA*, are well suited for identifying a putative sRNA binding site in a particular candidate mRNA sequence (14). It is essential to experimentally validate the pairing predicted by the genetic and computational methods by altering both the RNA sequences that participate in duplex formation and then measuring the consequences on sRNA-mediated regulation.

3.1. Construction of GFP-Translational Fusions to sRNA Targets Using Gene Splicing by Overlap Extension

1. A familiarity with bacterial culturing, standard and colony polymerase chain reaction (PCR), and cloning procedures is assumed for this and subsequent method sections. Gene *s*plicing by *o*verlap *e*xtension (SOE) (15) is used to construct the GFP-translational fusion reporters using a high-fidelity polymerase and a *gfp* gene template. *V. cholerae* chromosomal DNA serves as the template for the target gene PCR. Alternatively, luciferase-transcriptional fusions have also been used and are mentioned in Note 1.
2. To construct a translational gene fusion of a target gene to *gfp*, three rounds of PCR are performed as described below. We describe the design of a HapR–GFP fusion, which is illustrated in Fig. 1. Adjustments in primer length and location

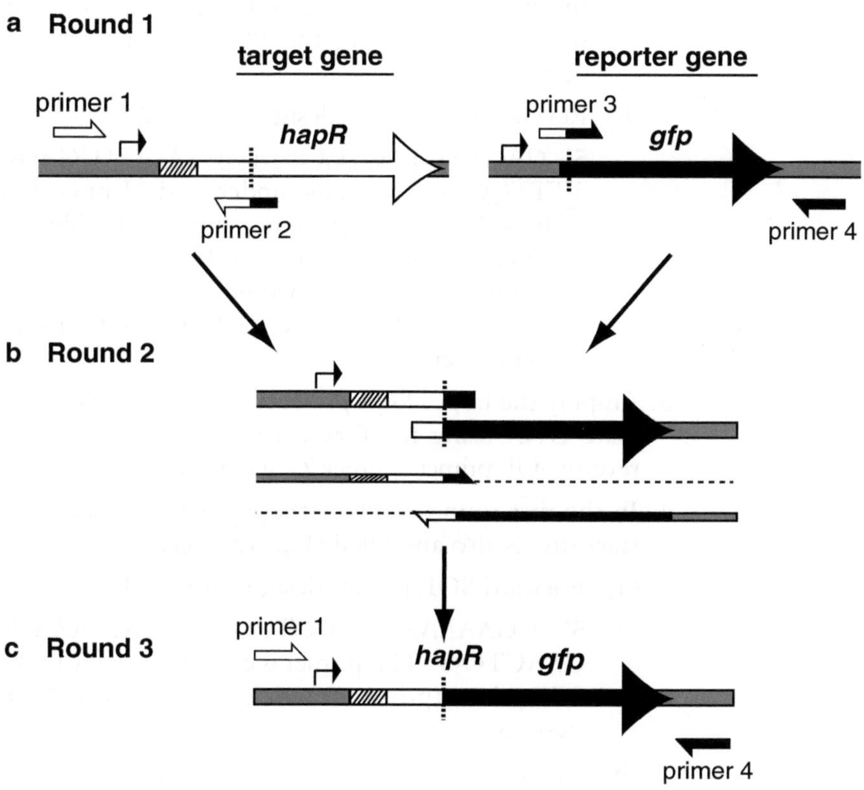

Fig. 1. SOE PCR for GFP-translational fusions. (**a**) During the first round of PCR, the *hapR* target gene product is amplified with the upstream target primer (primer 1) and reverse SOE primer (primer 2) to include the *hapR* promoter region (*small black arrow*), putative sRNA binding site in the 5′ UTR (*hashed rectangle*), and the region encoding the first 10 amino acids of HapR (*white*). The reporter gene, *gfp* (*black*), is amplified with a forward SOE primer (primer 3) that is the reverse complement of primer 2 and a reverse reporter primer (primer 4), and does not include the ATG initiation codon. The *vertical dotted lines* denote the junctions between codons 10 and 11 of *hapR* (*left*), and the initiation codon and codon 2 of *gfp* (*right*). (**b**) During the second round of PCR, the products from the first round serve as megaprimers. The *vertical dotted lines* indicate the junctions between codon 10 of *hapR* and codon 2 of *gfp* created by the design of the SOE primers. (**c**) During the third round of PCR, primers 1 and 4 amplify the translational gene fusion of *hapR* in-frame (*dotted line*) with *gfp*.

are described in Note 2. In the first round, a region is amplified that includes the promoter, 5′ UTR, and the first 10 codons of the target gene (Fig. 1a, *left*).

(a) Forward target primer design (primer 1):

5′-AA**OOOOOO**XXXXXXXXXXXXXXXXXXXX-3′. The Xs indicates sequence approximately 400 nt upstream of the target mRNA gene (in this example *hapR*) to be amplified. The exact location of the primer is not critical but we attempt to avoid the inclusion of a complete upstream open reading frame. The precise placement of this primer depends on obtaining a good forward/reverse primer pair with similar annealing temperatures, since there is no flexibility in the location of reverse SOE primer 2. In this and subsequent primer design sections, the bold underlined Os indicate an appropriate restriction site for cloning the PCR product into a suitable vector.

(b) Reverse SOE primer design (primer 2):

5′- GAGTTCTTCTCCTTTGCTAGCTCGAGGGCGTTTTTCG-3′. The single underlined 21 nt correspond to codons 2–7 of the *gfp* gene, and the double underlined 16 nt correspond to codons 5–10 of the *hapR* gene. The junction between these two underlined regions maintains the correct reading frame for HapR and for GFP in the fusion protein.

3. Amplify the target PCR product from the chromosomal template DNA using the forward target primer (primer 1) and reverse SOE primer (primer 2) under standard PCR conditions.

4. In the first round of PCR, the *gfp* gene, excluding the ATG start site, is also amplified (Fig. 1a, *right*).

(a) Forward SOE primer design (primer 3):

5′- CGAAAAACGCCCTCGAGCTAGCAAAGGAGAAGAACTC-3′. This primer is complementary to the reverse SOE primer (primer 2) and described as in Subheading 3.1, step 2b.

(b) Reverse *gfp* primer design (primer 4):

5′-AAOOOOOOXXXXXXXXXXXXXXXXXXXX-3′.

The Xs indicate sequence downstream of the *gfp* gene to be amplified. As long as the PCR product includes the *gfp* stop codon, the precise placement of the reverse *gfp* primer, as with the forward target primer (primer 1) in Subheading 3.1, step 2a, depends on obtaining an annealing temperature that matches forward SOE primer 3.

5. Amplify the *gfp* PCR product from the plasmid template using the forward SOE primer (primer 3) and the reverse *gfp* primer (primer 4).
6. Verify the size of both products from the first round of PCR by running each on an agarose gel, staining, and visualizing by brief UV exposure at 365 nm (to minimize DNA damage); and then purify the products by gel extraction (eluting into 10 μL of elution buffer).
7. Amplify a second round PCR product for 20–25 cycles, by mixing 3 μL of each purified PCR product from the previous step in a 50 μL reaction that lacks primers. The PCR products from Subheading 3.1, steps 3 and 5, each serves as a "mega primer" in this second round of amplification (Fig. 1b).
8. Use 2 μL of the second round reaction as a template in a 50 μL reaction and amplify for 20–25 cycles the third round PCR product with primers 1 and 4 (Fig. 1c).
9. The PCR product at this step is analyzed and purified as in Subheading 3.1, step 6, cloned into a suitable plasmid and transformed into an *E. coli* donor strain as described below.
   (a) Digest the PCR product with restriction enzymes recognizing the sites engineered into primers 1 and 4. Clean with a DNA clean-up kit, and then ligate into a suitable conjugative vector (for example, pEVS143, which carries the gene required for kanamycin resistance) that is also digested with the corresponding restriction enzymes.
   (b) Transform the ligation reaction into an *E. coli* strain (such as DH5α) and plate for transformants on LB medium containing kanamycin (100 μg/mL).
   (c) Analyze multiple candidate colonies by colony PCR or by restriction mapping of plasmid DNA preparations to identify clones that contain the PCR product insertion. Sequence several plasmids bearing inserts using primers that anneal to DNA flanking the insertion site to verify the junctions between the vector and the PCR product, and the absence of unwanted DNA mutations.
   (d) Transform an *E. coli* Dap$^-$ auxotroph strain with plasmid DNA obtained from a successful clone identified in the previous step, and plate the transformants on LB medium containing kanamycin (100 μg/mL) and diaminopimelic acid (dap) (300 μg/mL). A restreaked single colony serves as an *E. coli* donor for mating into *V. cholerae* strains in the later steps.

3.2. Manipulation of sRNA Levels

Once a reporter fusion has been constructed for the target gene (Subheading 3.1), testing whether the gene is regulated by a small RNA, and characterizing the interaction, requires the ability to manipulate sRNA levels. Qrr sRNA levels can be controlled indirectly by modulating extracellular AI levels, or by constructing *V. cholerae* mutant strains which constitutively repress or activate *qrr* expression.

1. *AI-mediated control of sRNA levels*: Qrr sRNA production in *V. cholerae* is controlled in response to cell density by the concentration of AIs in the growth medium. Hence, control of sRNA levels can be obtained by adding known amounts of autoinducers to a *V. cholerae* mutant strain deleted for both autoinducer synthase genes (Δ*cqsA*, Δ*luxS*). Methods for constructing chromosomal gene deletions and point mutations have been described elsewhere (16).
   (a) To transfer the target gene–GFP fusion from Subheading 3.1 into *V. cholerae* Δ*cqsA*, Δ*luxS*, mate the *E. coli* Dap^- donor carrying the target gene–GFP reporter plasmid with *V. cholerae* Δ*cqsA*, Δ*luxS* on LB medium containing dap (300 μg/mL). On the next day, select for transconjugants on LB lacking dap and containing the antibiotic for which the plasmid carries a resistance gene. On the next day, restreak an isolated colony on the same medium and grow an overnight culture in LB with the appropriate antibiotic at 37°C.
   (b) AIs can be supplied exogenously to the AI^- strain by collecting cell-free culture supernatants from an AI-producing strain (wild-type *V. cholerae*) grown to high cell density. Cell-free supernatant from a Δ*cqsA*, Δ*luxS V. cholerae* culture is used as a negative control. Alternatively, purified AIs can be used if available (see Note 3). Obtain cell-free supernatants by growing *V. cholerae* wild-type and Δ*cqsA*, Δ*luxS* to high cell density, centrifuging the culture to remove cells, and filter sterilizing the supernatant through a 0.22-μm filter.
   (c) Dilute the overnight culture of *V. cholerae* Δ*cqsA*, Δ*luxS* 1:200 in LB with the appropriate antibiotic, and dispense 100 μL of the diluted culture into the wells of a microtiter plate.
   (d) Prepare serial dilutions from 10^0 to 10^{-5} of the cell-free supernatants from wild-type and synthase mutant *V. cholerae* cultures in LB with antibiotic, and add 50 μL of the appropriate dilution to triplicate wells of the microtiter plate. This results in a 1:300 dilution of the overnight culture in a final volume of 150 μL, containing 33% supernatant. Incubate the microtiter plate at 37°C with agitation until the desired cell density is reached.
   (e) Measure the fluorescence and absorbance (at 600 nm) for each well in a microplate reader. The "mean relative

fluorescence ± standard deviation" (fluorescence/mL)/ (OD_{600}/mL) is calculated for the triplicate cultures and is plotted against the concentration of cell-free supernatant to determine if AI concentration and, by extrapolation, Qrr sRNA levels alter the expression of the target gene.

2. *Constitutive control of sRNA levels*: Production of Qrr sRNAs is eliminated in *V. cholerae* by constructing strains with deletions of the genes encoding the known activators of the sRNAs (*luxO* or *rpoN*), or by deleting the four sRNA genes (Δ*qrr*1–4). Conversely, the Qrr sRNAs are expressed constitutively in a *V. cholerae luxO* D47E mutant. This point mutation in the *luxO* gene results in a mutated LuxO protein, which mimics the phosphorylated, active state of LuxO (7). Finally, sRNA expression is uncoupled from QS regulation by introducing a Qrr over-expression plasmid into the Δ*qrr*1–4 *V. cholerae* strain.

   (a) *V. cholerae locked QS mutant strains*: The target gene–GFP fusion plasmid is introduced into the *V. cholerae* Δ*luxO* and *luxO* D47E mutant strains as described in Subheading 3.2, step 1a, and GFP fluorescence is measured as described in Subheading 3.2, step 1e.

   (b) *Qrr over-expression plasmid*: The constitutive promoter is added upstream of the sRNA gene by PCR amplification of the sRNA from the *V. cholerae* chromosome using a forward primer that includes the promoter sequence at the 5′ end. We use the P_{tac} promoter lacking a Lac operator (17).

      i. Forward sRNA primer design: 5′-AAOOOOOOGAGCTG**TTGACA**ATTAATCATCGGCTCG**TATAAT**GTGTGGXXXXXXXXXXXXXXXXXXXXX-3′. The Xs indicate the first 20 nt of the sRNA gene to be amplified. The −35 and −10 promoter recognition sequences of P_{tac} are indicated *in bold*.

      ii. Reverse sRNA primer design: 5′-AAOOOOOOXXXXXXXXXXXXXXXXXXXX-3′. The Xs indicate sequence downstream of the sRNA gene to be amplified. Since sRNA genes normally encode a strong terminator sequence, which ensures that transcription terminates correctly, the exact location of the reverse sRNA primer is not critical. Refer to Subheading 3.1, step 2a, regarding the annealing temperature considerations for amplification with the forward sRNA primer.

      iii. Amplify the sRNA gene from *V. cholerae* chromosomal DNA using the forward/reverse sRNA primer pair and standard PCR conditions.

iv. Analyze and purify the PCR product as described in Subheading 3.1, step 9.

v. Mate the plasmid into a Δ*qrr*1–4 *V. cholerae* strain that contains the target gene–GFP fusion plasmid from Subheading 3.1, and select transconjugants on the appropriate combination of antibiotics. Simultaneously set up an additional mating transferring just the empty vector into the same *V. cholerae* strain to serve as a negative control. Grow the two *V. cholerae* strains in parallel and measure fluorescence and absorbance as described in Subheading 3.2, step 1e (see Note 4).

3.3. Northern Blot Analysis of RNA Kinetics

1. The half-life of RNA is measured by treating a growing culture with rifampicin to block transcription initiation, and then monitoring the fate of the existing RNA pool by removing aliquots for Northern blotting at various times after rifampicin addition. To determine whether an sRNA affects the half-life of an mRNA, cells should be grown under conditions where the sRNA is expressed, and the experiment should be carried out both in wild-type cells and in mutant cells deleted for the sRNA gene(s). Figure 2 shows a Northern blot of *hapR* mRNA in wild-type and Δ*qrr*1–4 mutant cells. In addition, a mutant over-expressing the sRNA can be included in the experiment. This is particularly useful if the sRNA-mediated change in mRNA half-life is modest. See Note 5 for additional applications of Northern blot. This protocol assumes familiarity with proper handling of radioactive materials.

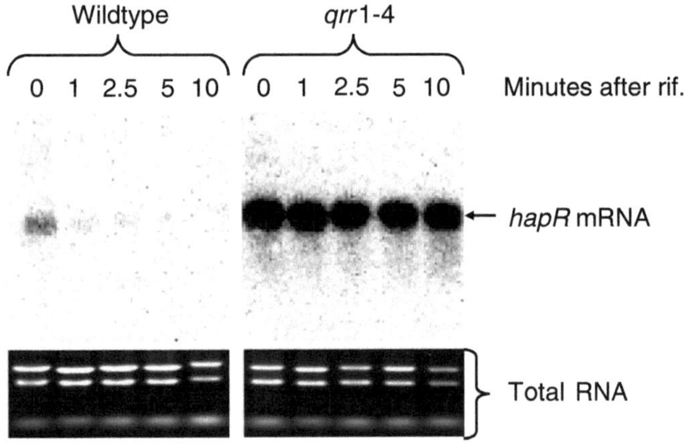

Fig. 2. Northern blot of *hapR* mRNA in wild-type and Δ*qrr*1–4 *V. cholerae*. Rifampicin was added to *V. cholerae* cultures at $OD_{600} = 0.1$, and aliquots were removed at the times indicated above the blots and treated as described in Subheading 3.3. Total RNA was visualized with ethidium bromide and is shown as the loading control.

2. *Preparation*: To obtain sufficient RNA for a Northern blot, each collected aliquot should contain the equivalent of 1 mL of *V. cholerae* culture at $OD_{600} = 1$ (e.g., 10 mL culture at $OD_{600} = 0.1$). For each time point, prepare a labeled collection tube of appropriate size containing 0.2 vol. stop solution (95% ethanol, 5% phenol) and place it on ice. Rifampicin stock solution (~100 mg/mL) is prepared fresh daily in methanol.
3. Grow cell culture to the desired cell density. The first aliquot should be collected immediately *before* the addition of rifampicin. Transfer cells to a collection tube on ice, invert the tube to mix with stop solution, and keep the collected aliquots on ice at all times. Add rifampicin (100 μg/mL) to the culture, start timer, and then collect aliquots at 1, 2.5, 5, and 10 min.
4. Centrifuge the collected aliquots, transfer to microfuge tubes, spin again, remove supernatant, and store pellets at −80°C overnight.
5. Steps 6–14 should be carried out in an RNase-free environment: Make all solutions with DEPC-treated water, clean surfaces and glassware with RNase Away, use filter pipette tips, and wear gloves.
6. Thaw frozen cells on ice. While cells are still frozen, add 1 mL trizol per tube, and re-suspend by pipetting. Incubate at room temperature (RT) for 5 min.
7. Add 200 μL of chloroform per tube. Shake vigorously for 15 s, and then incubate at RT for 5 min. Centrifuge at 15,000 RCF for 15 min at 4°C.
8. Carefully transfer upper (colorless) phase to a new microfuge tube. It is essential to avoid mixing of the phases. Best results will be obtained if the upper phase is recovered immediately after centrifugation.
9. Add 500 μL isopropanol per tube. Mix by inverting several times and then incubate at RT for 10 min. Spin at maximum speed for 15 min at 4°C and discard supernatant.
10. Add 1 mL 75% ethanol and invert gently to wash the pellet. Spin at 6000 RCF for 5 min at 4°C. Remove the supernatant. Air-dry pellet for 5 min, and then dissolve the pellet in 50 μL DEPC-treated water by incubating at 55°C for 10–15 min.
11. Measure RNA concentration with a Nanodrop or other spectrophotometer. We routinely obtain 500–1500 ng/μL of RNA using this method. Store RNA at −80°C or use immediately.
12. *Prepare gel*: Mix 1.2 g agar with 135 mL DEPC water and dissolve in microwave. Then add 15 mL 10× MOPS buffer and 8.1 mL formaldehyde (150 mL is appropriate for a

15 cm × 15 cm gel, adjust volume if using different size gel tray). In a fume hood, pour gel in gel tray treated with RNase Away. Running buffer is 1× MOPS.

13. Use 10 μg total RNA per lane for the Northern blot. Adjust volumes with DEPC water so that all the wells are loaded with similar volumes. Add the same volume of 2× RNA loading dye. Prepare RNA ladder as described by manufacturer. Heat all the samples at 70°C for 10 min and move to ice.
14. Load the gel and run it at 85 V for 2.5–3 h (adjust the voltage to equal ~5 V/cm if running gel of different sizes), visualize gel with UV light, and photograph. Equal amounts of total RNA should be visible in each lane. See Note 6 for another, more sensitive, loading control.
15. Wash the gel twice for 20 min in 10× SSC at RT.
16. *Set up transfer at RT*: Fill large tray with 10× SSC (minimum 1 L) and cover partly with glass plate. Wet a piece of absorbent paper in 10× SSC and lay it across the glass plate so that it reaches into the liquid in the tray on both the sides. For this and the following steps, roll a glass pipette across the surface to smooth out any trapped air bubbles between the layers. Place gel upside down on top of paper. Slice off the wells of the gel with a razor blade. Cut a gel-sized piece of membrane, label with a pencil in one corner, soak it in 10× SSC, and then lay it on top of the gel. Soak two pieces of gel-sized absorbent paper in 10× SSC and lay on top of the membrane. Seal the tray area around the gel "sandwich" with parafilm to avoid evaporation. Put two stacks (~15 cm in height) of paper towels on top of the sandwich and top it off with another glass plate, followed by a heavy item, such as a text book. Leave the sandwich overnight to allow sufficient time for the RNA to transfer from the gel to the membrane by capillary forces.
17. On the next day, take the sandwich apart, dry the membrane between two fresh, dry sheets of absorbent paper at 30°C for 30 min, and then crosslink the RNA to the membrane using the "autocrosslink" setting on a Stratalinker.
18. Prepare 30 mL of prehybridization solution per membrane (100 mL: 30 mL 20× SSC, 4 mL 50× Denhardt solution, 1 mL 10% SDS, and 65 mL water). Heat hybridization oven to 65°C, place membrane (RNA side in) in hybridization tube, and add prehybridization solution. Incubate for 3–6 h. See Note 7 on hybridization temperature and probe length.
19. Prepare the probe by asymmetric PCR using ~100 ng of a purified PCR product of the gene in question as template. Set up a standard Taq polymerase PCR, but use only one primer (the one that is complementary to the transcribed target RNA),

add 2 μL ^{32}P α-dATP, and use nonradioactive nucleotides at the following final concentrations: 0.2 mM dGTP/dCTP/dTTP and 0.0012 mM dATP.

20. Purify the radioactive PCR product using a standard PCR purification kit (discard collection tube at each step to avoid radioactive spills), elute in 50 μL elution buffer, and count 1 μL of the probe in a scintillation counter. Probe counts should be ~5×10^5 cpm/μL. Add the entire probe to the hybridization tube containing the membrane and continue incubation at 65°C overnight.
21. On the next day, wash the membrane for 20 min at 65°C with 1× SSC+0.1% SDS, followed by three times for 20 min washes at 65°C with 0.2× SSC+0.1% SDS.
22. Wrap the membrane in plastic wrap and scan the membrane in a phosphoimager to quantify radioactivity and calculate mRNA half-life. Finally, expose the wrapped membrane to autoradiographic film in a cassette with intensifying screens. Put the cassette at −80°C overnight and develop the film the next day. We find that publication quality images are more readily obtained using film than by scanning the membrane. See Note 8 for re-use of the membrane.

3.4. Computational Prediction of a Putative sRNA Binding Site in a Candidate mRNA

1. These instructions utilize the *TargetRNA* algorithm developed by Tjaden (14) and assume that the bacterial genome sequence of interest is available in the "Genome" menu of the website. Other resources are available for similar analyses when one has the sequence of the suspected mRNA target despite the lack of a sequenced genome, see Note 9. This procedure describes a method for predicting the putative sRNA binding site in the mRNA of a candidate target gene controlled by the *V. cholerae* Qrr sRNAs. A region of the target mRNA to search is defined by the user relative to the annotated translational start codon, so the transcriptional start site (+1) of a candidate mRNA is not required. The specific parameters below successfully identify the Qrr sRNA binding sites in the 5′ UTR of *V. cholerae* genes that have been validated by experimental data, but each parameter may be adjusted as necessary.
2. Open the *TargetRNA* web interface at the following URL: http://snowwhite.wellesley.edu/*targetRNA*/index.html.
3. Select the Advanced Search option, and then choose *V. cholerae* in the genome menu under "sRNA options".
4. Input the sequence of the sRNA in the 5′ to 3′ direction, remembering to replace each T with a U if you are starting with a DNA sequence. We do not check the "Remove terminator" box to exclude the rho-independent terminator from

base-pairing consideration when using the small 32 nt probe.

5. Under "target options" the default setting focuses the search around the start codon. We constrain the search options to a region that extends from −100 nt upstream to 20 nt downstream of the target gene start codon, which typically contains the sRNA binding site.
6. Select a "hybridization seed" of "0" to eliminate constraints on minimal consecutive base pairings, and also allow G:U pairing, which is possible in RNA/RNA, but not DNA/DNA duplexes.
7. To scan for an sRNA binding site in a particular target, enter the annotation number (with capital letters; VC0583, for example) in the "single target" window. Alternatively, the gene name may be used (the first letter in lower case; hapR, for example). To use this method to scan the entire *V. cholerae* genome for putative sRNA target genes, see Note 10.
8. In "Scoring Options" the *P* value threshold is set to "1.0" and thermodynamic energy scoring is not used.
9. For *V. cholerae*, we exclude from consideration any predicted interactions that indicate pairing between the mRNA and a region of the sRNA that participates in the formation of the three largest stem-loops because these sRNA regions are predicted to participate in Hfq interactions and not mRNA binding.
10. Submit the request using the Search button. Figure 3 illustrates the predicted pairing of the 32 nt sRNA region with *hapR*, *vca0939*, and *vc1021* (*luxO*) mRNA.

3.5. Experimental Validation by Site-Directed Mutagenesis

1. The base pairs of an mRNA that are predicted to interact with an sRNA are altered to the complementary base to disrupt pairing; for example, a G in the mRNA that is predicted to pair with a C in the sRNA is mutated to a C. Site-directed mutagenesis is performed on the plasmid carrying the target gene–GFP fusion constructed in Subheading 3.1. We describe mutagenesis of single nucleotides, but it may be necessary to alter more than one base to disrupt sRNA/mRNA pairing.
2. The Stratagene Quikchange mutagenesis kit is used for this procedure. The desired mutation is incorporated into two complementary oligonucleotides, which are subsequently used as primers for PCR amplification of the entire plasmid. This leads to mutagenesis of the selected residue(s) in the target gene–GFP fusion.
3. *Primer design*: Two complementary oligonucleotides are designed which carry the desired mutation in the middle, flanked by 10–15 nt of unmodified nucleotide sequence on either side.

```
sRNA            7  CCUAGCCAACUGACGUUGUU-AGUGA  31
                   ||| ||||| ||| |||||| |: |||
hapR          -11  GGAACGGUUAACU-CAACAACUAACU -35

sRNA      1  GGGUCACCUAGCCAACUGACGUUGUUAGUGAA  32
             |||   |:|||||||||  |    ||||:|| : |:||
vca0939 -17  CCCCUUGGAUCGGUU-A--GCAAUAAAAAUUU -45

sRNA            7  CCUAGCCA  14
                   ||||||||
vc1021         -3  GGAUCGGU -10
```

Fig. 3. Base pairing predicted by *TargetRNA* between a conserved 32 nt region of the Qrr sRNAs and the mRNA of *hapR*, *vca0939*, and *vc1021* (*luxO*). Base pairings that extend into the sRNAs stem loops that are predicted to be structural, and not participate in sRNA interactions, are *outside the thin vertical dotted lines*. The location of the pairing within the 5′ UTR of each mRNA is indicated relative to the initiation codon (+1) by the numbers flanking each mRNA sequence. The relative location of the sRNA bases within the 32 nt region is also indicated. *Solid lines* between bases indicate Watson–Crick pairing, *two dots* show predicted G:U pairing. *Annotation* of *luxO* is as described in (5).

The total length of the primer should be between 25 and 45 nt, and the melting temperature should be >78°C. Ideally, each primer should have a minimum GC content of 40% and there should be a G or C residue at the 3′ end of each primer.

4. PCR, *Dpn*I digestion, and transformation of XL1-Blue supercompetent cells is performed as described in the Quikchange kit manual, using the materials provided by the manufacturer. Plate cells on LB with kanamycin (100 μg/ml).
5. To verify the mutagenesis products, restreak 4–12 candidate colonies, grow overnight cultures in liquid LB medium with kanamycin, miniprep the plasmid DNA, and submit samples for sequencing with primers 1 and 4 from Subheading 3.1.
6. Transform the target gene–GFP reporter plasmid carrying the correct mutation into an *E. coli* dap⁻ auxotroph donor strain as described in Subheading 3.1, step 9, and then mate the plasmid into *V. cholerae* recipient strains described in Subheading 3.2 for analysis.
7. The identical site-directed mutagenesis is performed on the plasmid that carries the sRNA gene under control of a constitutive promoter (see Subheading 3.2). For example, a C in the sRNA that is predicted to pair with a G in the mRNA is mutated to a G. This plasmid is then mated into *V. cholerae* carrying the GFP reporter plasmid for analysis, as described in Subheading 3.2, step 2.b.v.
8. Finally, the mutated sRNA expression plasmid (for example, with the C to G mutation) and the mutated GFP reporter

plasmid (for example, with the G to C mutation) are introduced into the Qrr1–4⁻ *V. cholerae* strain to validate whether restoration of the base pairing by the compensatory mutations on either side of the sRNA/mRNA duplex (for example, the C–G pairing is now a G–C pairing) restores proper sRNA control of the mRNA target gene.

4. Notes

1. Target PCR products cloned into the *Bam*HI site of a luciferase vector, pBBRlux, create Lux transcriptional fusions that have been used successfully to monitor regulation of a QS target gene via the Qrr sRNAs (9). Luciferase levels are measured in a BioTek Synergy 4 plate reader by the microtiter plate method described above.
2. The location and length of the upstream target primer (primer 1) and downstream *gfp* primer (primer 4) can be adjusted to obtain a suitable annealing temperature, assuming that the promoter of the target gene and stop codon of *gfp*, respectively, are still included in the PCR products generated. The internal primers (primers 2 and 3) can also be increased in length and adjusted to include additional codons, however, it is essential that the junction preserves the codons in frame. Including an inordinate number of codons from the reporter gene may also impair the folding (and fluorescence) of GFP in the protein fusion, so we typically design the primers to include 10 codons and only increase the length if necessary.
3. Purified autoinducers are preferred over cell-free supernatants, if they are available to the researcher. Purified AI-2 is described in (20) and purified CAI-1 is described in (21). Carry out experiment as described in Subheading 3.2, but omit Subheading 3.2, step 3b, and instead prepare a microtiter plate with serial dilutions of purified AI-2 and CAI-1 from 50 pM to 5 μM final concentrations in a final volume of 150 μL LB with kanamycin and transfer 1.5 μL culture to each well to obtain a 1:100 dilution.
4. To stably maintain the P_{tac}-*qrr4* vector in cells that already harbor the target gene–GFP vector, two compatible cloning vectors must be chosen. For example, if the target gene–GFP construct is cloned in pEVS141 (p15A replication origin, kanamycin resistance), the P_{tac}-*qrr4* vector must be cloned into a vector which uses a different antibiotics cassette and an origin from a different compatibility group (e.g., pBBRMCS1 (pBBR origin, chloramphenicol resistance) or pLAFR (pRK2 origin, tetracycline resistance)).

5. This protocol describes how to measure the half-life of an mRNA. Because RNA levels are visualized directly by this method without the need for synthesis of reporter proteins, Northern blots are useful tools for measuring changes in mRNA and sRNA levels upon induction, changes in cell density, addition of external AIs, etc., with great time resolution. To do so, carry out protocol as described in Subheading 3.3, but omit rifampicin in Subheading 3.3, step 3, and add external AIs or inducer at time 0 instead.
6. Total RNA stained with ethidium bromide and visualized under UV light gives a rough measure of how much total RNA was loaded in each lane. For precise calculations of mRNA half-life, we recommend reprobing the membrane (see Note 7) with a probe for 5S RNA. In our experience, 5S RNA levels are sufficiently stable after treatment with rifampicin that 5S RNA serves as a reliable loading control. A template for 5S RNA probes is amplified from *V. cholerae* genomic DNA using primers "5S RNA up": 5′-CTTGGCGACCATAGCGTTTTG-3′ and "5S RNA dn": 5′-GCCTGGCGATGTTCTACTCTCAC-3′. The probe is obtained from the template DNA as described in Subheading 3.3, step 18.
7. We carry out hybridization and the following washes at the high temperature of 65°C to minimize background signal. However, if using a probe which is shorter than ~50 nt, the probe may not anneal well to the membrane at 65°C, and hybridization and washes at 42°C are recommended. Our probes are generally designed to be 100–500 nt long.
8. Northern blot membranes can be stripped and reused with additional probes. To strip a membrane, wash it in 0.5% SDS for 30 min at 95°C before initiating a new round of hybridization. It is only recommended to reprobe membranes for RNA species of a different size than the initial RNA, because a weak signal may be obtained from the original probe even after stripping the blot, and this could interfere with the interpretation of the new hybridization signal if the RNAs are of a similar size.
9. The Mfold algorithm (18) can be used to predict specific interactions between a user-defined concatamer of a Qrr sRNA and an mRNA target sequence as described in (9). The RNAhybrid algorithm (19) also allows the users to input both target mRNA and microRNA sequences at the following URL: http://bibiserv.techfak.uni-bielefeld.de/rnahybrid/submission.html.
10. To scan the entire *V. cholerae* genome, the "single target" window is left blank. We typically use the 32 nt probe to eliminate the possibility of interactions with the sRNA regions that are not conserved between Vibrios, and with the three stable sRNA helices that are predicted to participate in Hfq interactions (2).

Acknowledgment

This material is based upon the work supported by the National Science Foundation under Grant No. 0919821. Any opinions, findings, and conclusions or recommendations expressed in this material are those of the author(s) and do not necessarily reflect the views of the National Science Foundation.

References

1. Hammer, B. K., and Bassler, B. L. (2008) *Signal integration in the Vibrio harveyi and Vibrio cholerae quorum sensing circuits*, Washington, DC Stephen C Winans and Bonnie L. Bassler 323–332.
2. Lenz, D. H., Mok, K. C., Lilley, B. N., Kulkarni, R. V., Wingreen, N. S., and Bassler, B. L. (2004) The small RNA chaperone Hfq and multiple small RNAs control quorum sensing in *Vibrio harveyi* and *Vibrio cholerae*, *Cell 118*, 69–82.
3. Miller, M. B., Skorupski, K., Lenz, D. H., Taylor, R. K., and Bassler, B. L. (2002) Parallel quorum sensing systems converge to regulate virulence in *Vibrio cholerae*, *Cell 110*, 303–314.
4. Zhu, J., Miller, M. B., Vance, R. E., Dziejman, M., Bassler, B. L., and Mekalanos, J. J. (2002) Quorum-sensing regulators control virulence gene expression in *Vibrio cholerae*, *Proc Natl Acad Sci USA 99*, 3129–3134.
5. Svenningsen, S. L., Tu, K. C., and Bassler, B. L. (2009) Gene dosage compensation calibrates four regulatory RNAs to control *Vibrio cholerae* quorum sensing, *EMBO J 28*, 429–439.
6. Hammer, B. K., and Bassler, B. L. (2009) Distinct sensory pathways in *Vibrio cholerae* El Tor and classical biotypes modulate cyclic dimeric GMP levels to control biofilm formation, *J Bacteriol 191*, 169–177.
7. Hammer, B. K., and Bassler, B. L. (2003) Quorum sensing controls biofilm formation in *Vibrio cholerae*, *Mol Microbiol 50*, 101–104.
8. Zhu, J., and Mekalanos, J. J. (2003) Quorum sensing-dependent biofilms enhance colonization in *Vibrio cholerae*, *Dev Cell 5*, 647–656.
9. Hammer, B. K., and Bassler, B. L. (2007) Regulatory small RNAs circumvent the conventional quorum sensing pathway in pandemic *Vibrio cholerae*, *Proc Natl Acad Sci USA 104*, 11145–11149.
10. Dunn, A. K., Millikan, D. S., Adin, D. M., Bose, J. L., and Stabb, E. V. (2006) New rfp- and pES213-derived tools for analyzing symbiotic *Vibrio fischeri* reveal patterns of infection and lux expression in situ, *Appl Environ Microbiol 72*, 802–810.
11. Dehio, C., and Meyer, M. (1997) Maintenance of broad-host-range incompatibility group P and group Q plasmids and transposition of Tn5 in *Bartonella henselae* following conjugal plasmid transfer from *Escherichia coli*, *J Bacteriol 179*, 538–540.
12. Vogel, J., and Wagner, E. G. (2007) Target identification of small noncoding RNAs in bacteria, *Curr Opin Microbiol 10*, 262–270.
13. Aiba, H. (2007) Mechanism of RNA silencing by Hfq-binding small RNAs, *Curr Opin Microbiol 10*, 134–139.
14. Tjaden, B. (2008) *TargetRNA*: a tool for predicting targets of small RNA action in bacteria, *Nucleic Acids Res 36*, W109–W113.
15. Horton, R. M., Hunt, H. D., Ho, S. N., Pullen, J. K., and Pease, L. R. (1989) Engineering hybrid genes without the use of restriction enzymes: gene splicing by overlap extension, *Gene 77*, 61–68.
16. Skorupski, K., and Taylor, R. K. (1996) Positive selection vectors for allelic exchange, *Gene 169*, 47–52.
17. de Boer, H. A., Comstock, L. J., and Vasser, M. (1983) The tac promoter: a functional hybrid derived from the trp and lac promoters, *Proc Natl Acad Sci USA 80*, 21–25.
18. Zuker, M. (2003) Mfold web server for nucleic acid folding and hybridization prediction, *Nucleic Acids Res 31*, 3406–3415.
19. Rehmsmeier, M., Steffen, P., Hochsmann, M., and Giegerich, R. (2004) Fast and effective prediction of microRNA/target duplexes, *RNA 10*, 1507–1517.
20. Schauder, S, Shokat, K., Surette, M. G., and Bassler, B.L. (2001) The LuxS family of bacterial autoinducers: biosynthesis of a novel quorum-sensing signal molecule, Mol Microbiol 41, 463–476.
21. Higgins, D. A., Pomianek, M. E., Kraml, C. M., Taylor, R. K., Semmelhack, M. F., and Bassler, B. L. (2007) The major Vibrio cholerae autoinducer and its role in virulence factor production, Nature 450, 883–886.

Chapter 15

Quantifying *Pseudomonas aeruginosa* Quinolones and Examining Their Interactions with Lipids

Gregory C. Palmer, Jeffrey W. Schertzer, Lauren Mashburn-Warren, and Marvin Whiteley

Abstract

Pseudomonas aeruginosa produces a quorum sensing molecule termed the *Pseudomonas* Quinolone Signal (2-heptyl-3-hydroxy-4-quinolone; PQS) that regulates an array of genes involved in virulence. This chapter addresses four related techniques useful for detecting and quantifying PQS. First, extraction of PQS from complex mixtures (e.g. cell cultures) is described. Separation of PQS from extracts by Thin-Layer Chromatography (TLC) is used in combination with the natural fluorescence of the molecule for quantification. A second separation technique for the PQS precursor HHQ using High-Performance Liquid Chromatography (HPLC) is also described, and this assay exploits the molecule's characteristic absorbance for quantification. A third method for quantification of PQS from simple mixtures (e.g. enzyme assays) using fluorescence is outlined. Finally, a protocol for determining PQS interactions with membrane lipids through Fluorescence Resonance Energy Transfer (FRET) is presented. These techniques allow for quantification and characterization of PQS from diverse environments, a prerequisite to understanding the biological functions of QS molecules.

Key words: *Pseudomonas* quinolone signal, Thin-layer chromatography, High-pressure liquid chromatography, Fluorescence resonance energy transfer

1. Introduction

The Gram-negative opportunistic pathogen *Pseudomonas aeruginosa* synthesizes several quorum sensing (QS) molecules, including two acylhomoserine lactones (3-oxo-dodecanoyl and butyryl homoserine lactone; AHLs), which are synthesized and sensed by the *las* and *rhl* QS systems, respectively (1). A third QS molecule is a hydrophobic quinolone (2-heptyl-3-hydroxy-4-quinolone) termed the *Pseudomonas* Quinolone Signal (PQS) (2). PQS, bound to its transcriptional regulator PqsR (MvfR), regulates an

Kendra P. Rumbaugh (ed.), *Quorum Sensing: Methods and Protocols*, Methods in Molecular Biology, vol. 692,
DOI 10.1007/978-1-60761-971-0_15, © Springer Science+Business Media, LLC 2011

array of genes involved in the production of important virulence factors like hydrogen cyanide and pyocyanin (3). While the AHL autoinducers are soluble and can readily diffuse within the extracellular milieu, the hydrophobic PQS molecule is packaged into outer membrane vesicles (MV), and PQS is required for MV formation in *P. aeruginosa* (4). PQS has also been shown to stimulate MV formation in other Gram-negative bacteria (5). Additionally, other Gram-negative bacteria such as *Rhodobacter capsulatus* (6) and *Sinorhizobium meliloti* (7) produce hydrophobic quorum sensing signals that may also interact with outer membrane lipids.

PQS biosynthesis is affected by genetic and environmental factors. The *las* QS system with its aforementioned autoinducer 3-oxo-dodecanoyl homoserine lactone regulates transcription of *pqsH*, which encodes the terminal enzyme in the PQS biosynthetic pathway. PqsH catalyzes the addition of the 3-hydroxyl group to the immediate PQS precursor molecule 2-heptyl-4-quinolone (HHQ) (8, 9). In addition, nutritional cues also affect PQS production. The amino acids phenylalanine and tyrosine significantly enhance PQS production (10), and it is hypothesized that this is due to flux through a precursor pathway shared by PQS production and aromatic amino acid biosynthesis (10).

The study of PQS as a QS molecule demands accurate determination of the concentration of signal present in both complex and simple mixtures. In this chapter, two methods for separating PQS and HHQ from complex mixtures, Thin-Layer Chromatography (TLC) and High-Performance Liquid Chromatography (HPLC), are described and used in conjunction with the natural fluorescence/absorbance of PQS and HHQ to determine their concentrations. We also describe a fluorescence assay for quantifying PQS in simple mixtures and the use of Fluorescence Resonance Energy Transfer (FRET) to determine interactions between PQS (or other hydrophobic signaling molecules) and membrane lipids. Through the use of these techniques, PQS and HHQ can be quantified from diverse environments including biological samples and whole bacterial cultures, and insights into interactions between quinolones and membrane lipids can be gained.

2. Materials

2.1. TLC for Separation and Quantification of PQS

1. Acidified ethyl acetate ACS grade (acidified with 0.15 mL/L glacial acetic acid).
2. Separation funnel.
3. Rotary evaporator.
4. Nitrogen gas and dry down apparatus.

5. Glass extraction vials (15 mL and 5 mL).
6. Monobasic potassium phosphate (KH_2PO_4) ACS grade.
7. 20×20 cm Silica 60 F_{254} pre-coated plates for thin-layer chromatography.
8. Drying oven 100°C.
9. Methanol ACS grade.
10. Dichloromethane ACS grade.
11. Glass TLC chamber.
12. Fluorescence imager and spot densitometry analysis software (e.g. Syngene G:Box with GeneTools software).
13. 2 mM PQS standard (MW 259) in methanol.

2.2. HPLC for Separation and Quantification of HHQ

1. HPLC with Photo Diode Array (PDA) detector (e.g. Varian ProStar).
2. Reverse phase C8 column 140×4.5 mm 1 mL/min flow rate (Varian).
3. Acidified methanol HPLC grade (acidified with 1% glacial acetic acid by volume).
4. Acidified water HPLC grade (acidified with 1% glacial acetic acid by volume).
5. 2 mM HHQ standard (MW 243) in acidified methanol.

2.3. PQS Quantification from Simple Mixtures

1. Glass tubes or bottles (depending on sample volume).
2. Ethyl acetate (ACS grade acidified with 0.1 mL/L glacial acetic acid).
3. Nitrogen gas and dry down apparatus.
4. High-grade methanol minimal fluorescent impurities.
5. Fluorometer.

2.4. FRET to Detect PQS Interactions with Membrane Lipids

2.4.1. Preparation of Soluble Lipopolysaccharide and Phospholipids

1. Purified lipopolysaccharide (LPS).
2. Phosphatidylethanolamine (PE) from *E. coli* (Sigma) and phosphatidyl glycerol (PG) from phosphatidylcholine (Sigma).
3. Chloroform HPLC grade.
4. Methanol HPLC grade.
5. Ultrasonic Bath.

2.4.2. Labeling of LPS and Phospholipids with Donor and Acceptor Dyes

1. *Donor conjugated dye*: 4-Nitro-benz-2-oxa-phosphatidylethanolamine (NBD-PE), in chloroform (Molecular Probes).
2. *Acceptor conjugated dye*: Rhodamine-phosphatidylethanolamine (Rh-PE), in chloroform (Molecular Probes).
3. Chloroform.

4. Nitrogen gas.
5. 20 mM HEPES, pH 7.0.
6. Tip Sonicator.

2.4.3. FRET Measurements

1. Quinolones, in methanol.
2. Labeled LPS or phospholipids, 1 mM in HEPES buffer (see Subheading 3 for labeling protocol).
3. Fluorometer, heated at a constant 37°C with stirring.

3. Methods

3.1. TLC for Separation and Quantification of PQS

3.1.1. PQS Extraction

1. Using a sterile inoculation loop, streak relevant bacterial strains onto LB agar plates. Wild-type PA14 *P. aeruginosa* as well as *pqsA*⁻ (no quinolone control) and *pqsH*⁻ (no PQS control) mutants are recommended. Incubate at 37°C overnight.
2. The next day, inoculate 10 mL of LB or other desired medium from the plate and incubate at 37°C overnight.
3. The next day, determine the OD_{600} of overnight cultures and inoculate 10 mL of LB or other desired medium to $OD_{600}=0.05$. Incubate at 37°C shaking (250 rpm) until the culture has reached $OD_{600}=0.5$ (mid-exponential phase).
4. Once the culture has reached mid-exponential phase pellet cells by spinning cultures at 7000 ×*g* for 5 min. Re-suspend cells in 1 mL of sterile medium (see Note 1). Determine the OD_{600} of the concentrated culture, and inoculate 55 mL of LB or other desired medium to $OD_{600}=0.01$. Incubate newly inoculated medium at 37°C shaking (250 rpm) until the culture desired density for extraction of PQS. In this example, a mid-exponential phase culture will be examined ($OD_{600}=0.5$).
5. Remove 50 mL of culture to a 500 mL flask washed with acidified ethyl acetate. Add 50 mL acidified ethyl acetate, vortex mixture for 30 s, and place mixture in a separation flask for 5–10 min or until aqueous and organic phases have completely separated.
6. Collect the clear organic phase top layer in a clean glass container.
7. Repeat steps 5 and 6 twice to fully extract PQS from 50 mL culture.
8. Use rotary evaporator to completely dry down ethyl acetate extract. Add 10 mL to rotary evaporator flask and twist to ensure that the entire flask contacts the ethyl acetate. Completely dry down the remaining ethyl acetate under a constant stream of nitrogen gas.

9. Extracts may be re-suspended in methanol for spotting on TLC plates (see next section) or stored in the dark until needed.

3.1.2. PQS Quantification with TLC

1. Prepare standards by diluting 2 mM PQS stock in methanol. As an example, the low range of standard visible on the plate is 100 ng; thus, standards in which 100, 200, 300, 400, and 500 ng/10 μL are spotted on the plate are an appropriate range.
2. Pre-heat the drying oven to 100°C.
3. Prepare TLC plate by incubating in 5% KH_2PO_4 solution at room temperature for 30 min. Transfer the plate to the pre-heated drying oven for a minimum of 1 h (see Note 2).
4. Concentrate PQS samples by completely drying down extracts and re-suspending them in 50 μL methanol. The extent to which samples should be concentrated depends on the range of PQS standards and the optical density of the culture extracted.
5. Prepare TLC chamber with 95:5 mixture of dichloromethane: methanol. Allow 10 min for solvent system to equilibrate within chamber.
6. Remove the plate from the oven and lightly draw a line in pencil 1–2 cm from the bottom of the plate. Sample and standard PQS will be spotted on this line; tick marks denoting the intended location of a spot may also be added (see Note 3).
7. Pre-run the marked plate by placing the plate in TLC chamber with equilibrated solvent. The solvent should not be above the line drawn in the previous step. Allow the solvent to migrate until the entire plate is covered. Allow the excess solvent to evaporate from the plate; pre-run plates should be returned to the drying oven until ready to be spotted.
8. Spot 10 μL of samples leaving at least 1 cm between the samples. Spot standards on either side of the samples. Spotting standards on either side of the samples (and between samples if space is available) allows for the standardization of any disproportionate illumination that may affect quantification.
9. Place the plate in TLC chamber and allow the solvent to migrate until the entire plate is covered. Be sure that the spotting solvents have completely evaporated before placing the plate in the TLC chamber (see Note 4).
10. Remove the plate from TLC chamber and image PQS spots using fluorescence imager. Excite the plate with UV light and observe fluorescent spots (see Note 5). Sample PQS should co-migrate with standard on the TLC plate (Fig. 1).
11. Perform spot densitometry analysis on imaged TLC. Plot fluorescence units against the amount of PQS standard to

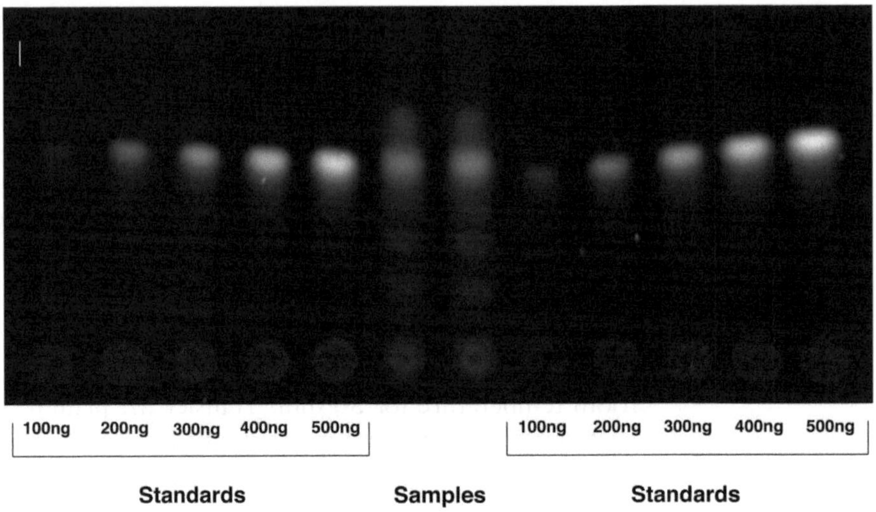

Fig. 1. Thin-Layer Chromatography separating PQS extracted from whole *P. aeruginosa* cultures (samples). PQS standards (100–500 ng) are shown on either side of the samples.

generate a standard curve. The standard curve should be linear, and PQS concentrations in samples can be calculated using the standard curve slope equation. Samples outside the standard curve must be diluted and re-run to be within the range of the standard.

3.2. HPLC for Separation and Quantification of HHQ

1. Re-suspend HHQ sample to be analyzed in 200 μL acidified methanol and remove to HPLC vials. HHQ samples may be obtained from the PQS extraction protocol detailed in the previous Subheading 3.1.
2. Prepare HHQ standards by diluting 2 mM stocks to 50, 100, 200, 300, 400, and 500 μM in acidified methanol. The final volume of each stock should be 200 μL, and the stocks should be removed to HPLC vials.
3. The mobile phase for the HPLC is acidified methanol and acidified water with the following program: 60% acidified methanol for 10 min, ramp up to 100% in 5 min, hold at 100% for 5 min, drop to 60% in 1 min, and hold at 60% for 3 min.
4. Analyze 50 μL of standards and samples sequentially over HPLC using the solvent program above.
5. Extract the chromatogram at 314 nm to view HHQ peak. HHQ elutes at 6 min (Fig. 2). Both HHQ and PQS may be viewed by extracting the chromatogram at 325 nm (see Note 6).
6. Plot peak absorbance values against the concentration of HHQ standard to generate a standard curve. The standard curve should be linear, and HHQ concentrations in samples

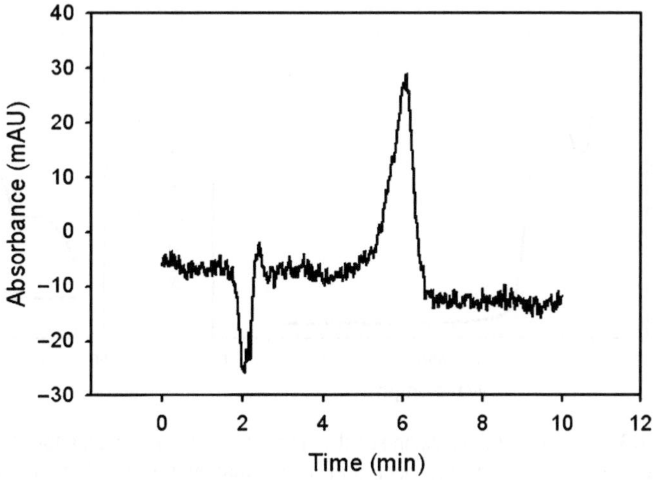

Fig. 2. Representative HPLC chromatogram. The relevant portion of an HPLC chromatogram showing HHQ elutes at approximately 6 min.

can be calculated using the standard curve slope equation. Samples outside the standard curve must be diluted and re-run to be within the range of the standard.

3.3. PQS Quantification from Simple Mixtures

1. This method is suitable for detection and quantification of PQS in simple mixtures of known composition (i.e. enzymatic assays, defined media, etc.). It is important that no ethyl acetate-extractible contaminant fluoresces at the wavelengths described.
2. The simple mixture containing PQS is extracted using three times the volume of acidified ethyl acetate. For example, add 3 mL acidified ethyl acetate to 1 mL of the mixture.
3. Agitate the extraction mixture. Several short bursts using a vortexer are recommended for small volumes; manual mixing may be required for larger volumes.
4. If the separation vessel can be accommodated by a centrifuge, a short low-speed spin (2 min at 2000 ×*g*) will result in a sharp phase separation.
5. Recover as much as possible of the upper organic phase (the ethyl acetate) into a new glass vessel by pipetting. Record the recovered volume as this value will be important in back-calculating PQS concentration in the original mixture.
6. Place the recovered ethyl acetate sample under a stream of nitrogen gas until the liquid has completely evaporated.
7. Once all the samples have been evaporated, re-suspend each in a known volume of high-grade methanol. Once again record this volume for back-calculation.

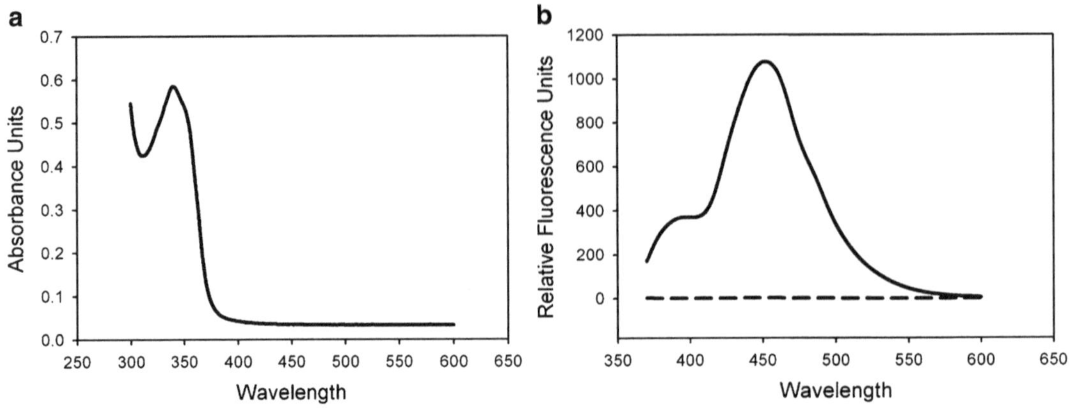

Fig. 3. PQS absorption and emission spectra. The maximum absorbance for PQS (**a**) occurs at approximately 340 nm. The maximum emission of PQS occurs at approximately 450 nm. A *dotted line* indicates HHQ emission, which is 1000-fold less than PQS.

8. Transfer the methanol samples into a black 96-well plate (or other appropriate sample holder required for use in the fluorometer) and place in the fluorometer. Excite at 342 nm and read fluorescence at 450 nm (Fig. 3). Plot fluorescence values against the concentration of PQS standards to generate a standard curve. The standard curve should be linear, and PQS concentrations in samples can be calculated using the standard curve slope equation. Samples outside the standard curve must be diluted and re-run to be within the range of the standard (see Note 7).

3.4. Preparation of Soluble LPS and Phospholipids

1. Weigh out LPS so that the final concentration is 1 mM in 1 mL.
2. Add 100 μL of chloroform to the LPS and incubate at 90°C for 5 min.
3. Add 10 μL of methanol and incubate at 90°C for 5 min.
4. Sonicate the LPS/chloroform/methanol mixture for 20 min in the ultrasonic bath.
5. Add 10 μL methanol and incubate at 90°C for 5 min.
6. Repeat step 5 three times for a total volume of 150 μL (50 μL total methanol).
7. Sonicate the LPS/chloroform/methanol mixture for 15 min in the ultrasonic bath. After sonication the LPS should be in solution and appear clear. At this point the LPS is ready for labeling.
8. For the phospholipids PE and PG, separately weigh out for a final concentration of 1 mM in 1 mL final volume.
9. Re-suspend PE and PG separately in 150 μL chloroform.

10. Sonicate the phospholipid/chloroform mixture in the ultrasonic bath for 20 min. The mixture should be clear and is now ready for labeling.

3.5. Labeling of LPS Aggregates and Phospholipid Liposomes

1. To the LPS/chloroform/methanol mixture add NBD-PE and Rh-PE (both in chloroform) to a final molar ratio of 100:1:1 LPS:NBD-PE:Rh-PE. Use the same molar ratio when labeling PE and PG.
2. Evaporate the solvent with a continuous stream of N_2 gas, until no liquid is present.
3. Add 1 mL of 20 mM HEPES, pH 7.0, to the labeled LPS or phospholipids.
4. Sonicate with the tip sonicator for around 5 min. Make sure that the LPS/phospholipids do not get too warm during sonication.
5. For preparing labeled LPS aggregates and phospholipid liposomes, LPS and phospholipids are temperature cycled for a total of four cycles. After sonication incubate the aqueous mixture at 4°C for 30 min and then transfer to a heat block set at 55°C for 30 min. Repeat the temperature cycling three more times.
6. Store at 4°C overnight. Labeled samples are then ready for FRET measurements and last about 2 weeks at 4°C.

3.6. FRET Measurements

1. Allow the fluorometer sample chamber to equilibrate to a constant 37°C with stirring.
2. Set the excitation wavelength to 470 nm and the emission wavelengths to 531 nm and 593 nm.
3. Dilute the labeled LPS aggregates or the phospholipid liposomes in 20 mM HEPES buffer, pH 7.0, so that the final concentration is 10 μM.
4. Add the labeled sample to the cuvette and let equilibrate at 37°C with stirring. The volume will vary depending on the size of your cuvette.
5. Adjust the emission intensities so that the ratio of the donor intensity and the acceptor intensities are 1. Record these intensities for 50 s to establish a baseline.
6. After 50 s add the appropriate amount of quinolone (or methanol as a control) at a final molar ratio of 1:10 LPS:PQS and record the intensities for an additional 250 s for a total of 300 s. The molar ratio of LPS:PQS may need to be adjusted. If intercalculation has occurred, the intensity of the donor should increase, whereas the intensity of the acceptor should decrease.

4. Notes

1. If the effect of a specific medium on PQS production is of interest, washing the cell pellet twice and starving cells for 2 h in sterile buffer (e.g. phosphate buffered saline, PBS) is recommended before inoculation of 55 mL culture.
2. TLC plates may be rocked during KH_2PO_4 incubation.
3. TLC plates may be stored in a drying oven for longer than 1 h to prevent moisture from entering the plate before TLC run. The plates should be completely dry (directly out of oven) when spotting samples.
4. Sealing the TCL chamber with petroleum jelly can improve TLC quality.
5. When visualizing the TLC in the fluorescence imager, exciting the plate from above with UV light tends to result in a brighter image. Exciting from above and below is ideal.
6. PQS may also be separated and quantified by this protocol. To obtain a clear PQS peak it may be helpful to chelate iron from PQS samples before HPLC analysis.
7. The presence of HHQ in a simple mixture is not an issue in this protocol as fluorescence intensity from HHQ is 1000-fold less than PQS (Fig. 3b).

References

1. Parsek, M. R., and Greenberg, E. P. (2000) Acyl-homoserine lactone quorum sensing in gram-negative bacteria: a signaling mechanism involved in associations with higher organisms, *Proc Natl Acad Sci USA* **97**, 8789–8793.
2. Pesci, E. C., Milbank, J. B., Pearson, J. P., McKnight, S., Kende, A. S., Greenberg, E. P., and Iglewski, B. H. (1999) Quinolone signaling in the cell-to-cell communication system of *Pseudomonas aeruginosa*, *Proc Natl Acad Sci USA* **96**, 11229–11234.
3. Deziel, E., Gopalan, S., Tampakaki, A. P., Lepine, F., Padfield, K. E., Saucier, M., Xiao, G., and Rahme, L. G. (2005) The contribution of MvfR to *Pseudomonas aeruginosa* pathogenesis and quorum sensing circuitry regulation: multiple quorum sensing-regulated genes are modulated without affecting *lasRI*, *rhlRI* or the production of *N*-acyl-L-homoserine lactones, *Mol Microbiol* **55**, 998–1014.
4. Mashburn, L. M., and Whiteley, M. (2005) Membrane vesicles traffic signals and facilitate group activities in a prokaryote, *Nature* **437**, 422–425.
5. Mashburn-Warren, L., Howe, J., Garidel, P., Richter, W., Steiniger, F., Roessle, M., Brandenburg, K., and Whiteley, M. (2008) Interaction of quorum signals with outer membrane lipids: insights into prokaryotic membrane vesicle formation, *Mol Microbiol* **69**, 491–502.
6. Schaefer, A. L., Taylor, T. A., Beatty, J. T., and Greenberg, E. P. (2002) Long-chain acyl-homoserine lactone quorum-sensing regulation of *Rhodobacter capsulatus* gene transfer agent production, *J Bacteriol* **184**, 6515–6521.
7. Marketon, M. M., Gronquist, M. R., Eberhard, A., and Gonzalez, J. E. (2002) Characterization of the *Sinorhizobium meliloti sinR/sinI* locus and the production of novel *N*-acyl homoserine lactones, *J Bacteriol* **184**, 5686–5695.
8. Deziel, E., Lepine, F., Milot, S., He, J., Mindrinos, M. N., Tompkins, R. G., and Rahme, L. G. (2004) Analysis of *Pseudomonas aeruginosa* 4-hydroxy-2-alkylquinolines

(HAQs) reveals a role for 4-hydroxy-2-heptylquinoline in cell-to-cell communication, *Proc Natl Acad Sci USA* **101**, 1339–1344.

9. Gallagher, L. A., McKnight, S. L., Kuznetsova, M. S., Pesci, E. C., and Manoil, C. (2002) Functions required for extracellular quinolone signaling by *Pseudomonas aeruginosa*, *J Bacteriol* **184**, 6472–6480.
10. Palmer, K. L., Aye, L. M., and Whiteley, M. (2007) Nutritional cues control *Pseudomonas aeruginosa* multicellular behavior in cystic fibrosis sputum, *J Bacteriol* **189**, 8079–8087.

Chapter 16

Linking Quorum Sensing Regulation and Biofilm Formation by *Candida albicans*

Aurélie Deveau and Deborah A. Hogan

Abstract

Candida albicans biofilms are surface-associated, structured communities composed of yeast, hyphal, and pseudohyphal cells surrounded by an extracellular matrix. *C. albicans* biofilms often lead to life-threatening systemic infections and are particularly difficult to eradicate because of their high levels of resistance to antibiotics. Farnesol, an autoregulatory molecule secreted by *C. albicans*, inhibits hyphal growth and the expression of a number of morphology-specific genes that are necessary for robust biofilm formation. Many stages of biofilm development are impacted by farnesol including the adherence of cells to the substratum, the architecture of mature biofilms, and the dispersal of cells from biofilms. For these reasons, understanding the mechanisms of action of farnesol could lead to the development of new antifungal compounds that target *C. albicans* biofilm cells, perhaps rendering biofilms more sensitive to antibiotics. Here, we describe several methods for the analysis of the effects of farnesol on biofilm formation and function.

Key words: Farnesol, Biofilm, *Candida albicans*, Morphology

1. Introduction

Biofilms develop when cells attach to a surface and then grow in a manner that leads to organized communities of differentiated microbial cells encased in a matrix of exopolymeric materials. In in vitro models, *Candida albicans* biofilm formation involves three main steps (1, 2): the initial colonization of the substratum by yeast cells, growth and hypha formation, and the production of an extracellular matrix primarily composed of β-1,3-glucan (3, 4). Mature biofilms typically contain a mixture of yeast, hyphae, and pseudohyphae. Ultimately, yeast cells disperse from the biofilm.

Kendra P. Rumbaugh (ed.), *Quorum Sensing: Methods and Protocols*, Methods in Molecular Biology, vol. 692, DOI 10.1007/978-1-60761-971-0_16, © Springer Science+Business Media, LLC 2011

As has been found for many microbial species, *C. albicans* quorum sensing regulation has been shown to modulate all stages of biofilm life cycles (attachment, maturation, and dispersal) (5, 6). The best characterized quorum sensing molecule produced by *C. albicans* is farnesol, an autoregulatory sesquiterpene that regulates the interconversions between its yeast and filamentous forms (7). Farnesol leads to a reduction of the size of biofilms, regardless of whether it is added to cells prior to attachment to the surface or to preformed mature biofilms (8). Interestingly, farnesol does not appear to impact the size of biofilms when farnesol is added during the initial stages of biofilm growth that follow initial attachment (8), indicating that there may be biofilm developmental steps that are insensitive to farnesol. Nickerson et al. have observed that cells do not alter their morphology in response to farnesol during early hyphal growth (9), but it is not yet known how farnesol insensitivity occurs and if the same phenomenon explains resistance of young biofilms to the effects of farnesol. While effects on biofilm morphology have not been observed when farnesol is added to developing biofilms, other physiological parameters may be affected. Indeed, farnesol has been shown to impact properties other than the morphological switch, such as resistance to oxidative stress (10, 11) or apoptosis (12) in planktonic conditions. Other *C. albicans*-produced quorum sensing molecules, such as tyrosol (13), have may also alter biofilm development (14).

Because of the impact of farnesol on morphology (7) and of the importance of morphology for biofilm formation (15–18), it has been hypothesized that exogenous farnesol affects biofilm development by repressing hypha formation and the expression of hypha-specific genes (9). Little is known, however, about the effects of endogenous farnesol on the different stages of biofilm formation and biofilm phenotypes, such as drug resistance. We expect that its effects are complex. Davis-Hanna et al. (11) showed that farnesol represses hyphal growth by inhibiting the Ras1–adenylate cyclase–protein kinase A signaling pathway, which also impinges on many cellular processes including stress responses, metabolism, and drug resistance (19, 20). Other studies have found that farnesol affects other signaling pathways (21–23), and the connections, if any, between the different pathways affected by farnesol are not yet known. As mutants altered in the farnesol response and farnesol production (24) are discovered, the roles of farnesol in multicellular populations will be better understood.

A large variety of methods have been developed to grow and analyze *C. albicans* biofilms in the past years (25, 26). Four main parameters must be taken into account when designing biofilm experiments: the nature of the substrate (plastic microtiter dish, medically relevant materials, glass, etc.), nutrient availability, flow, if any, and incubation conditions (in vitro vs. in vivo).

Each of these parameters has been shown to modify biofilm size and has advantages and disadvantages (for an extensive review, see ref. 23). Here, we provide a detailed method for the growth of static biofilms in microtiter plates. This method has the advantage of being easy to set up and allowing one to test a wide variety of conditions and strains. Complementary approaches are also described. Finally, we suggest several techniques to explore the role of farnesol on biofilm formation and function in vitro. We propose methods to follow attachment, maturation, and dispersion of biofilms, as well as three-dimensional architecture. These approaches can be used in a variety of ways including the analysis of mutants altered in their responses to farnesol or monitoring of the expression of farnesol regulated genes in situ.

2. Materials

2.1. Cell Culture

1. Agar.
2. Anhydrous glucose.
3. Peptone.
4. 1× Phosphate-buffered saline (PBS).
5. Yeast extract.
6. Yeast nitrogen base (YNB) without amino acids (Research Products International Corp.).
7. 30 and 37°C Incubators.
8. Borosilicate culture tubes 16×150 mm with caps.
9. Plastic petri dishes 100×15 mm.
10. Tissue culture roller drum (New Brunswick Scientific).
11. Centrifuge for washing cells.

2.2. Farnesol Preparation

1. Ethyl acetate (see Note 1).
2. Glacial acetic acid (see Note 1).
3. Farnesol solution: Prepare a fresh 50 mM stock solution of *trans,trans*-Farnesol (Sigma–Aldrich) in a screw thread glass vial by diluting 6.2 μL of stock farnesol in 492 μL acidified ethyl acetate (see Note 2) (0.01% glacial acetic acid). *Caution*: Farnesol is highly sensitive to oxygen. We advise aliquoting farnesol stock solution upon reception in screw thread glass vials under nitrogen flow. The 50 mM stock solution should also be prepared under nitrogen flow to limit the loss of activity.
4. Glass vials, screw thread, w/polyvinyl-faced pulp lined closure (Fisherbrand).
5. Nitrogen or argon gas supply.

2.3. Biofilm Growth

1. 15-mL Centrifuge tubes.
2. Glass flasks.
3. Microtiter plates – microtest flat bottom 96-, 24-, 12-, 6-well plates, polystyrene (Falcon).
4. Plastic box to create a humidified chamber.

2.4. Metabolic Activity Assay Reagents

1. Alamar Blue (AbD Serotec).
2. 1× PBS.
3. Spectrophotometer.

2.5. Physical Quantitation of Biofilm

1. Crystal violet stain: 0.02% crystal violet (w/v) in water (see Note 1). Refer to MSDS for proper handling and disposal of this chemical.
2. Triton X-100.
3. Glass beads (Fisher Scientific).
4. Orbital table shaker (Boeckel Scientific).
5. Trays for washing plates.

2.6. Biofilm Imaging

1. Glass bottom culture dishes – 35-mm Petri dish (MatTek Corporation).
2. Calcofluor white M2R (Sigma–Aldrich) is dissolved 0.2 mg/mL in MilliQ water. One normal NaOH is added to solubilize the dye, which is then filter-sterilized. Store away from light at room temperature for 3–4 weeks.
3. Concanavalin Alexa Fluor 488 conjugate (CON-A; Invitrogen) is dissolved at 5 mg/mL in PBS. Fifty microliters of aliquots should be stored at −20°C and should not be refrozen once thawed. CON-A can be stored for 6 months at −20°C. CON-A can be viewed at the 488 nm excitation wavelength and 505 nm emission with a long-pass filter.
4. FUN-1 dye (2-chloro-4-(2,3-dihydro-3-methyl-(benzo-1, 3-thiazol-2-yl)-methylidene-1-phenylquinolinium iodide); Invitrogen) is commercially available as a 10 mM solution in dimethylsulfoxide. Store at −20°C. FUN-1 can be viewed at the 543 nm excitation wavelength and 560 nm emission with a long-pass filter.
5. Epifluorescence microscope (e.g., Zeiss Axiovert inverted microscope equipped with a 63× objective and Axiovision software).
6. Confocal microscope.
7. Volocity 3.5.1 (Improvision Inc.; Lexington, MA).
8. COMSTAT image analysis software package (27).

3. Methods

3.1. Plate Biofilm Microtiter Assay

The microtiter plate biofilm assay has already been successfully used to analyze biofilm susceptibility to antibiotics (28, 29), the effects of farnesol on biofilm size (8), and to compare farnesol production and biofilm development by different *Candida* species (30, 31). The protocol presented below is adapted from refs. 8, 28, 32. In brief, cells are incubated in microtiter dishes for a short period of time to allow attachment to surfaces; the adherence phase is followed by a wash step to remove planktonic cells. The adherent cells are then allowed to grow and form biofilms for a desired period of time (usually 24–48 h). Biofilms can be quantified by measuring metabolic activity or physical properties. To study the effects of farnesol on *C. albicans* biofilm development, farnesol can be added at different times depending on whether one is studying the effects of farnesol on attachment, biofilm development, or dispersal. We recommend including at least three to four replicates per condition. The basic biofilm assay protocol is as follows:

1. Prepare yeast peptone glucose (YPD) broth and agar. Dissolve 10 g of yeast extract with 20 g of peptone in 900 mL of MilliQ water. Prepare separately a 20% w/v glucose solution by dissolving 20 g of glucose in 100 mL of MilliQ water. Autoclave the resulting suspensions at 121°C for 45 min. Cool to 60°C and then add 100 mL of glucose solution to YP medium. To make YPD plates, add agar (2% w/v final concentration) to the yeast extract–peptone solution prior to autoclaving. Pour ~20–25 mL into Petri dishes.
2. Prepare low glucose minimal medium (YNB 0.2%). Dissolve 6.7 g of YNB in 100 mL of MilliQ water to obtain 10× YNB. Filter-sterilize this solution using sterile 0.22-μm-diameter vacuum steritop system. Keep at 4°C. *Caution*: Do not autoclave YNB as autoclaving can degrade media constituents. Mix 10 mL of filter-sterilised 10× YNB with 1 mL of sterile 20% glucose to a final volume of 100 mL with sterile MilliQ water. If using auxotrophic strains, add appropriate supplements. See Note 3 below for a description of other media that can be used.
3. Inoculate *C. albicans* strains of interest from frozen stock onto YPD 2% plate and incubate overnight at 30°C, then store at room temperature. Strains should be freshly streaked from freezer stocks after 1 week.
4. Inoculate cells from several colonies in 5 mL of YPD 2% broth in borosillicate culture tubes. Grow at 30°C in a roller drum for 14–16 h.

5. Harvest cells by centrifugation and wash once with fresh medium. Dilute culture in YNB 0.2% glucose (see Note 3) to an OD_{600nm} of 0.2 [≈10^6 cells/mL (see Note 4)] into glass flasks.
6. Pipette 100 µL of the diluted culture into each well of a 96-well (see Note 5) microtiter plate (see Note 6). Cover plate and incubate at 37°C for 90 min.
7. Remove the medium and planktonic cells with a pipette and wash the well gently with water, buffer, or fresh medium two times.
8. Pipette 100 µL of fresh medium into each well.
9. Put the plate into a plastic box containing a wet paper towels to prevent excessive evaporation. Incubate at 37°C for 2–48 h.
10. Quantify biofilm size following the methods described in Subheading 3.4.

3.2. Alternative Methods for Biofilm Growth

The plate microtiter assay allows one to rapidly test a large number of conditions or strains, but it may not accurately reflect in vivo conditions. Therefore, one may consider using this method for a first-step analysis, followed by methods using medically relevant surfaces such as monitoring biofilm formation on silicone sheets (18) or polymethylmethacrylate (33), or using nutrient supply regimes that can be modeled using a flow cell (34). The silicone sheet system is designed to mimic biofilm formation on implanted biomedical devices such as catheters or heart valves (25). Briefly, sheets of silicone are coated with bovine serum to enhance initial attachment of cells (35), then washed and incubated in a suspension of *C. albicans* cells at 37°C for 90 min to allow for fungal attachment to the sheets. The silicone sheets are then washed to remove unattached cells, and biofilms are grown for 48 h. The extent of biofilm formation can be assessed by comparing the population of cells in the planktonic phase to those associated with the biofilm.

Flow cell models incorporate the conditions of sheer stress and nutrient influx that fungal cells may encounter in the host when biofilms develop on tubing or host tissues. *C. albicans* biofilms grown under flow conditions are thicker and show increased architectural complexity compared to biofilms grown under static conditions (34). In flow cell models, cells are allowed to attach to a surface for a short period of time prior to the initiation of flow. Flow cell systems can be run for several days to accumulate large amounts of biofilm biomass.

3.3. Assessing the Effects of Farnesol on Biofilm Formation

3.3.1. Farnesol Effects on Initial Attachment

To assess the effects of farnesol on initial attachment, begin by following steps 1–5 from the biofilm growth protocol described above (see Subheading 3.1). Add freshly prepared farnesol (from a 50 mM stock solution) to one flask containing the suspension of *C. albicans* cells (10^6 cells/mL) to reach the desired final concentration (see Note 7) (see Subheading 3.1, step 4). Add the same volume of acidified ethyl acetate to the other flask to create a vehicle control. *Caution*: It is important to mix farnesol or acidified ethyl acetate in glass tubes or flasks before inoculating polystyrene plates instead of adding directly to the wells because at high concentrations, ethyl acetate can have localized effects on plastic that may disturb the settlement of cells. The farnesol-treated cells can be analyzed using the standard biofilm assay described above (see Subheading 3.1).

3.3.2. Farnesol Effects on Biofilm Development

Follow up to step 6 in Subheading 3.1 and incubate plates at 37°C for 1–24 h to allow biofilm growth. Incubation times will depend on the biofilm development stage that you want to study (hyphal growth, maturation or late maturation, and dispersal). Prepare two aliquots of fresh medium amended with either farnesol, at the desired concentration, or the vehicle alone. Remove the supernatant from microtiter plate with a pipette, wash twice with sterile water or buffer, and add the fresh solution containing farnesol. Incubate plate for additional time and quantify biofilm size following the methods described in Subheading 3.4.

3.3.3. Farnesol Effects on Biofilm Dispersal

To assess the effects of farnesol on cell dispersal, first grow mature biofilms for 48 h as described in 3.1." Biofilms can be "refed" to increase biofilm size by exchanging the spent supernatant for fresh medium. Prepare two aliquots of YNB 0.2% glucose containing either farnesol at the desired concentration or ethyl acetate. Remove the supernatant from microtiter plate by pipetting, wash twice with sterile water, and add the fresh solution containing farnesol. Incubate at 37°C for another 24 h. Quantify biofilm size following the methods described in Subheading 3.4. Biofilm released cells can be recovered from the supernatant if necessary.

3.4. Monitoring the Effects of Farnesol on Biofilm Growth

Several methods can be used to monitor the effects of farnesol on biofilm formation. The choice of method will depend on whether adhesion, cell morphology, biofilm size, or biofilm architecture is being examined. Light microscopy using an inverted microscope is a powerful way to follow the initial attachment steps and to study and quantify the consequences of the inhibition of hyphal growth by farnesol on the structures of early biofilms. However, the high density of cells and the three-dimensional structure of the biofilm impedes direct microscopic visualization of more mature biofilms. At later stages of biofilm development, imaging by confocal microscopy allows for the precise monitoring of the variation in biofilm size and architecture (18, 26).

While microscopy is a powerful tool for biofilm analysis, it can be limited by the time necessary to process each sample, and it may not allow for the analysis of multiple samples simultaneously. When higher throughput analyses are required, alternative quantitative assays are recommended. Metabolic assays that rely on reduction of a redox reactive dye (XTT or Alamar Blue) are the most frequently used (8, 26). Alternatively, biofilm size can be evaluated by quantifying biomass, either by measuring the dry weight or by staining with crystal violet (36). Crystal violet is a basic dye that binds to negatively charged surface molecules and polysaccharides in the extracellular matrix. One advantage of crystal violet is that one can process samples in situ and times analysis is faster than that can be achieved using dry weight measurements. When performing biofilm studies in microtiter dishes, dry weight measurement requires one to scrape each biofilm from the bottom of each well and to filter each sample individually which can introduce error particularly when biofilms are small. In contrast, dry weight measurement is well adapted when using other substrates such as silicone sheets, tubing, or dental fragments to grow biofilms.

Metabolic assays, crystal violet staining, or biomass measurements provide different information. While metabolic assays such as XTT and Alamar Blue only assess the metabolic activity of living cells, crystal violet staining and biomass measurements quantify cells and matrix material in aggregate. Variation in biofilm biomass may not always correlate with variation in metabolic activity. The use of different techniques in parallel can provide complementary information. Furthermore, these assays have different ranges of sensitivity. In our hands, metabolic assays allow for the detection of smaller differences in comparison to crystal violet staining.

3.4.1. Metabolic Quantitation: Alamar Blue

1. Remove supernatant from biofilm-containing well with a pipette.
2. Wash wells gently with 100 μL of H_2O twice using a pipette to remove unattached cells.
3. Add 0.1 mL of PBS (see Note 8) + 10% of Alamar Blue in each well. A negative control in an uninoculated well is necessary.
4. Incubate for an appropriate length of time (30 min–8 h) at 37°C on a rocker at 75 rpm. *Caution*: It is important to perform a kinetic study before starting the experiment to determine time points that are in the linear range of the assay. These time points will depend on the cell number and the activity of the cells within the biofilm.
5. Transfer liquid in new plate and measure absorbance at 570 and 600 nm. Alamar Blue changes color in response to the

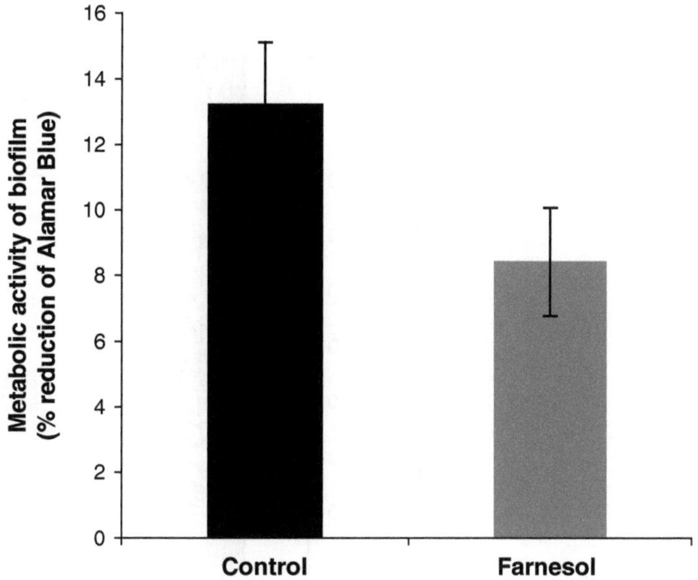

Fig. 1. Representative results obtained using Alamar Blue to quantify the effects of 50 μM farnesol on biofilm formation by *C. albicans* SC5314. Farnesol was added at time zero, and biofilms were grown for 24 h before measurement. For quantification, biofilms were incubated for 8 h with Alamar Blue before measuring the absorbance at 570 and 600 nm. The average of three independent replicates (±SE) is shown.

chemical reduction of the growth medium. Metabolic activity of cells leads to the reduction of Alamar Blue (see Note 9). Alamar Blue is blue in its oxidized state and turns red upon reduction. The percent reduction of Alamar Blue is calculated according to the manufacturer's instructions. An example of results obtained with Alamar Blue is provided in Fig. 1.

3.4.2. Quantitation by Crystal Violet Staining

1. Remove supernatant from biofilm-containing well with a pipette.
2. Wash wells gently with 100 μL of H_2O twice using a pipette to remove unattached cells.
3. Add 100 μL of 0.02% crystal violet solution to each well and incubate at room temperature for 10 min.
4. Shake microtiter dish out over a waste tray to remove the crystal violet solution (see Note 10). Wash the biofilm dish by submerging in water three times and shake out as much liquid as possible after each wash by tapping the inverted plate on paper towels.
5. Add 100 μL of 0.1% Triton X-100 and three glass beads and shake gently on a orbital table shaker until the full biofilm is released from the plastic and disrupted.
6. Homogenize by pipetting and measure the optical density of each sample at a wavelength of 590 nm.

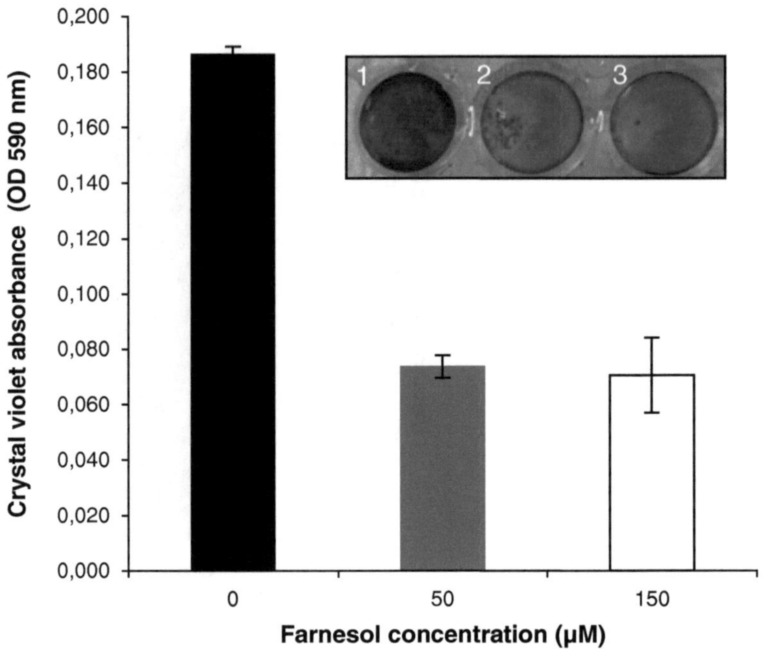

Fig. 2. Representative results obtained using crystal violet to quantify the effects of farnesol on biofilm formation of *C. albicans* SC5314. Farnesol was added at time zero, and biofilms were grown for 48 h before measurement. The stained wells are shown prior to solubilization of the dye. Control biofilms (1), biofilms grown with 50 μM (2), and 150 μM (3) farnesol are shown. The average of three independent replicates (±SE) is presented.

An example of results obtained with crystal violet is provided in Fig. 2.

3.4.3. Biofilm Imaging

Light microscopy can be used in the early steps of the biofilm formation to follow the effects of farnesol on adhesion to the surface and on morphology of cells. For analysis of more mature biofilms, confocal scanning laser microscopy (CSLM) should be performed. CSLM gives useful information on the three-dimensional architecture of biofilms, but only at low resolution. It is necessary to use scanning electron microscopy to obtain high-resolution images.

3.4.3.1. Imaging Biofilms with Light Microscopy

1. Grow biofilm as described in Subheading 3.1. The assessment of attachment is most easily performed in the larger wells of 12- or 6-well plates. Analysis is facilitated by the use of a long-working-distance objective. Instead of using regular microplates, glass bottom culture dishes can also be used to enhance quality of images. Instead of using an OD_{600nm} of 0.2 for inoculating plates, a 1/50th dilution of 0.1 OD_{600nm} is preferred to achieve a density that will allow for counting during the attachment and hyphal induction stages.

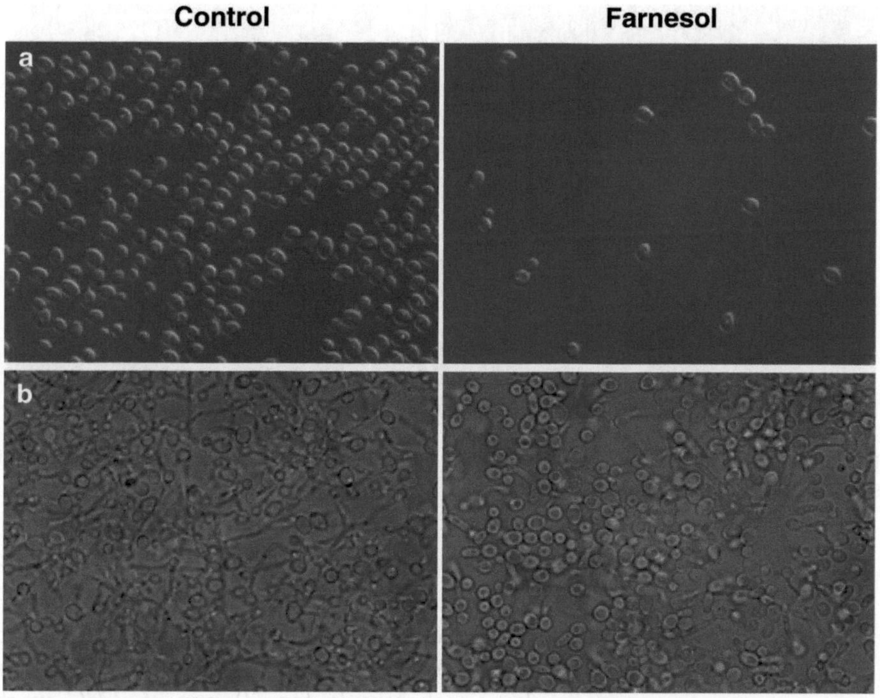

Fig. 3. Effects of farnesol on adhesion of *C. albicans* on glass bottom slide Petri dishes (**a**) and on cell morphology (**b**) as observed with light transmission microscopy. Farnesol (100 μM) was added at time zero, and cells were allowed to attach for 90 min before removing unattached cells. Observations were made immediately after removing unattached cells (**a**) or after an 8-h incubation period (**b**).

2. After incubation with farnesol at 37°C for the desired length of time, remove the supernatant by pipetting and wash wells gently with sterile water or buffer (see Note 11); remove the wash solution and repeat once. Add 2 mL of 0.2% YNB and count the number of cells that remain attached to the bottom of the well. Multiple fields, randomly chosen, per treatment will need to be analyzed to perform the appropriate statistical analyses. An example of observations obtained with light microscopy is provided in Fig. 3.

3.4.3.2. Imaging Biofilms with Confocal Microscopy

The quantification of biofilm structures requires the use of fluorescent dyes. Calcofluor White, CON-A, and FUN-1 are the most common dyes used to analyze biofilm architecture with CSLM (18, 26). Calcofluor White binds to chitin and β-glucan of fungal cell walls, while CON-A binds to mannose and polysaccharides. Therefore, Calcofluor and CON-A allow visualization of fungal cells and matrix plus fungal cells, respectively. FUN-1 is converted into an orange fluorescent dye by metabolically active cells. Thus, it permits the analysis of localized metabolic activity within biofilms:

1. Grow biofilm as described in Subheading 3.1, but instead of using regular microplates, use glass bottom culture dishes.

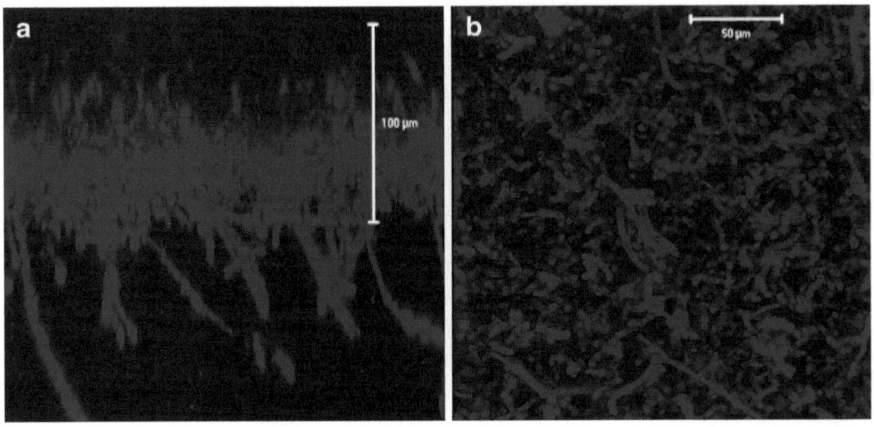

Fig. 4. Architecture of biofilm of strain DAY185 on silicone substrate as visualized by CSLM after staining with CON-A. (**a**) Assembly side view. (**b**) Top view. Images kindly provided by Dr. A. Mitchell.

2. Incubate biofilm for desired time, then remove liquid from each well by pipetting and gentle washing with sterile H_2O twice.
3. Add 2 mL of dye mix (2 μL of FUN-1 working solution and 10 μL of CON-A in PBS or 5 μL of Calcofluor white in water) and incubate each sample for 30 min at 37°C in the dark. Caution: FUN-1 degrades after 1 h and forms yellow artifacts after prolonged exposure. Stain and analyze the sample quickly.
4. Gently remove the dye mix solution and wash twice with PBS. *Caution*: Work in the dark as much as possible to avoid bleaching of the dyes.
5. Examine the biofilm using an inverted CSLM and acquire images at 1 μm intervals across the depth of the biofilm to create z-stacks. The thickness of each optical slice in a z-stack should be the same for both channels.
3. Deconvolve volumes by iterative restoration using the microscope 3D reconstruction software or Volocity 3.5.1 software. Biofilm structures (volume, roughness, maximum height, etc.) can be analyzed using the COMSTAT image analysis software package (27). Examples of results obtained using CSLM are provided in Fig. 4.

4. Notes

1. *Caution*: toxic, use proper precautions.
2. Farnesol can also be diluted in methanol or ethanol. In our growth conditions, acidified ethyl acetate does not impact the growth rate of *C. albicans*, while methanol and ethanol do have effects on growth. However, laboratories using other

media reported the opposite result (K. Nickerson, personal communication). These contradictory results are probably due to differences in the composition of nutrient media used.

3. Composition of the medium strongly influences cell morphology, cell proliferation, and extracellular matrix production. Other media such as Spider medium (37), RPMI 1640 medium (2, 8), or phosphate buffer with serum (2) have been used for *C. albicans* biofilm formation. The low glucose medium described here allows for striking differences to be observed upon the addition of farnesol; other media with potent inducers of hyphal growth, such as YNBNP (38) or medium with serum, can also be used if large, robust biofilms are required.
4. We have found it important to start with cell densities no greater than 10^6 cells/mL, as farnesol accumulation may inhibit hyphal growth and biofilm formation at higher cell densities.
5. Microtiter plates with different well sizes can also be used. Biofilms from wells in 6-well and 12-well plates often have a sufficient number of cells for easy analysis of specific transcript or protein levels. When 6-, 12-, or 24-well plates are used, the volume of medium should be adjusted to 3, 1, or 0.5 mL, respectively.
6. The composition of the surface of the microtiter plate or Petri dish can affect the adherence of cells and biofilm development. For example, *C. albicans* sticks poorly to polyvinyl or glass. Coating with serum or saliva can be used to stimulate adhesion and to mimic in vivo conditions of biofilm formation. Plastics that model denture material (polymethylmethacrylate) (33) or catheter material (silicone elastomer) (18) are also commonly used in the study of *C. albicans* biofilm formation.
7. We routinely use farnesol at a final concentration ranging between 50 and 100 μM which results in a decrease in biofilm formation from 25 to 70%. However, Ramage et al. described a significant effect of farnesol on initiation step and mature biofilm with concentration of 3 and 30 μM, respectively.
8. Alamar Blue only works in pH range between 6.8 and 7.4.
9. A higher metabolic activity of biofilms can be the consequence of either a higher metabolic activity of the cells or a larger number of cells.
10. Crystal violet is a hazardous compound, and its degradation products are as toxic as the native compound. Waste should be disposed of as a biohazard waste.
11. *C. albicans* cells may attach weakly to glass in comparison to polystyrene plastics. Therefore, care must be taken when

holding the plates and washing the biofilms. Plates should not be shaken violently, and pipette should not be aimed at the biofilm when washing it but rather on the side of the Petri dish. To increase adherence, cover glass Petri dishes (MatTek) can be coated with serum or saliva.

Acknowledgment

We thank Dr. Aaron Mitchell (Carnegre mellon University, PA, USA) for providing CSLM pictures. The data described in this chapter were generated in studies funded by NIH (K22 DE016542, D.A.H.).

References

1. Ramage, G., Saville, S. P., Thomas, D. P., and Lopez-Ribot, J. L. (2005) *Candida* biofilms: an update. *Eukaryot Cell 4*, 633–8.
2. Chandra, J., Kuhn, D. M., Mukherjee, P. K., Hoyer, L. L., McCormick, T., and Ghannoum, M. A. (2001) Biofilm formation by the fungal pathogen *Candida albicans*: development, architecture, and drug resistance. *J Bacteriol 183*, 5385–94.
3. Al-Fattani, M. A., and Douglas, L. J. (2006) Biofilm matrix of *Candida albicans* and *Candida tropicalis*: chemical composition and role in drug resistance. *J Med Microbiol 55*, 999–1008.
4. Nobile, C. J., Nett, J. E., Hernday, A. D., Homann, O. R., Deneault, J. S., Nantel, A., Andes, D. R., Johnson, A. D., and Mitchell, A. P. (2009) Biofilm matrix regulation by *Candida albicans* Zap1. *PLoS Biol 7*, e1000133.
5. Parsek, M. R., and Greenberg, E. P. (2005) Sociomicrobiology: the connections between quorum sensing and biofilms. *Trends Microbiol 13*, 27–33.
6. Blankenship, J. R., and Mitchell, A. P. (2006) How to build a biofilm: a fungal perspective. *Curr Opin Microbiol 9*, 588–94.
7. Hornby, J. M., Jensen, E. C., Lisec, A. D., Tasto, J. J., Jahnke, B., Shoemaker, R., Dussault, P., and Nickerson, K. W. (2001) Quorum sensing in the dimorphic fungus *Candida albicans* is mediated by farnesol. *Appl Environ Microbiol 67*, 2982–92.
8. Ramage, G., Saville, S. P., Wickes, B. L., and Lopez-Ribot, J. L. (2002) Inhibition of *Candida albicans* biofilm formation by farnesol, a quorum-sensing molecule. *Appl Environ Microbiol 68*, 5459–63.
9. Nickerson, K. W., Atkin, A. L., and Hornby, J. M. (2006) Quorum sensing in dimorphic fungi: farnesol and beyond. *Appl Environ Microbiol 72*, 3805–13.
10. Westwater, C., Balish, E., and Schofield, D. A. (2005) *Candida albicans*-conditioned medium protects yeast cells from oxidative stress: a possible link between quorum sensing and oxidative stress resistance. *Eukaryot Cell 4*, 1654–61.
11. Davis-Hanna, A., Piispanen, A. E., Stateva, L. I., and Hogan, D. A. (2008) Farnesol and dodecanol effects on the *Candida albicans* Ras1-cAMP signalling pathway and the regulation of morphogenesis. *Mol Microbiol 67*, 47–62.
12. Shirtliff, M. E., Krom, B. P., Meijering, R. A., Peters, B. M., Zhu, J., Scheper, M. A., Harris, M. L., and Jabra-Rizk, M. A. (2009) Farnesol-induced apoptosis in *Candida albicans*. *Antimicrob Agents Chemother 53*, 2392–401.
13. Chen, H., Fujita, M., Feng, Q., Clardy, J., and Fink, G. R. (2004) Tyrosol is a quorum-sensing molecule in *Candida albicans*. *Proc Natl Acad Sci USA 101*, 5048–52.
14. Alem, M. A., Oteef, M. D., Flowers, T. H., and Douglas, L. J. (2006) Production of tyrosol by *Candida albicans* biofilms and its role in quorum sensing and biofilm development. *Eukaryot Cell 5*, 1770–9.
15. Baillie, G. S., and Douglas, L. J. (1999) Role of dimorphism in the development of *Candida albicans* biofilms. *J Med Microbiol 48*, 671–9.
16. Ramage, G., VandeWalle, K., Lopez-Ribot, J. L., and Wickes, B. L. (2002) The filamentation pathway controlled by the Efg1 regulator protein is required for normal biofilm formation and development in *Candida albicans*. *FEMS Microbiol Lett 214*, 95–100.

17. Richard, M. L., Nobile, C. J., Bruno, V. M., and Mitchell, A. P. (2005) *Candida albicans* biofilm-defective mutants. *Eukaryot Cell 4*, 1493–502.
18. Nobile, C. J., Andes, D. R., Nett, J. E., Smith, F. J., Yue, F., Phan, Q. T., Edwards, J. E., Filler, S. G., and Mitchell, A. P. (2006) Critical role of Bcr1-dependent adhesins in *Candida albicans* biofilm formation in vitro and in vivo. *PLoS Pathog 2*, e63.
19. Harcus, D., Nantel, A., Marcil, A., Rigby, T., and Whiteway, M. (2004) Transcription profiling of cyclic AMP signaling in *Candida albicans. Mol Biol Cell 15*, 4490–9.
20. Bahn, Y. S., Molenda, M., Staab, J. F., Lyman, C. A., Gordon, L. J., and Sundstrom, P. (2007) Genome-wide transcriptional profiling of the cyclic AMP-dependent signaling pathway during morphogenic transitions of *Candida albicans. Eukaryot Cell 6*, 2376–90.
21. Smith, D. A., Nicholls, S., Morgan, B. A., Brown, A. J., and Quinn, J. (2004) A conserved stress-activated protein kinase regulates a core stress response in the human pathogen *Candida albicans. Mol Biol Cell 15*, 4179–90.
22. Kruppa, M., Krom, B. P., Chauhan, N., Bambach, A. V., Cihlar, R. L., and Calderone, R. A. (2004) The two-component signal transduction protein Chk1p regulates quorum sensing in *Candida albicans. Eukaryot Cell 3*, 1062–5.
23. Kebaara, B. W., Langford, M. L., Navarathna, D. H., Dumitru, R., Nickerson, K. W., and Atkin, A. L. (2008) *Candida albicans* Tup1 is involved in farnesol-mediated inhibition of filamentous-growth induction. *Eukaryot Cell 7*, 980–7.
24. Navarathna, D. H., Hornby, J. M., Krishnan, N., Parkhurst, A., Duhamel, G. E., and Nickerson, K. W. (2007) Effect of farnesol on a mouse model of systemic candidiasis, determined by use of a DPP3 knockout mutant of *Candida albicans. Infect Immun 75*, 1609–18.
25. Nett, J., and Andes, D. (2006) *Candida albicans* biofilm development, modeling a host–pathogen interaction. *Curr Opin Microbiol 9*, 340–5.
26. Chandra, J., Mukherjee, P. K., and Ghannoum, M. A. (2008) In vitro growth and analysis of *Candida* biofilms. *Nat Protoc 3*, 1909–24.
27. Heydorn, A., Nielsen, T. A., Hentzer, M., Sternberg, C., Givskov, M., Ersboll, B., and Molin, S. (2000) Quantification of biofilm structures by the novel computer program COMSTAT. *Microbiology 146*, 2395–2407.
28. Ramage, G., Vande Walle, K., Wickes, B. L., and Lopez-Ribot, J. L. (2001) Standardized method for in vitro antifungal susceptibility testing of *Candida albicans* biofilms. *Antimicrob Agents Chemother 45*, 2475–9.
29. Pierce, C. G., Thomas, D. P., and Lopez-Ribot, J. L. (2009) Effect of tunicamycin on *Candida albicans* biofilm formation and maintenance. *J Antimicrob Chemother 63*, 473–9.
30. Martins, M., Henriques, M., Azeredo, J., Rocha, S. M., Coimbra, M. A., and Oliveira, R. (2007) Morphogenesis control in *Candida albicans* and *Candida dubliniensis* through signaling molecules produced by planktonic and biofilm cells. *Eukaryot Cell 6*, 2429–36.
31. Weber, K., Sohr, R., Schulz, B., Fleischhacker, M., and Ruhnke, M. (2008) Secretion of E,E-farnesol and biofilm formation in eight different Candida species. *Antimicrob Agents Chemother 52*, 1859–61.
32. Krom, B. P., Cohen, J. B., McElhaney Feser, G. E., and Cihlar, R. L. (2007) Optimized candidal biofilm microtiter assay. *J Microbiol Methods 68*, 421–23.
33. Chandra, J., Mukherjee, P. K., Leidich, S. D., Faddoul, F. F., Hoyer, L. L., Douglas, L. J., and Ghannoum, M. A. (2001) Antifungal resistance of candidal biofilms formed on denture acrylic in vitro. *J Dent Res 80*, 903–8.
34. Uppuluri, P., Chaturvedi, A. K., and Lopez-Ribot, J. L. (2009) Design of a simple model of *Candida albicans* biofilms formed under conditions of flow: development, architecture, and drug resistance. *Mycopathologia 168*, 101–9.
35. Nikawa, H., Jin, C., Hamada, T., Makihira, S., Kumagai, H., and Murata, H. (2000) Interactions between thermal cycled resilient denture lining materials, salivary and serum pellicles and *Candida albicans* in vitro. Part II. Effects on fungal colonization. *J Oral Rehabil 27*, 124–30.
36. Peeters, E., Nelis, H. J., and Coenye, T. (2008) Comparison of multiple methods for quantification of microbial biofilms grown in microtiter plates. *J Microbiol Methods 72*, 157–65.
37. Nobile, C. J., and Mitchell, A. P. (2005) Regulation of cell-surface genes and biofilm formation by the *Candida albicans* transcription factor Bcr1p. *Curr Biol 15*, 1150–5.
38. Hogan, D. A., Vik, A., and Kolter, R. (2004) A *Pseudomonas aeruginosa* quorum-sensing molecule influences *Candida albicans* morphology. *Mol Microbiol 54*, 1212–23.

Chapter 17

Design of Synthetic Mammalian Quorum-Sensing Systems

Wilfried Weber and Martin Fussenegger

Abstract

Synthetic quorum-sensing systems in mammalian cells has enabled the implementation of time- and distance-dependent bioprocesses, as well as the design of synthetic ecosystems emulating clinically important host–parasite interactions. In this chapter, we provide a detailed protocol of the design of a mammalian cell-to-cell signaling device and its integration into a mammalian quorum-sensing system for cell density-induced expression product genes. Cell-to-cell signaling is based on a sender cell, metabolically engineered for expression of alcohol dehydrogenase converting ethanol into acetaldehyde, and a receiver cell line for the dose-dependent translation of the acetaldehyde concentration into transgene expression by an acetaldehyde-responsive promoter. This protocol can be adapted easily to various cell types and transgenes for the design of versatile mammalian cell-based quorum-sensing systems.

Key words: Acetaldehyde, AIR, Alcohol dehydrogenase, Cell phone, Synthetic biology, Synthetic ecosystem

1. Introduction

While bacterial quorum-sensing components have been applied for the construction of self-regulated control of population density (1), pattern formation (2) or the implementation of synthetic predator–prey ecosystems (3), the design of mammalian cell-based quorum-sensing systems has, thus far, been hampered by the lack of synthetic cell-to-cell communication devices. A recent study (4) has described the first synthetic mammalian intercell signaling system (a molecular "cell phone"), enabling the design of quorum sensing in mammalian cells as well as the design of multispecies synthetic ecosystems emulating clinically important host–parasite interactions (4).

The cell-to-cell signaling system comprised a sender and a receiver cell line. The sender cell line was genetically engineered

Kendra P. Rumbaugh (ed.), *Quorum Sensing: Methods and Protocols*, Methods in Molecular Biology, vol. 692, DOI 10.1007/978-1-60761-971-0_17,

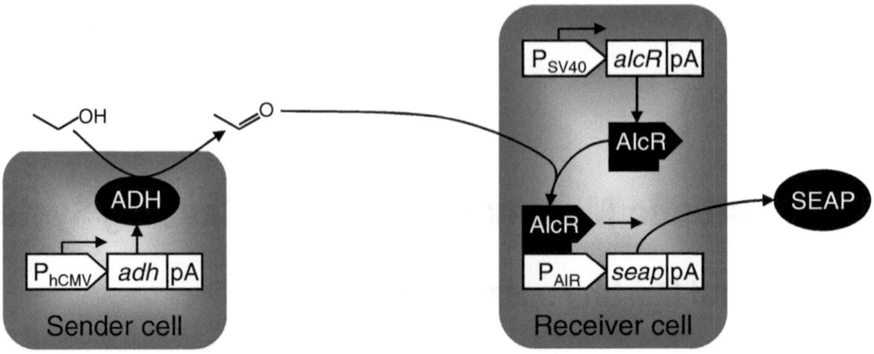

Fig. 1. Mammalian cell quorum-sensing system. The sender cell is genetically engineered for production of alcohol dehydrogenase (*adh*) under the control of the human cytomegalovirus promoter (P_{hCMV}). The enzyme ADH converts ethanol spiked into the cell culture medium into acetaldehyde, which diffuses by gas or liquid phase to the receiver cells. The receiver cells are engineered for the expression of the acetaldehyde-dependent transactivator AlcR, which, in the presence of acetaldehyde, activates the P_{AIR} promoter which drives expression of the reporter gene *seap* (human placental secreted alkaline phosphatase) in an acetaldehyde-responsive and dose-dependent manner. P_{SV40}, simian virus 40 promoter; *pA* polyadenylation signal.

for the expression of mouse liver alcohol dehydrogenase converting ethanol supplemented to the cell culture medium into acetaldehyde, which, with a boiling point of 21°C, diffuses by gas or the liquid phase to the receiver cells. The receiver cells harbor the transactivator AlcR, which, in the presence of acetaldehyde, activates its cognate promoter P_{AIR}, which controls expression of the target gene (Fig. 1) (5). Since the transfer of acetaldehyde from the sender to the receiver cell depends on time, distance, and concentration, the molecular "cell phone" enables the design of processes like mammalian cell-based quorum sensing. Quorum sensing can, for example, be applied to the design of cell density-dependent bioprocesses, during which target gene production is autonomously induced as soon as the optimum cell density has been reached (4).

We provide a detailed step-by-step protocol of the construction of a mammalian sender and receiver cell line for the design of a cell-to-cell signaling as well as a synthetic quorum-sensing system.

2. Materials

2.1. Reagents/ Solutions

1. 2× SEAP buffer : 20 mM homoarginine, 1 mM $MgCl_2$, 21% (w/v) diethanolamine/HCl, pH 9.8, store in the dark at 4°C for up to 6 months.
2. Acetaldehyde, store at 4°C.
3. Dimethyl sulfoxide (DMSO), cell culture tested.

4. Ethanol 96%, nondenatured.
5. Fetal calf serum.
6. HTS complete medium: HTS medium containing 5% (v/v) fetal calf serum and 1% (v/v) penicillin–streptomycin solution.
7. HTS medium.
8. Neomycin stock solution; 100 mg/ml ddH_2O, store in aliquots at −20°C for up to 6 months.
9. Phosphate solution: 50 mM HEPES, 280 mM NaCl, 1.5 mM Na_2HPO_4, pH 7.05, filter-sterilize and store at 4°C for up to 6 months.
10. Phosphate-buffered saline solution without magnesium and calcium, adjust to pH 7.2.
11. pNPP solution: 120 mM para-nitrophenylphosphate (Sigma 104® Phosphatase Substrate, Sigma, St. Louis, MO, USA) in 2× SEAP buffer. Store in single-use aliquots at −20°C (thaw once only).
12. Puromycin (Sigma, St. Louis, MO) stock solution: 10 mg/ml in ddH_2O, store at −20°C for up to 6 months.
13. Sterile 1 M $CaCl_2$ stock solution.
14. Sterile glycerol, sterilized by autoclaving.
15. Trypsin solution: Trypsin–EDTA (1×) in Hank's balanced salt solution (HBSS) without calcium and magnesium (Invitrogen, Carlsbad, CA, USA).

2.2. Other Material/Hardware

1. Air-tight plastic boxes with a volume of 1 l (e.g., Tupperware-like boxes).
2. Cell counting device (Casy1® counter, Innovatis, Reutlingen, Germany) or a standard hemocytometer can also be used.
3. Cell culture-certified disposable plasticware: T-25 flasks, 6-well plates, 10-cm Petri dishes, 35-mm Petri dishes, 96-well plates, serological pipettes, 15- and 50-ml Falcon tubes, and cryotubes.
4. Centrifuge to spin microtiter plates.
5. CHO-K1 cells (ATCC CCL 61).
6. Incubator for cultivation of mammalian cells at 37°C in a humidified atmosphere containing 5% CO_2.
7. Microplate reader to measure absorbance kinetics at 405 nm in 96-well plates.
8. Multidispenser pipette.
9. Plasmid pPUR (Clontech, Palo Alto, CA, USA) conferring puromycin resistance.

10. Plasmid pSV2*neo* (Clontech, Palo Alto, CA, USA) conferring neomycin resistance.
11. Plasmids: pIVL-2 (P_{hCMV}-*adh*-pA, (6)) encoding the mouse liver alcohol dehydrogenase under the control of the human cytomegalovirus promoter P_{hCMV}; pWW195 (P_{SV40}-*alcR*-pA, (5)) encoding the acetaldehyde-inducible transactivator AlcR; pWW192 (P_{AIR}-*seap*-pA, (5)), encoding the acetaldehyde-inducible promoter P_{AIR} controlling expression of the human placental secreted alkaline phosphatase SEAP (Fig. 1).
12. Silica-based anion-exchange DNA purification kits (Genomed Jetstar 2.0 Midiprep, Genomed AG, Bad Oeynhausen, Germany).
13. StrataCooler cryopreservation module (Stratagene, La Jolla, CA, USA).
14. Water bath, 65°C.

3. Methods

3.1. Construction of the Receiver Cell Line CHO-pWW195-pWW192

The receiver cell line harbors the acetaldehyde-dependent transactivator AlcR as well as the AlcR-responsive promoter P_{AIR} (5), which controls expression of the reporter gene *seap* (human placental secreted alkaline phosphatase). As an expression plasmid for AlcR, pWW195 (5) harbors the AlcR gene under the control of the simian virus 40-derived promoter P_{SV40} (P_{SV40}-*alcR*-pA; pA, polyadenylation signal). The acetaldehyde-inducible promoter P_{AIR} driving expression of *seap* is encoded on plasmid pWW192 (P_{AIR}-*seap*-pA, (5)). The receiver cell line will be constructed in two steps: (1) the chromosomal integration of pWW195 and (2) the integration of pWW192 followed by screening for clones with optimal regulation performance (see Fig. 1 and Notes 1 and 2). Basic cell culture techniques including sterile cultivation and passaging of cells are described elsewhere (7).

3.1.1. Stable Transfection of pWW195 into CHO-K1 Cells

CHO-K1 cells are stably cotransfected with pWW195 and the neomycin resistance-conferring plasmid pSV2*neo* according to the following protocol. The resulting cell line will be referred to as CHO-pWW195.

1. Isolate pWW192, pWW195, pSV2*neo*, and pPUR plasmid DNA using ion exchange-based DNA purification kits according to the manufacturer's protocol (Genomed Jetstar 2.0 midiprep kit).
2. Seed 2×10^5 CHO-K1 cells, resuspended in 2 ml of HTS complete medium per well, into two wells of a 6-well plate 15 h before transfection.

3. Transfect the cells using the following optimized calcium phosphate-based transfection protocol for CHO-K1 cells (scaled to one well of a 6-well plate).
   (a) Prepare 60 μl of a 500 mM $CaCl_2$ solution containing 6 μg of total DNA (5.75 μg of pWW195, 0.25 μg of pSV2*neo*) in a 50-ml Falcon tube.
   (b) Add 60 μl of phosphate solution within 5 s during vortexing, to allow the formation of calcium phosphate–DNA complexes.
   (c) Add 2 ml of HTS medium supplemented with 2% (v/v) FCS after exactly 25 s, to stop the formation of calcium phosphate–DNA complexes.
   (d) Aspirate the medium from the CHO-K1-containing well.
   (e) Add the calcium phosphate–DNA complex of step (c) to the cells.
   (f) Centrifuge the cell culture plate at 450×*g* for 5 min which facilitates the interaction of calcium phosphate–DNA precipitates with the cell monolayer.
   (g) Place the cells in the incubator for 3 h.
   (h) Aspirate the medium and add 2 ml of HTS supplemented with 2% (v/v) FCS and 15% (v/v) glycerol.
   (i) After exactly 30 s, replace the glycerol-containing medium with 2 ml of HTS medium and tap the plate gently to remove residual glycerol.
   (j) Replace the washing medium with 2 ml of HTS complete medium.
   (k) Place the plate in incubator.
4. Simultaneously carry out the same transfection procedure (steps a–k) for the second well using 6 μg pWW195 to provide a negative selection control in the absence of pSV2*neo*.
5. 48 h posttransfection, aspirate the medium, add 0.5 ml of trypsin solution to each well, and place the plate in an incubator for 5–10 min. Tap the 6-well plate gently to ensure that the cells are completely detached from it.
6. Transfer the cells from each well to a separate 15-ml Falcon tube containing 7 ml of HTS complete medium. Centrifuge for 3 min at 300×*g*.
7. Aspirate and discard the supernatant, resuspend the cell pellet in 10 ml of HTS complete medium, and transfer to a 10-cm Petri dish. Add neomycin to a final concentration of 400 μg/ml (see Note 3).
8. Change the medium every 2–3 days; make sure that the medium always contains 400 μg/ml neomycin. Repeat until

all the cells in the control population (without pSV2*neo*) are dead (typically after 10–12 days).

9. Cultivate the pWW195 + pSV2*neo*-transfected cell population in HTS complete medium supplemented with 400 µg/ml neomycin until colonies of at least 2 mm in diameter are visible. Change the medium when it starts to turn yellow/orange.

3.1.2. Single-Cell Cloning of CHO-pWW195 Cells

Transgenic monoclonal cell lines are generated from the stable mixed population constructed above (Subheading 3.1.1) by limiting dilution (see Note 4).

1. Aspirate the medium from the mixed stable cells (Subheading 3.1.1, step 9). Wash cells once with 5 ml of sterile and prewarmed (37°C) PBS.
2. Add 2 ml of trypsin solution to each Petri dish and place in incubator for 5–10 min.
3. Gently tap the plate until all cells are detached.
4. Aspirate cells and transfer to a 15-ml Falcon tube containing 7 ml of HTS complete medium. Centrifuge the cells for 3 min at 300 × *g*, aspirate the supernatant, and resuspend the cell pellet in 5 ml of HTS complete medium.
5. Count cells with a cell counting device (e.g., Casy® Cell Counter).
6. Dilute the cell suspension to three cells/ml in HTS complete medium with 400 µg/ml neomycin to give a final volume of 120 ml (depending on the initial cell concentration, serial dilutions might be necessary).
7. Resuspend the diluted cells and pipette the cell suspension into six 96-well plates (200 µl/well) with a multidispenser pipette.
8. Place the plates in an incubator. Check cell growth after 5–7 days. Mark wells with a single cell clone on the lid of the plate.
9. Let the cells grow until 30–50% of the bottom of the well is covered by the growing clone.
10. The remaining cells of the mixed cell population can be frozen for later use according to the following protocol:
    (a) Centrifuge the cells (3 min at 300 × *g*), aspirate, and discard supernatant.
    (b) Resuspend the cell pellet in 5 ml of ice-cold FCS containing 10% (v/v) DMSO.
    (c) Transfer the cell suspension to cryotubes (1 ml/tube), put into a StrataCooler cryopreservation module, and freeze at –80°C overnight.

(d) Transfer the cryotubes to a liquid-nitrogen tank or a freezer (−140°C) for long-term preservation.

3.1.3. Selection of Transgenic CHO-pWW195 Cell Clones

The monoclonal CHO-pWW195 cell lines (Subheading 3.1.2) are then analyzed for the expression of AlcR. Therefore, the individual cell clones will be transiently transfected with pWW192 (Fig. 1) and analyzed for SEAP production in the presence and absence of acetaldehyde.

1. Pour 2 ml of HTS complete medium supplemented with 400 μg/ml neomycin into each well of ten 6-well plates.
2. Select 60 cell clones for analysis of AlcR expression from the 96-well plates.
3. Aspirate the medium from the 60 clones. Take care not to aspirate the cells. It might be better to use a fine pipette tip (e.g., for a 200-μl pipette) than a Pasteur pipette.
4. Add 150 μl of trypsin solution to each clone and place in incubator for 5–10 min.
5. Detach the cells by pipetting and transfer the cells of each well to a separate well of the 6-well plates from step 1.
6. Place the cells in the incubator until near 90% confluence. Since initial densities and growth rates of the different cell clones might vary, the cells reach confluence at different points in time; therefore, the following selection procedure (steps 7–18) must probably be performed on different days, i.e., whenever a clone reaches confluence.
7. For each cell clone to be analyzed, add 2 ml of HTS complete medium supplemented with 400 μg/ml neomycin to one well of a 6-well plate and 0.5 ml of the same medium to one well each of two 24-well plates.
8. Aspirate the medium of the cell clone to be analyzed, add 500 μl of trypsin solution to each well, place in an incubator for 5–10 min, and gently tap the plate to detach the cells.
9. Transfer the cells to a 15-ml Falcon tube containing 3 ml of HTS complete medium.
10. Count the cells in each tube and pipette 50,000 cells per clone into one well of the 6-well plate and into the two wells of the two 24-well plates (step 7 above). The 6-well plate will serve as the maintenance culture, while the 24-well plates are for screening. Place the cells in an incubator overnight.
11. The next day, transfect the cells in the 24-well plates with pWW192, according to the above protocol (Subheading 3.1.1, step 3), with the exception that all volumes and amounts of reagents are scaled down from the 6-well plate to the 24-well plate (all volumes and amounts divided by 4).

12. Place the 24-well plates in 1-L plastic boxes without lids in the incubator. Wait for 10 min until CO_2 and humidity reach equilibrium in the incubator and the boxes, and then rapidly put the lids on the boxes (see Note 5).
13. Prepare a 10% (v/v) solution of ice-cold acetaldehyde in ice-cold ethanol, lift one corner of the lid, and add 2 μl of the dilution to one box. Close lid immediately. Avoid direct contact of the acetaldehyde with the plastic box by placing a piece of aluminum foil in the box and pipetting the acetaldehyde onto the foil to avoid possible damage to the box. Put the boxes back into the incubator for another 48 h. The acetaldehyde will evaporate, dissolve in the cell culture medium, and induce the P_{AIR} promoter in the presence of AlcR. The box prevents acetaldehyde from circulating within the incubator and cross-inducing other acetaldehyde-responsive cell populations. Use cooled (−20°C) pipette tips to pipette the acetaldehyde.
14. Analyze the cells for acetaldehyde-responsive SEAP production.
    (a) Remove 200 μl of medium and transfer the sample to a well of a 96-well plate.
    (b) Seal the plate with adhesive tape or foil and place in a water bath (65°C) for 30 min. Avoid contact of the water with the tape to prevent water from seeping into the plate. This heat treatment inactivates nonspecific phosphatases produced by CHO cells; SEAP is not affected by temperature. If heat inactivation is too short, then an unspecific SEAP signal will result.
    (c) Centrifuge the sealed 96-well plate for 2 min at $500 \times g$ to pellet potential cell debris and to spin the condensate water on the tape into the well.
    (d) Pipette 80 μl of supernatant into a well of a new 96-well plate. Add 100 μl of 2× SEAP buffer. Add one well containing medium only as a negative control.
    (e) Program the multiwell-plate reader for absorbance at 405 nm in kinetic mode with one measurement every 30 s for 1 h and heat the reader to 37°C.
    (f) Pipette 20 μl of pNPP solution per sample-containing well of the 96-well plate and immediately start the measurements in the multiwell-plate reader. pNPP will be cleaved by SEAP to *p*-nitrophenolate (pNP) and absorbs light at 405 nm (8).
    (g) Determine the increase in optical density per minute within the linear part of the absorbance time course and calculate SEAP according to Lambert–Beer's law $c = E/(a \times d)$, where $a = 18{,}600$/M/cm; c = increase in pNP concentration/min [M/min], d = length of the light path in the liquid [cm] (typically 0.5 cm in this 96-well

set-up), and E is the increase in optical density per minute. The activity in units per liter [U/L; µmol/min/L] is calculated as follows: Activity [U/L] = $c \times 10^6 \times 200/80$ (dilution factor). In addition to SEAP, a variety of other mammalian cell-compatible reporter genes is available (8–10).

(h) Figure 2a shows a typical distribution of acetaldehyde-responsive SEAP production.

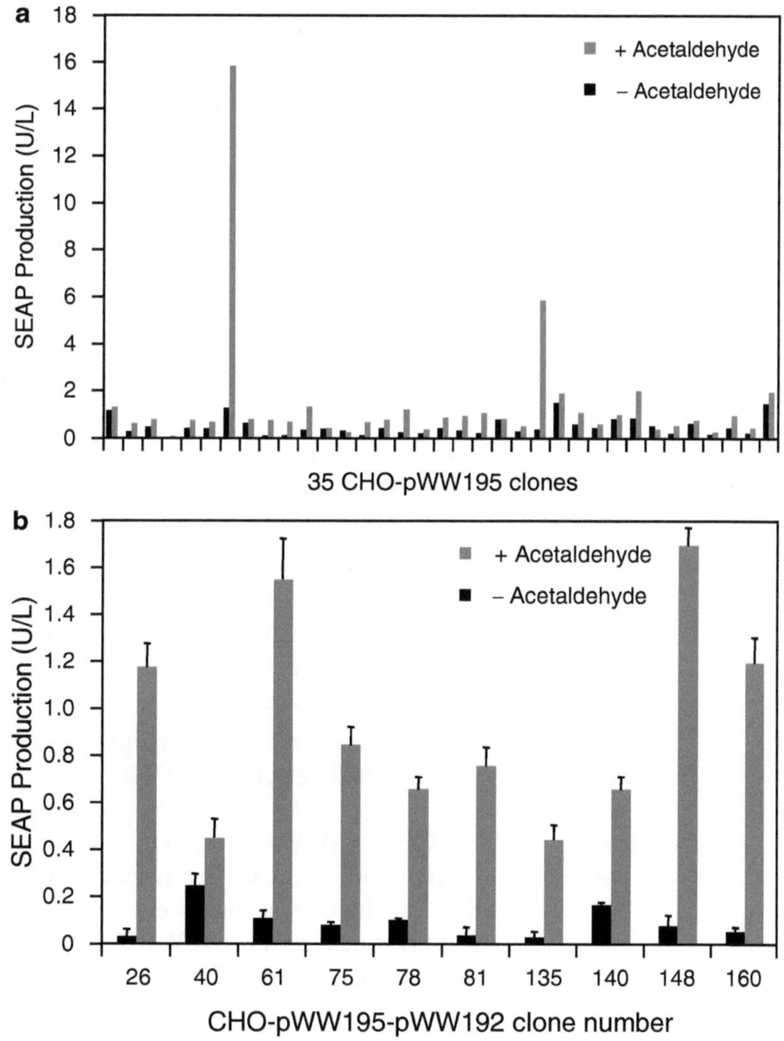

Fig. 2. Screening for receiver cell clones. (**a**) Screening for transactivator cell clones CHO-pWW195. 35 CHO-pWW195 clones were seeded in 24-well plates (50,000 cells per well according to Subheading 3.1.3, steps 1–14), transfected with plasmid pWW192 and cultivated in the presence or absence of acetaldehyde (Subheading 3.1.3, step 13) for 48 h before quantification of SEAP production. (**b**) Validation of receiver cells CHO-pWW195-pWW192. Ten CHO-pWW195-pWW192 clones were seeded in 24-well plates (50,000 cells per well) and cultivated in the presence or absence of acetaldehyde for 48 h before the quantification of SEAP production. The error bars indicate the standard deviation of triplicate cultures (see Subheading 3.1.4, steps 1–6).

15. Select five clones with high SEAP expression in the presence of acetaldehyde and low/no expression in its absence.
16. Expand the five clones, starting from the corresponding maintenance cultures in the 6-well plate (step 10 above) in 10-cm Petri dishes. Aspirate the medium, wash once with 2 ml of prewarmed PBS, add 300 μl of trypsin solution to each well, incubate for 5–10 min, gently tap the plate to detach the cells, and transfer the cells to a 10-cm Petri dish containing 10 ml of HTS complete medium supplemented with 400 μg/ml neomycin. Incubate the plates until the cells reach near 90% confluence.
17. Validation of the cell clones.
    (a) Detach and count the cells from each clone. Aspirate the medium, wash once with 5 ml of prewarmed PBS, add 2 ml of trypsin solution, place in an incubator for 5–10 min, tap the plate gently to detach the cells, transfer to a 15-ml Falcon tube with 5 ml of HTS complete medium, centrifuge for 3 min at 300 × *g*, aspirate and discard the supernatant, resuspend in 5 ml of complete HTS medium supplemented with 400 μg/ml neomycin, and count the cells.
    (b) Seed 100,000 cells per clone in a 10-cm Petri dish in 10 ml of HTS complete medium supplemented with 400 μg/ml neomycin as the maintenance culture.
    (c) Per cell clone seed three wells each of two 24-well plates and perform transfection with pWW192 and subsequent SEAP analysis as described above (steps 10–14) to confirm the acetaldehyde-inducible expression characteristics in triplicate.
18. Cryopreserve the remaining cells of the cell clones (Subheading 3.1.2, step 10).

3.1.4. Stable Transfection of CHO-pWW195 Cells with pWW192

The CHO-pWW195 cell clone with the best induction ratio (SEAP production in the presence of acetaldehyde divided by SEAP production in the absence of the inducer) will subsequently be stably transfected with pWW192 harboring the AlcR-responsive promoter to control SEAP production. Puromycin (encoded on plasmid pPUR) will be used as the second selection marker.

1. Perform transfection of one CHO-pWW195 cell clone with plasmids pWW192 and pPUR according to Subheading 3.1.1, steps 1–4.
2. After transfection, wait for 72 h (see Note 6).
3. Perform selection in complete HTS medium supplemented with 400 μg/ml neomycin and 10 μg/ml puromycin as described above (Subheading 3.1.1). Perform single-cell

cloning, cell expansion, and seed in a 6- and a 24-well plate as described above (Subheading 3.1.1, step 5 to Subheading 3.1.3, step 7), with the exception that the medium is additionally supplemented with 10 μg/ml puromycin.

4. Place 24-well plates in plastic boxes, and one box is supplemented with acetaldehyde as described in Subheading 3.1.3, step 13. After 48 h, SEAP production is quantified in the supernatant of cultures grown in the presence or absence of acetaldehyde (Subheading 3.1.3, step 14). Expand the five clones with the best induction factors in 10-cm Petri dishes.
5. The acetaldehyde-inducible SEAP production profiles are validated in triplicate and frozen for long-term storage (Subheading 3.1.3, steps 17 and 18). The resulting receiver cell line is referred to as CHO-pWW195-pWW192.
6. Figure 2b gives a typical distribution of the regulation characteristics.

3.2. Construction of the Sender Cell Line CHO-ADH

The sender cell line is metabolically engineered for the expression of mouse liver alcohol dehydrogenase (ADH) for the enzymatic conversion of ethanol into acetaldehyde. Therefore, the plasmids pIVL-2 (P_{hCMV}-*adh*-pA) providing constitutive ADH expression will stably be transfected into CHO-K1 cells. As selection marker, the puromycin resistance-conferring plasmid pPUR will be cotransfected. The resulting cell line is referred to as CHO-ADH.

3.2.1. Stable Transfection of CHO-K1 Cells with pIVL-2 and Clonal Expansion

1. Perform transfection of CHO-K1 cells with pIVL-2 and pPUR according to Subheading 3.1.1, steps 1–4, by replacing pWW195 with pIVL-2 and pSV2*neo* with pPUR. Wait 72 h after transfection (see Note 6) and then start selection and subsequent cultivation in complete HTS medium supplemented with 10 μg/ml puromycin instead of 400 μg/ml neomycin according to the protocol described above (Subheading 3.1.1, step 5 to Subheading 3.1.2, step 10).
2. Select 30 clones and expand them in 6-well plates to near 90% confluence (see Subheading 3.1.3, steps 1–6). Thirty clones will probably be enough, since screening is for expression only, not to determine regulation performance.

3.2.2. Screening for ADH-Producing Cell Clones

Screen for ADH production by cultivating the CHO-ADH clones in the presence of CHO-pWW195-pWW192 cells in ethanol-containing medium. ADH production will convert ethanol into acetaldehyde, which will subsequently activate SEAP production (see Note 7 for alternative screening methods).

1. Seed CHO-ADH cell clones from the 6-well plates (Subheading 3.2.1, step 2).
   (a) Aspirate the medium and wash the cells with 2 ml of prewarmed PBS.
   (b) Add 500 μl of trypsin and place the plates in an incubator for 5–10 min; gently tap the plates to detach the cells.
   (c) Transfer the cells to a 15-ml Falcon tube containing 5 ml of complete HTS medium supplemented with 10 μg/ml puromycin, centrifuge for 3 min at 300 × *g*, aspirate, and discard the supernatant.
   (d) Resuspend the cells in 5 ml of complete HTS medium supplemented with 10 μg/ml puromycin and count the cells.
   (e) For each clone, seed 50,000 cells per well in 2 ml of complete HTS medium supplemented with 10 μg/ml puromycin in one well of a 6-well plate (as maintenance culture) and in one 35-mm cell culture dish.
2. Add 50,000 CHO-pWW195-pWW192 cells to each 35-mm cell culture dish containing the CHO-ADH clones (according to step 1 above). Complete total volume to 3 ml.
3. Add 3 μl ethanol to each 35-mm cell culture dish.
4. Place the 35-mm cell culture dishes in the incubator and cultivate for 48 h. The dishes should be at least 10 cm apart to ensure that acetaldehyde that is produced in one does not diffuse to the next dish and induce SEAP production.
5. Perform the SEAP test on medium from the 35-mm cell culture dishes as described in Subheading 3.1.3, step 14. Select the three CHO-ADH clones with the highest SEAP production among the CHO-pWW195-pWW192 cells.
6. Expand the selected CHO-ADH clones in 10-cm Petri dishes starting from the maintenance culture in the 6-well plate (Subheading 3.2.2, step 1e); freeze the cells for long-term storage (Subheading 3.1.2, step 10).

3.3. Mammalian Cell-Based Quorum-Sensing

The sender and receiver cell lines are now ready to set up the cell-to-cell communication configurations, controlled by cell density, time, or distance (4). As an example, we describe here a protocol for the design of a mammalian cell-based quorum-sensing system that mediates cell density-dependent transgene expression.

1. Add 5×10^5 CHO-pWW195-pWW192 cells in 5 ml of complete HTS medium to each of eight cell culture flasks (seed according to Subheading 3.2.2, step 1).
2. Add 0, 5×10^3, 10^4, 2×10^4, 5×10^4, 10^5, 2×10^5, and 5×10^5 CHO-ADH to each flask. Add complete HTS medium to adjust the total volume of each flask to 10 ml.

3. Add 10 μl of ethanol to each flask. Adjust the lid to the "vented" position. Close lid after 3–4 h. This ensures that the flask is equilibrated with the 5% CO_2-containing humid atmosphere of the incubator and avoids seepage of acetaldehyde into the incubator.
4. Place the flasks in the incubator and take samples of the medium (150 μl per flask) twice a day for 72 h.
5. After sampling, freeze the samples at −20°C until analysis.
6. After the end of the experiment, quantify SEAP in all samples (Subheading 3.1.3, step 14).
7. Blot SEAP production in U/L versus time. In this configuration, the onset of gene expression will depend on the cell density of the CHO-ADH sender cells. A smaller number of CHO-ADH cells will result in the later onset of gene expression, since the accumulation of expression-inducing acetaldehyde is slower than when the density of the CHO-ADH cells is high resulting in the almost immediate onset of SEAP production. Figure 3 shows a representative graph of cell density-dependent gene expression from a coculture of increasing densities of CHO-ADH cells with a receiver cell engineered for acetaldehyde-inducible expression of β-interferon instead of SEAP (4).

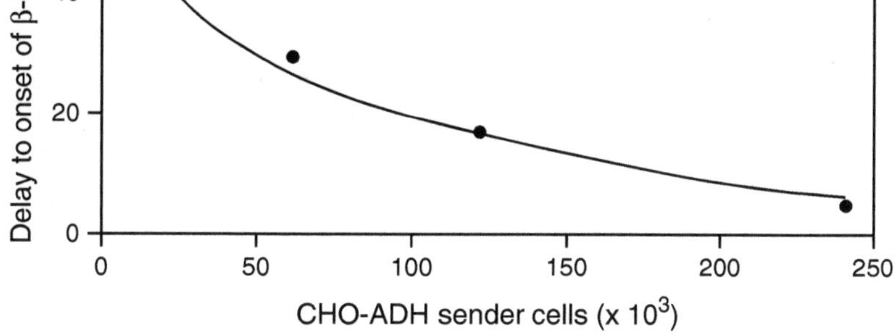

Fig. 3. Mammalian cell quorum sensing. 500,000 $_{AIR}$CHO-IFN cells, transgenic for acetaldehyde-inducible expression of β-interferon (12), were cultivated in 10 ml of medium containing 1‰ ethanol and supplemented with increasing numbers of CHO-ADH sender cells analog to Subheading 3.3. β-interferon production was followed over time, and the onset of β-interferon production was determined and plotted against the number of CHO-ADH sender cells.

4. Notes

1. The acetaldehyde-inducible promoter P_{AIR} was incorporated into flexible vector backgrounds, enabling the simultaneous expression of multiple genes by bidirectional or multicistronic vectors in a conventional plasmid-based format or as lentiviral vectors, thereby enabling the straightforward construction of receiver cell lines (5, 11, 12).
2. In principle, the transactivator-(pWW195) and promoter-(pWW192) containing vectors can be transfected together with a resistance plasmid and selected for simultaneous integration into the chromosome. However, based on our experience with related inducible expression systems, the induction factor of the resulting cell clones is lower than when sequential transfection is performed. A possible explanation of this phenomenon is that, upon cotransfection, both plasmids probably integrate into the same chromosomal locus (13). The strong constitutive promoter, driving expression of the transactivator, then cross-activates the minimal promoter in the inducible expression construct and causes rather high leaky expression.
3. The protocols provided here can be adapted to other cell types and other selection markers. Since different cell lines show different susceptibility to antibiotic selection, it is advisable to perform an antibiotic susceptibility test before starting construction of stable cell lines. Testing antibiotic susceptibility can be performed by cultivating 5×10^4 cells per well of a 6-well plate in 2 ml of medium supplemented with increasing concentrations of the selective agent (e.g., for neomycin: 0, 10, 20, 40, 80, 160, 320, 640 μg/ml). The medium is changed every 3 days (or earlier if the medium turns orange) for 7–14 days. The selective antibiotic concentration is the one that is lethal to all the cells.
4. As well as by limiting dilution (see Subheading 3.1.2), single-cell clones can be generated by fluorescence-activated cell sorting. This is of special interest if the target gene encodes a fluorescent protein, thereby enabling the rapid selection of positive clones. A detailed protocol for single-cell cloning by FACS is described elsewhere (14).
5. Given the gaseous nature of the signaling molecule acetaldehyde, the intensity of the signal depends on the cultivation volume. Therefore, all experiments should be conducted in closed polypropylene (or a material that is resistant to acetaldehyde) boxes of defined volume. Acetaldehyde concentrations in the gas phase can be monitored using the Gastec gas analyzing system (Gastec, Kanagawa, Japan, cat. no. 92M).

6. Based on our experience, selection with puromycin should be started 72 h posttransfection so that the cells have enough time to produce the resistance protein. Earlier onset of puromycin selection led to significantly lower number of clone numbers. For neomycin, selection can be initiated 48 h after transfection.
7. Screening for alcohol dehydrogenase expression can also be performed by cultivating the different clones in the presence of 25 μM allyl alcohol. Oxidation of allyl alcohol by ADH results in the production of toxic acrolein; therefore, cytotoxicity indicates that a given clone is ADH positive (6).

Acknowledgment

We thank Marcia Schoenberg for critical comments on the manuscript. This work was supported by the Swiss National Science Foundation (grant no. 3100A0-112549).

References

1. You, L., Cox, R. S., III, Weiss, R., and Arnold, F. H. (2004) Programmed population control by cell–cell communication and regulated killing. *Nature* **428**, 868–71.
2. Basu, S., Gerchman, Y., Collins, C. H., Arnold, F. H., and Weiss, R. (2005) A synthetic multicellular system for programmed pattern formation. *Nature* **434**, 1130–4.
3. Balagadde, F. K., Song, H., Ozaki, J., Collins, C. H., Barnet, M., Arnold, F. H., Quake, S. R., and You, L. (2008) A synthetic *Escherichia coli* predator-prey ecosystem. *Mol Syst Biol* **4**, 187.
4. Weber, W., Daoud-El Baba, M., and Fussenegger, M. (2007) Synthetic ecosystems based on airborne inter- and intrakingdom communication. *Proc Natl Acad Sci U S A* **104**, 10435–40.
5. Weber, W., Rimann, M., Spielmann, M., Keller, B., Daoud-El Baba, M., Aubel, D., Weber, C. C., and Fussenegger, M. (2004) Gas-inducible transgene expression in mammalian cells and mice. *Nat Biotechnol* **22**, 1440–4.
6. Clemens, D. L., Forman, A., Jerrells, T. R., Sorrell, M. F., and Tuma, D. J. (2002) Relationship between acetaldehyde levels and cell survival in ethanol-metabolizing hepatoma cells. *Hepatology* **35**, 1196–204.
7. Doyle, A., Griffiths, J. B., and Newell, D. G. (1995) *Cell & Tissue Culture: Laboratory Procedures*, Vol. Update 9, Wiley, Hoboken, NJ.
8. Schlatter, S., Rimann, M., Kelm, J., and Fussenegger, M. (2002) SAMY, a novel mammalian reporter gene derived from *Bacillus stearothermophilus* alpha-amylase. *Gene* **282**, 19–31.
9. Fluri, D. A., Kelm, J. M., Lesage, G., Baba, M. D., and Fussenegger, M. (2007) InXy and SeXy, compact heterologous reporter proteins for mammalian cells. *Biotechnol Bioeng* **98**, 655–67.
10. Fluri, D. A., Kemmer, C., Daoud-El Baba, M., and Fussenegger, M. (2008) A novel system for trigger-controlled drug release from polymer capsules. *J Control Release* **131**, 211–9.
11. Hartenbach, S., and Fussenegger, M. (2005) Autoregulated, bidirectional and multicistronic gas-inducible mammalian as well as lentiviral expression vectors. *J Biotechnol* **120**, 83–98.
12. Weber, W., Rimann, M., de Glutz, F. N., Weber, E., Memmert, K., and Fussenegger, M. (2005) Gas-inducible product gene expression in bioreactors. *Metab Eng* **7**, 174–81.
13. Derouazi, M., Martinet, D., Besuchet Schmutz, N., Flaction, R., Wicht, M., Bertschinger, M., Hacker, D. L., Beckmann, J. S., and Wurm, F. M. (2006) Genetic characterization of CHO production host DG44 and derivative recombinant cell lines. *Biochem Biophys Res Commun* **340**, 1069–77.
14. Weber, W., and Fussenegger, M. (2004) Inducible gene expression in mammalian cells and mice. *Methods Mol Biol* **267**, 451–66.

Part III

Quorum Sensing Disruption Strategies

Chapter 18

Qualitative and Quantitative Determination of Quorum Sensing Inhibition In Vitro

Tim Holm Jakobsen, Maria van Gennip, Louise Dahl Christensen, Thomas Bjarnsholt, and Michael Givskov

Abstract

The formation of biofilms in conjunction with quorum sensing (QS)-regulated expression of virulence by opportunistic pathogens contributes significantly to immune evasion and tolerance to a variety of antimicrobial treatments. The present protocol describes methods to determine the in vitro efficacy of potential quorum sensing inhibitors (QSIs). Work on *Pseudomonas aeruginosa* has shown that chemical blockage of QS is a promising new antimicrobial strategy. Several live bacterial reporter systems been developed to screen extracts and pure compounds for QSI activity. Here we describe the usage of reporter strains consisting of a *lasB-gfp* or *rhlA-gfp* fusion in *P. aeruginosa* for qualitative and quantitative evaluation of the inhibition of the two major QS pathways, monitored as reduced expression of green fluorescence. By the use of an in vitro flow cell system it is possible to study the QSI activity by monitoring its ability to interfere with the protective functions of bacterial biofilm. For evaluation of the global effects of QSI compounds, we present a protocol for the DNA microarray-based transcriptomics. Using these in vitro methods it is possible to evaluate the potential of various QSI compounds.

Key words: Quorum sensing inhibitor, Halogenated furanones, QSI monitor screen, DNA microarray, In vitro continuous-culture biofilm flow cell system, Confocal Scanning Laser Microscopy

1. Introduction

Treatment of infectious diseases is becoming increasingly more difficult as bacterial resistance to antibiotics evolves with an alarming rate. Several opportunistic pathogens rely on biofilm formation and quorum sensing (QS)-controlled expression of virulence factors in the process of establishing persistent infections in humans and animals. Both processes help the bacteria against the host defense and otherwise detrimental effects of antimicrobial drug treatments (1–4). However, research has shown that

Kendra P. Rumbaugh (ed.), *Quorum Sensing: Methods and Protocols*, Methods in Molecular Biology, vol. 692,
DOI 10.1007/978-1-60761-971-0_18,

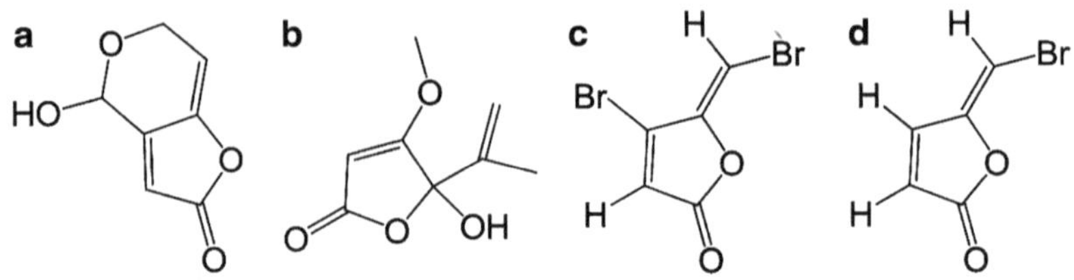

Fig. 1. Chemical structures of QS inhibitors produced by *Penicillium* species (**a** and **b**) and synthetic halogenated furanones (**c** and **d**). (**a**) Patulin, (**b**) penicillic acid, (**c**) furonone C30, and (**d**) furanone C56.

administration of QS inhibitors (QSIs) can block QS-controlled phenotypes including production of virulence factors (5, 6) and thereby function as antimicrobials (6, 7).

The first discovery of environmental QSIs came from the Australian red macroalga *Delisea pulchra*, which was found to produce an antifouling cocktail of halogenated furanone compounds of which some were able to inhibit QS (8) (see Fig. 1). To increase the effect against the opportunistic human pathogen *Pseudomonas aeruginosa*, synthetic derivatives were synthesized. Two such halogenated furanone compounds, C30 (6) and C56 (5, 7), showed significant activity against QS in *P.aeruginosa*. The discovery of the ability of halogenated furanones to inhibit QS gave rise to the idea that other compounds with similar capability could be found in a variety of ecological niches. Our further search revealed that QSIs can be extracted from many different natural sources including fungi (9), herbs (10, 11), corals, and sponges (12). A multitude of synthetic QSI compounds have been described in the following articles (13–19).

The present chapter describes methods we have invented and routinely use to identify and test the potency of QSIs for their ability to attenuate transcription of the bulk of QS-controlled genes in *P. aeruginosa*. We present simple, quantitative LasR- and RhlR-controlled monitor systems (5, 20) as well as a qualitative method to identify interference or disruption of important biofilm phenotypes. Finally, the chapter includes a description of how to evaluate the target specificity of QSI compounds using DNA microarray-based transcriptomics.

2. Materials

2.1. QS Monitor Screens

1. Bacterial monitor strains: LasB-GFP, *P. aeruginosa* (PAO1-ATCC) harboring a *lasB-gfp*(ASV) fusion together with P_{lac}-*lasR*-mini-Tn5 inserted upstream to enhance the sensitivity (5), and RhlA-GFP, *P. aeruginosa* (PAO1-ATCC) harboring

a *rhlA-gfp*(ASV) fusion together with P_{lac}-*lasR*-mini-Tn5 inserted upstream to enhance the sensitivity (20).

2. Black 96-well microtiter plates (Nunc): Black microtiter plates are used to hinder interference between the wells.
3. To measure fluorescence and growth of monitor strains, a VICTOR™ X multi label plate reader (Perkin Elmer) is used. Growth is measured as an optical density at 450 nm. Fluorescence is measured as excitation and emission wavelength at 485 and 535 nm, respectively.
4. AB minimal growth medium: Add 1 mM $MgCl_2$, 0.1 mM $CaCl_2$, and 0.01 mM $FeCl_3$ to Milli-Q water and sterilize the solution by autoclaving. Thereafter, add 10% A10 (see description below). Store at 4°C.
5. A10: Add 20 g of $(NH_4)_2SO_4$, 60 g of Na_2HPO_4, 30 g of KH_2PO_4, and 30 g of NaCl to 1 L Milli-Q water. Adjust pH to 6.4 and sterilize solution by autoclaving. Store at 4°C for several months.
6. Glucose.
7. Bacto™ Casamino Acids (Becton Dickinson).

2.2. In Vitro Continuous-Culture Biofilm Flow Cell System

1. Bacterial strains: *P. aeruginosa* (PAO1-ATCC) obtained from the *Pseudomonas* Genetic Stock Center (http://www.pseudomonas.med.ecu.edu, strain PAO0001), and *P. aeruginosa* (PAO1-ATCC) constitutively expressing a stable GFP (21).
2. AB trace minimal growth medium: Add 1 mM $MgCl_2$, 0.1 mM $CaCl_2$, and 100 μL/L trace metals (*see* description below) to Milli-Q water and sterilize the solution by autoclaving. Thereafter, add 10% A10 (*see* description below). Store at 4°C.
3. Trace metal solution: 200 mg/L $CaSO_4 \times 2H_2O$, 200 mg/L $FeSO_4 \times 7H_2O$, 20 mg/L, $MnSO_4 \times H_2O$, 20 mg/L $CuSO_4 \times 5\ H_2O$, 20 mg/L $ZnSO_4 \times 7H_2O$, 10 mg/L $CoSO_4 \times 7H_2O$, 12 mg/L $Na_2MoO_4 \times H_2O$, and 5 mg/L H_3BO_3. It can be stored at room temperature for several months. Remember to mix before use.
4. A10: Add 20 g of $(NH_4)_2SO_4$, 60 g of Na_2HPO_4, 30 g of KH_2PO_4, and 30 g of NaCl to 1 L Milli-Q water. Adjust pH to 6.4 and sterilize solution by autoclaving. Store at 4°C.
5. 0.3 mM Glucose.
6. Bacto™ Casamino Acids (Becton Dickinson).
7. 16-channel Watson Marlow 205S peristaltic pump.
8. Equipment required to assemble the system:
   (a) Silicon tubing, int. diameter: 2 mm, ext. diameter: 4 mm (VWR) (labeled 'A' in Fig. 2).

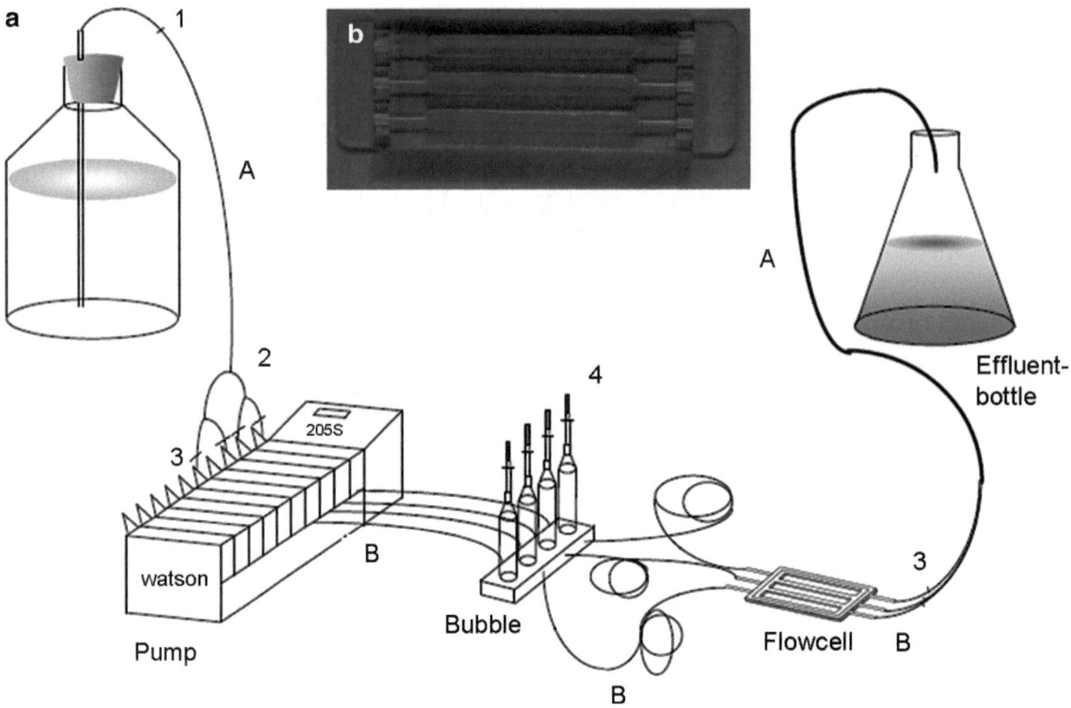

Fig. 2. (**a**) *In vitro* continuous-culture biofilm flow cell system setup. The digits and letters refer to the designation in the material section of the different connectors and tubes. (**b**) Three-channel flow cell, in which the biofilms are cultivated.

(b) Silicon tubing, int. diameter: 1 mm, ext. diameter: 3 mm (VWR) (labeled 'B' in Fig. 2).

(c) Straight connectors 1/8 in. (Buch Holm) (labeled '1' in Fig. 2).

(d) T-connectors 1/8 in. (Buch Holm) (labeled '2' in Fig. 2).

(e) Reducing connectors 1/8 in. × 1/16 in. (Buch Holm) (labeled '3' in Fig. 2).

(f) Barrel tip cap orange (Diatom) (labeled '4' in Fig. 2).

9. Silicone to assemble the coverslips with the flow chambers.
10. Surface-attached biofilms are cultivated in flow chambers with channel dimensions of 1 by 4 by 40 mm.
11. The substratum consists of a 24 mm by 50 mm microscope coverslip.
12. In order to sterilize biofilm system: Use Sodium hypochlorite (NaClO) solution (13%) – Bleach and dilute it in water to 0.5%. Be aware that NaClO is alkaline and may cause skin irritation; therefore, wear protective clothes, gloves, and safety glasses.

2.3. Confocal Scanning Laser Microscopy

Image acquisitions of biofilms are performed using a Confocal Scanning Laser Microscope (CSLM). We currently use a Leica SP5 (Leica Lasertechnik, GmbH, Heidelberg, Germany) equipped with a detector and a filter set for monitoring GFP and propidium iodide (PI). In addition, a reflection detector for acquiring bright-field images is installed. We recommend images to be obtained using a 40–100× oil objective. Image scanning is carried out with the 488 nm laser line from an Ar/Kr laser. Imaris software package (Bitplane, AG) can be used to generate pictures of the biofilm.

2.4. Culture Preparation for DNA Microarray

1. *Bacterial strain: P. aeruginosa* (PAO1-ATCC) obtained from the *Pseudomonas* genetic stock center (http://www.pseudomonas.med.ecu.edu, strain PAO0001).
2. AB minimal growth medium: Add 1 mM $MgCl_2$, 0.1 mM $CaCl_2$, and 0.01 mM $FeCl_3$ to Milli-Q water and sterilize the solution by autoclaving. Thereafter, add 10% A10 (see description below). Store at 4°C.
3. A10: Add 20 g of $(NH_4)_2SO_4$, 60 g of Na_2HPO_4, 30 g of KH_2PO_4, and 30 g of NaCl to 1 L Milli-Q water. Adjust pH to 6.4 and sterilize solution by autoclaving. Store at 4°C.
4. Bacto™ Casamino Acids (Becton Dickinson).
5. RNAlater is used to stabilize and protect RNA (Ambion).

3. Methods

We have developed a simple assay (5), which is able to detect the inhibition of either the *las-* or the *rhl*-encoded QS systems in *P. aeruginosa*. The monitor systems are constructed by fusing an unstable version of green fluorescence protein (GFP) (22) to the QS-controlled *lasB* and *rhlA* promoters in a wild-type background of *P. aeruginosa*. These monitors switch on expression of GFP in QS-dependent manner in batch cultures of *P. aeruginosa*, typical in late exponential or early stationary phases of growth. Hence, administration of a QSI compound to the growth medium will result in reduced expression of green fluorescence compared with the untreated batch culture. However, compounds inhibiting or reducing growth of the monitor strains will also affect the fluorescence. Therefore, to prevent scoring false positives, growth should be measured simultaneously. We calculate the specific activity of GFP expression as change in GFP expression per time unit divided by change in OD_{450} per unit time. Reduction in specific GFP expression and unaffected growth rate indicates the presence of a functional QSI compound. These screens are based on transcription of only two QS-regulated genes. There is a minimum of 170 QS-controlled genes. Consequently, only DNA microarray or "deep sequencing" technologies gives the opportunity to monitor changes in transcription of the entire bacterial genome and thereby gain a more specific knowledge of the target specificity. Furthermore, it is important to test the efficacy of a possible QSI in regard to bacteria living in a biofilm. The in vitro continuous-culture flow cell system (23) makes it possible to follow and investigate biofilm development receiving fresh nutrients continually. The continuous-culture biofilm flow cell system is perfect for visual inspection of formation, disruption, and killing of biofilms using CSLM. The biofilm is monitored using either GFP-tagged bacterial cells or SYTO9 as a stain. The killing of the bacteria is monitored using PI, which will stain the DNA of cells with impaired membrane, i.e., dead cells.

3.1. QS Monitor Assays

1. Inoculate the two monitor strains (*lasB*-GFP or *rhlA*-GFP) in 2× 10 ml ABT media supplemented with 0.5% glucose and 0.5% casamino acids.
2. Incubate cultures overnight at 30°C with shaking at 180 rpm.
3. Add 150 μL AB medium containing 0.5% glucose and 0.5% casamino acid to all wells in a black 96-well microtiter dish.
4. Add 150 μL of test sample (possible QSI) to the first row of microtiter dish (see Note 1).
5. Make a twofold serial dilution from row 1 to 11. No test sample/QSI are added to row number 12, which works as a reference to confirm the growth of the monitor strain and compare the fluorescence.
6. The overnight culture (monitor strain) is diluted in 0.9% NaCl to an optical density of 0.2 at 450 nm (OD_{450}).
7. Add 150 μL of diluted overnight culture to all wells to make a total volume of 300 μl in each well.
8. The microtiter dish is placed in the plate reader (VICTOR™ X multi label plate reader) and measurements are started. The fluorescence (GFP-expression) is measured with an excitation and emission wavelength at 485 nm and 535 nm, respectively, and growth of the bacteria is determined by measuring OD_{450}. Both GFP expression and growth is measured every 15 min for 15 h. The temperature is set at 34°C.

3.2. In Vitro Continuous-Culture Flow Cell System

1. The flow cells are assembled minimum 24 h prior to use by gluing (use silicone) a 24×50 mm glass coverslip onto the top of the flow cell (see Fig. 2b).
2. Assemble the rest of the system as shown in Fig. 2a.

3. The following description is based on a 16-channel pump from Watson Marlow. Sterilize the whole system using 1 L of a 0.5% NaClO solution in sterilized Milli-Q water. The pump is set at 90 rpm to fill the whole system. When the bubble traps are filled, the lids are put on and the pump is set at 12 rpm. Sterilize for approximately 2 h. Make sure that nothing is leaking, and use silicone to stop leaks. Remember to wear gloves and glasses when working with NaClO (see Note 2).
4. Empty the system at 90 rpm.
5. Wash the system with 2 × 1 L of sterilized Milli-Q water the same way as with the NaClO. Set the pump at 50 rpm when the system is filled. The system is emptied after one 1 L and filled up again with the new flask (see Note 2).
6. After the last wash, the system is emptied and filled up with AB trace minimal medium supplemented with 0.3 mM glucose. Place the system at 37°C overnight (see Note 2).
7. Use minimum two medium flasks. Hook for example channels 1–8 up to the medium flask, which contains the test QSI compound, and hook the growth medium flask up to channels 9–16 as the controls (see Note 3). The medium flasks (not the connecting tubings) can be kept on ice to avoid turn over (chemical instability) of the test compound, and flasks with freshly prepared contents can be prepared each day.
8. Grow an overnight culture with *P. aeruginosa* (PAO1) or a GFP-tagged PAO1.
9. Before inoculation with the overnight culture, the flow is stopped and the tubes are clamped off between the flow channels and the bubble traps.
10. Inoculate the flow chambers using 250 μL of the overnight culture diluted to an OD_{600} of 0.1 in 0.9% NaCl. Inject the diluted culture in the flow channels by using a syringe needle, which is inserted in the tubing next to the flow channel inlet. Close the injection hole with a thin layer of silicone.
11. The flow of media is arrested for 1 h to allow efficient colonization of the glass surface. Flip the flow cell upside down, placing it on the glass surface.
12. Flip the flow cells back in "upright" position (glass slides are facing upwards) and start the flow of media. Remove the clamps and let the medium perfuse the flow chambers at a constant rate of 1.75 for a 16-channel pump and 2.00 for a 12-channel pump which corresponds to approximately 3.3 mL/h using the peristaltic pumps.
13. We usually allow the biofilm to develop and mature in the flow chambers for 3 days.

14. At day three, the growth medium is changed to fresh growth medium containing antibiotic and antibiotic with QSI. The medium is changed by (1) stopping the flow, (2) clamp the tube off between the flow cells and bubble traps, (3) empty the bubble traps by pulling the syringe off and quickly again, (4) remove the barrel tip cap and fill the bubble traps with fresh growth medium containing for example antibiotics, e.g., at 90 rpm, (5) when the bubble traps are filled up, stop the system and remove the clamps, (6) start the system again on 1.75 rpm.
15. The biofilm is investigated after approximately 24 h.
16. For examination of the biofilm with CSLM, a viability staining kit is used:
    (a) If a non-GFP-tagged *P. aeruginosa* strain is used, LIVE/DEAD viability staining kit can be applied. SYTO 9 (Invitrogen) and PI are added at a concentration of 0.005 mM and 0.01 mM, respectively, 15 min before examination of the flow cells by injecting them the same way as the bacterial culture, described in step 3.2.10 (see Note 4).
    (b) If a GFP-tagged *P. aeruginosa* strain is used, only PI is added to the flow chambers (see Note 5).

3.3. Confocal Scanning Laser Microscopy

Biofilm formation in flow chambers is examined by CSLM. To visualize live and dead bacteria, SYTO 9 (live) and PI (dead) are used. Figure 3 is an example of 3-day-old biofilms of *P. aeruginosa* (PAO1) grown in flow chambers treated or untreated with tobramycin and garlic extract, which we have shown to inhibit QS in *P. aeruginosa* (11).

3.4. Culture Preparation for DNA Microarray

1. Inoculate *P. aeruginosa* (PAO1) in AB medium supplemented with 0.5% casamino acids.
2. Grow culture at 180 rpm at 37°C exponentially to an OD_{600} of approx. 0.5.
3. Dilute culture to an OD_{600} of 0.05 in 200 mL fresh medium.
4. Culture in a 1,000-mL conical flask at 180 rpm at 37°C.
5. At OD_{600} of 0.5, the culture is split into two 100-mL cultures in 500-mL flasks (see Note 6).
6. From a 100–1,000-fold concentrated stock solution (in the appropriate solvent), add QSI to one culture flask (the concentration must not affect growth rate of *P. aeruginosa*). A similar volume of the solvent used for preparing the stock solution is added to the other culture flask, which then serves as a reference (or no treatment) culture.

Fig. 3. Images of 3-day-old biofilms grown in flow chambers. Biofilms shown in picture **a**–**d** are grown with the following: (**a**) untreated, (**b**) tobramycin for 24 h, (**c**) garlic extract, and (**d**) tobramycin for 24 h + garlic extract. Live bacterial cells appear as *light gray*, whereas dead cells are *dark gray*. Reproduced from (11) with permission from Microbiology.

7. Grow cultures with shaking (180 rpm) at 37°C.
8. Retrieve samples at OD_{600} of 2.0 and immediately add RNAlater (1:2) (see Note 7).
9. Store samples at −20°C if not used the same day.
10. Isolate RNA by using the "RNeasy Mini Purification Kit" (Qiagen).

4. Notes

1. If the added test compound (QSI) is diluted in solvents that affect the growth of the monitor strain, dilute the solution in 0.9% NaCl to a concentration that is not affecting growth of the monitor strain.
2. It is important that there are no bubbles in the flow chambers. If bubbles are present, try to remove them by gently

knocking the flow chamber on the inlet side to the table. Remember to check for bubbles in the flow chamber during the entire experimental period, but after the flow cells are inoculated with bacteria, removal of bubbles should be avoided. If larger bubbles are generated, the biofilm development can be affected, resulting in unusable results. Smaller bubbles can be displaced by the bacteria and will therefore not affect the result.

3. The amount of growth media used is dependent on the number of flow channels used and therefore must be calculated before initiating the experiment to make sure that the system does not exhaust the media.
4. Syto 9 and PI are light-sensitive and have to be covered with aluminum foil. Remember to also cover the flow cells with aluminum foil during the 15 min of staining.
5. It is possible to add PI to the medium from the beginning of the experiment or when the media is changed to contain antibiotics and QSI. If this procedure is used, the final concentration of PI in the media has to be 0.0015 mM. Remember to cover every media-containing part with aluminum foil.
6. The flask has to be warmed in a 37°C incubator before use.
7. It is important that the growth of the treated and untreated culture is similar and that they reach the decided density (OD_{600} = 2.0) at the same time.

Acknowledgments

We acknowledge the contributions made by Drs. Morten Hentzer, Hong Wu, Thomas Bovbjerg Rasmussen, Jens Bo Andersen, and Allan Bech Christensen with regard to the initial developments of the model systems.

References

1. Lewenza, S., Conway, B., Greenberg, E. P. and Sokol, P. A. (1999) Quorum sensing in *Burkholderia cepacia*: identification of the LuxRI homologs CepRI, *J Bacteriol.* **181**, 748–756.
2. Bjarnsholt, T., Jensen, P. O., Burmolle, M., Hentzer, M., Haagensen, J. A., Hougen, H. P., et al. (2005) *Pseudomonas aeruginosa* tolerance to tobramycin, hydrogen peroxide and polymorphonuclear leukocytes is quorum-sensing dependent, *Microbiology.* **151**, 373–383.
3. Costerton, J. W., Stewart, P. S. and Greenberg, E. P. (1999) Bacterial biofilms: a common cause of persistent infections, *Science.* **284**, 1318–1322.
4. de Kievit, T. R. and Iglewski, B. H. (2000) Bacterial quorum sensing in pathogenic relationships, *Infect Immun.* **68**, 4839–4849.
5. Hentzer, M., Riedel, K., Rasmussen, T. B., Heydorn, A., Andersen, J. B., Parsek, M. R., et al. (2002) Inhibition of quorum sensing in *Pseudomonas aeruginosa* biofilm bacteria by a halogenated furanone compound, *Microbiology.* **148**, 87–102.
6. Hentzer, M., Wu, H., Andersen, J. B., Riedel, K., Rasmussen, T. B., Bagge, N., et al. (2003)

Attenuation of *Pseudomonas aeruginosa* virulence by quorum sensing inhibitors, *EMBO J.* **22**, 3803–3815.

7. Wu, H., Song, Z., Hentzer, M., Andersen, J. B., Molin, S., Givskov, M., et al. (2004) Synthetic furanones inhibit quorum-sensing and enhance bacterial clearance in *Pseudomonas aeruginosa* lung infection in mice, *J Antimicrob Chemother.* **53**, 1054–1061.
8. Givskov, M., de Nys, R., Manefield, M., Gram, L., Maximilien, R., Eberl, L., et al. (1996) Eukaryotic interference with homoserine lactone-mediated prokaryotic signalling, *J Bacteriol.* **178**, 6618–6622.
9. Rasmussen, T. B., Skindersoe, M. E., Bjarnsholt, T., Phipps, R. K., Christensen, K. B., Jensen, P. O., et al. (2005) Identity and effects of quorum-sensing inhibitors produced by *Penicillium* species, *Microbiology.* **151**, 1325–1340.
10. Bjarnsholt, T., Jensen, P. O., Rasmussen, T. B., Christophersen, L., Calum, H., Hentzer, M., et al. (2005) Garlic blocks quorum sensing and promotes rapid clearing of pulmonary *Pseudomonas aeruginosa* infections, *Microbiology.* **151**, 3873–3880.
11. Rasmussen, T. B., Bjarnsholt, T., Skindersoe, M. E., Hentzer, M., Kristoffersen, P., Kote, M., et al. (2005) Screening for quorum-sensing inhibitors (QSI) by use of a novel genetic system, the QSI selector, *J Bacteriol.* **187**, 1799–1814.
12. Skindersoe, M. E., Ettinger-Epstein, P., Rasmussen, T. B., Bjarnsholt, T., de Nys, R. and Givskov, M. (2008) Quorum sensing antagonism from marine organisms, *Mar Biotechnol (NY).* **10**, 56–63.
13. Muh, U., Schuster, M., Heim, R., Singh, A., Olson, E. R. and Greenberg, E. P. (2006) Novel *Pseudomonas aeruginosa* quorum-sensing inhibitors identified in an ultra-high-throughput screen, *Antimicrob Agents Chemother.* **50**, 3674–3679.
14. Amara, N., Mashiach, R., Amar, D., Krief, P., Spieser, S. A., Bottomley, M. J., et al. (2009) Covalent inhibition of bacterial quorum sensing, *J Am Chem Soc.* **131**, 10610–10619.
15. Riedel, K., Kothe, M., Kramer, B., Saeb, W., Gotschlich, A., Ammendola, A., et al. (2006) Computer-aided design of agents that inhibit the cep quorum-sensing system of *Burkholderia cenocepacia*, *Antimicrob Agents Chemother.* **50**, 318–323.
16. Persson, T., Hansen, T. H., Rasmussen, T. B., Skinderso, M. E., Givskov, M. and Nielsen, J. (2005) Rational design and synthesis of new quorum-sensing inhibitors derived from acylated homoserine lactones and natural products from garlic, *Org Biomol Chem.* **3**, 253–262.
17. Persson, T., Givskov, M. and Nielsen, J. (2005) Quorum sensing inhibition: targeting chemical communication in gram-negative bacteria, *Curr Med Chem.* **12**, 3103–3115.
18. Hjelmgaard, T., Persson, T., Rasmussen, T. B., Givskov, M. and Nielsen, J. (2003) Synthesis of furanone-based natural product analogues with quorum sensing antagonist activity, *Bioorg Med Chem.* **11**, 3261–3271.
19. Olsen, J. A., Severinsen, R., Rasmussen, T. B., Hentzer, M., Givskov, M. and Nielsen, J. (2002) Synthesis of new 3- and 4-substituted analogues of acyl homoserine lactone quorum sensing autoinducers, *Bioorg Med Chem Lett.* **12**, 325–328.
20. Yang, L., Rybtke, M. T., Jakobsen, T. H., Hentzer, M., Bjarnsholt, T., Givskov, M., et al. (2009) Computer-aided identification of recognized drugs as *Pseudomonas aeruginosa* quorum-sensing inhibitors, *Antimicrob Agents Chemother.* **53**, 2432–2443.
21. Bjarnsholt, T., Jensen, P. O., Burmølle, M., Hentzer, M., Haagensen, J. A., Hougen, H. P., et al. (2005) *Pseudomonas aeruginosa* tolerance to tobramycin, hydrogen peroxide and polymorphonuclear leukocytes is quorum-sensing dependent, *Microbiology.* **151**, 373–383.
22. Andersen, J. B., Sternberg, C., Poulsen, L. K., Bjorn, S. P., Givskov, M. and Molin, S. (1998) New unstable variants of green fluorescent protein for studies of transient gene expression in bacteria, *Appl Environ Microbiol.* **64**, 2240–2246.
23. Christensen, B. B., Sternberg, C., Andersen, J. B., Palmer, R. J., Jr., Nielsen, A. T., Givskov, M., et al. (1999) Molecular tools for study of biofilm physiology, *Methods Enzymol.* **310**, 20–42.

Chapter 19

Custom Synthesis of Autoinducers and Their Analogues

Jun Igarashi and Hiroaki Suga

Abstract

Bacterial quorum sensing (QS) system is a unique target for the development of a new class of drugs that potentially control pathogenicity and attenuate virulence. Thus, it has been of significant interest to discover small organic molecules that modulate QS circuits by competing with the signaling molecules, or so-called autoinducers (AIs), for binding to QS proteins. In this chapter, we summarize synthetic methodology for custom QS agonists and antagonists against the Lux system in Gram-negative bacteria.

Key words: AHL combinatorial libraries, QS agonists, QS antagonists, *gfp*-reporter strains

1. Introduction

In bacterial quorum sensing (QS) systems, the structure of autoinducers (AIs) dictates the signaling pathways used and coordinates gene expression. A furanosyl borate diester (3A-methyl-5,6-dihydro-furo[2,3-D][1,2,3]dioxaborole-2,2,6,6A-tetraol), known as AI-2 (Fig. 1) (1), acts as a common QS signal that plays a role in the cross talk of different species of Gram-negative and occasionally Gram-positive bacteria (2–4). On the other hand, acyl-homoserine lactones (acyl-HSLs, Fig. 1), known as AI-1 signals, have more structural diversity originating from their different acyl chain lengths and oxidative modifications at the C3 position (5, 6). The cognate receptor, a protein from the LuxR family, responds to the specific AI-1 and thus differentiates it from other bacterial AI-1 signals (7). The acyl length can be varied from C4 to C14 with even numbers, and the C3-modification can be varied from a saturated hydrocarbon to 3-(*S* or *R*)-hydroxy and 3-oxo (Fig. 1). It should be noted that since the acyl chain is derived from a part of the fatty acid

Kendra P. Rumbaugh (ed.), *Quorum Sensing: Methods and Protocols*, Methods in Molecular Biology, vol. 692,
DOI 10.1007/978-1-60761-971-0_19, © Springer Science+Business Media, LLC 2011

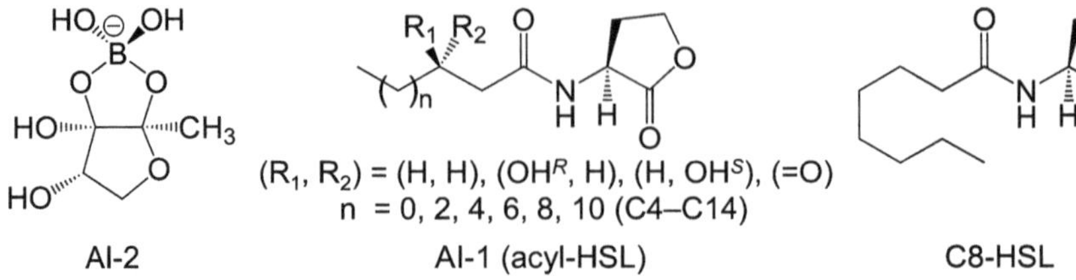

Fig. 1. Generic structures of QS signals. AI-2, AI-1, and C_8-HSL.

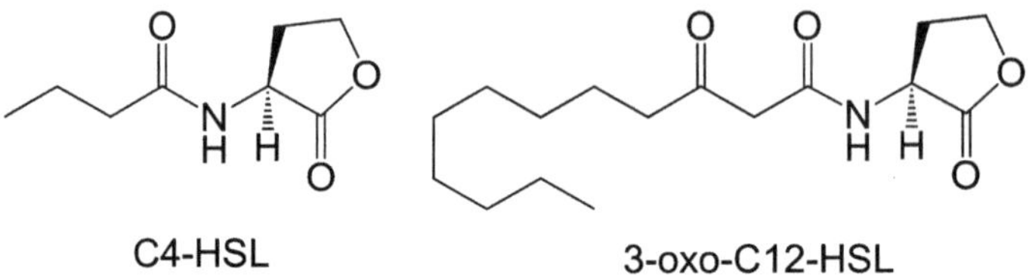

Fig. 2. *P. aeruginosa* QS signals. C_4-HSL and 3-oxo-C_{12}-HSL.

synthesis pathway, odd numbers of the chain length would not be generated (8, 9). It is also noteworthy that the acyl chain on HSLs with over C16 drastically changes its chemical properties, making them prone to aggregate with each other in water, or nonspecifically interact with hydrophobic sites in other proteins or membranes, and thus might not be used as a specific signal. As a result, 24 (6 times 4) variations of AI-1 are possible if all chemical combinations are considered (some additional complexity can be added to this combination when an unsaturated hydrocarbon is placed, but this is a rare case) (10). However, the fact is that less structural variations of AI-1 have been found thus far in bacterial communities and many Gram-negative bacteria share the same structure of AI-1. For instance, C8-HSL (Fig. 1) is a very typical AI-1 found in various species, such as *Burkholderia cepacia*, *Burkholderia glumae*, and *Ralstonia solanacearum*. Thus, some acyl-HSLs could be more widely found than others in certain bacterial communities, possibly acting as cross-species signals.

As aforementioned, the most typical AI-1 is the family of a simple alkanoyl-HSL in which the alkanoyl group is varied from C4 (butyl, Fig. 2) to C14 (myristoyl) (10–14). The collection of the series of synthetic alkanoyl-HSLs helps to elucidate the AI structure(s) when the individual ones are tested in a LuxI (synthase of alkanoyl-HSL)-deficient mutant reporter of interest. Fortunately, all molecules can be prepared by the same procedure involving a single step synthesis from the commercially available materials. On the other hand, a C3-modified AI-1, e.g., 3-oxo-C_{12}-HSL (Fig. 2) that is one of the *P. aeruginosa* AI-1 signals, requires more synthetic steps for the preparation of the acyl chain (2, 3, 15, 16). However, once researchers become familiar with

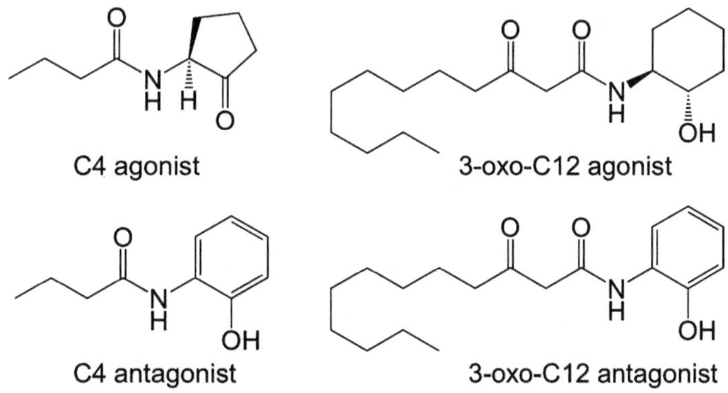

Fig. 3. Synthetic agonists and antagonists of *P. aeruginosa* QS.

the synthetic procedure of the 3-oxo-C_{12} chain, for instance, altering the chain length is fairly easy.

Some attempts have been made to discover synthetic analogs of AI-1 as agonists. Libraries of AI-1 analogs, where the hydrolytically labile HSL is substituted with other chemical substituents, was constructed and screened for specific agonistic activity. For instance, the analogs bearing 2-aminopentanone or 2-aminohexanol (Fig. 3) instead of HSL act as potent agonists (15–17) of *P. aeruginosa* AI-1 molecules. It should be noted that these HSL-substituents do not have the hydrolytically labile lactone structure, and therefore, they are chemically more stable than the natural AI-1 and also resistant against endogenous or exogenous lactonases. A virtue of the design of these synthetic agonists is their versatility, where the acyl chain can be readily altered to a desired length, like the AI-1 chemical synthesis, making custom synthetic agonists.

QS antagonists potentially provide a new class of antibiotics that do not yield resistant strains. Several different approaches have been used to discover antagonists: (1) random screening of natural product or chemical libraries (18, 19), (2) screening of combinatorial AI analog libraries altering the acyl chain to various acyl groups (20–23), and (3) screening of those with altered HSL groups to various amine groups (15–17). Although the respective approaches have advantages and disadvantages, in terms of the simplicity of design and synthesis of a biased library generated on the basis of the antagonist structures found in a bacterium, the third approach could be the most versatile to apply for screening against other bacterial QS systems. As an example, C4 antagonist and 3-oxo-C_{12} antagonist (Fig. 3) were discovered via the third approach (15–17). More recently, a biased library generated from the third approach has led to the discovery of potent antagonists against *B. glumae* (Hwang and Suga et al., unpublished results). Thus, knowledge of the basic synthetic procedures described below will allow researchers to synthesize their desired custom AI-1 analogs and screen them using appropriate reporter strains of target bacteria.

2. Materials (See Notes 1 and 2)

1. Acetyl chloride.
2. Propionyl chloride.
3. Butyryl chloride.
4. Hexanoyl chloride.
5. Heptanoyl chloride.
6. Octanoyl chloride.
7. Nonanoyl chloride.
8. Decanoyl chloride.
9. Lauryl chloride.
10. Myristoyl chloride.
11. p-TsOH, *p*-toluenesulfonic acid monohydrate.
12. EDC *N*-(3-dimethylaminopropyl)-*N'*-ethylcarbodiimide hydrochloride.
13. DMAP, 4-dimethylaminopyrideine.
14. DIPEA, *N*, *N*-diisopropylethylamine.
15. $MgSO_{4,}$ magnesium sulfate anhydrous (see Note 3).
16. $LiOH{\cdot}H_2O$, lithium hydroxide monohydrate.
17. TFA, trifluoroacetic acid.
18. $CDCl_3$, chloroform-d.
19. EM Reagent 0.25 mm silica gel 60-F plates.
20. EM silica gel 60 (230–400 mesh).
21. Finnigan MAT95 XL spectrometer (or other mass spectrometer).
22. Varian Gemini 300, Inova 400 or Inova 500 MHz spectrometer (for NMR).
23. Perkin-Elmer 1760 infrared spectrometer.

3. Methods

3.1. Synthesis of AI-1 Analogs, Acyl-HSL (Fig. 4)

1. Couple an acyl chloride (1.0 eq) with L-homoserine lactone hydrochloride (1.0 eq) and triethylamine (2.2 eq) in dichloromethane on a cool bath (see Note 4).
2. Wash the solution with 0.1 M hydrochloric acid (HCl) and brine.
3. Dry the organic layer over $MgSO_4$ and remove under reduced pressure.

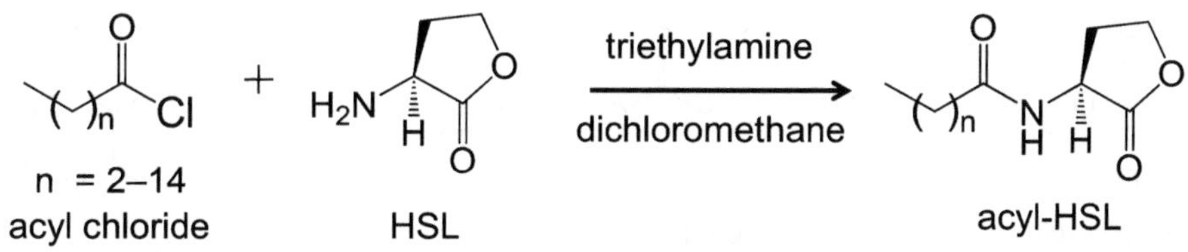

Fig. 4. Generic synthetic scheme for acyl-HSL.

4. Purify the residue by flash chromatography (hexane/ethyl acetate) to give the desired acyl homoserine lactone (acyl-HSL, n = 2–12, yield of 70–85%).
5. The NMR readings are as follows:

   C4-HSL; ^{1}H NMR (500 MHz, $CDCl_3$) δ 0.76 (t, *J* = 7.2 Hz, 3H), 1.46 (q, *J* = 7.2 Hz, 2H), 2.12 (t, *J* = 7.2 Hz, 2H), 2.18 (m, 1H), 2.45 (m, 1H), 4.24 (m, 1H), 4.38 (m, 1H), 4.49 (m, 1H).

   C6-HSL; ^{1}H NMR (500 MHz, $CDCl_3$) δ 0.90 (t, *J* = 7.2 Hz, 3H), 1.28–1.32 (m, 4H), 1.64 (m, 2H), 2.13 (m, 1H), 2.24 (t, *J* = 7.2 Hz, 2H), 2.87 (m, 1H), 4.31 (m, 1H), 4.51 (m, 1H), 4.55 (m, 1H), 6.00 (broad, 1H).

   C8-HSL; ^{1}H NMR (500 MHz, $CDCl_3$) δ 0.88 (t, 3H), 1.28–1.31 (m, 8H), 1.65 (m, 2H), 2.13 (m, 1H), 2.27 (t, 2H), 2.87 (m, 1H), 4.30 (m, 1H), 4.46 (m, 1H), 4.55 (m, 1H), 6.00 (broad, 1H).

   C10-HSL; ^{1}H NMR (500 MHz, $CDCl_3$) δ 0.87 (t, 3H), 1.24–1.28 (m, 12H), 1.61 (m, 2H), 2.11 (m, 1H), 2.24 (t, 2H), 2.86 (m, 1H), 4.28 (m, 1H), 4.46 (m, 1H), 4.52 (m, 1H), 5.97 (broad, 1H).

3.2. Synthesis of AI-1, 3-Oxo-C_{12}-HSL (Fig. 5)

1. Cool a mixture of Meldrum's acid (1.0 eq) and triethylamine (10 eq) in dichloromethane to 0°C.
2. Add decanoyl chloride (2.5 eq) dropwise via an air-tight syringe.
3. Raise to room temperature and continue stirring for 1 h.
4. Wash the solution with 0.1 M HCl, saturated aqueous sodium bicarbonate ($NaHCO_3$), and brine (saturated aqueous sodium chloride).
5. Dry the organic layer over $MgSO_4$ and concentrate (Fig. 5a). Use the resulting product for the next reaction without further purification.
6. Dissolve the residue in ethanol and reflux at 90°C for 3 h.
7. Remove the solvent under reduced pressure.
8. Purify the residue by flash chromatography (hexane/ethyl acetate) to give ethyl 3-oxo-dodecanoate in 70% yield (Fig. 5b).

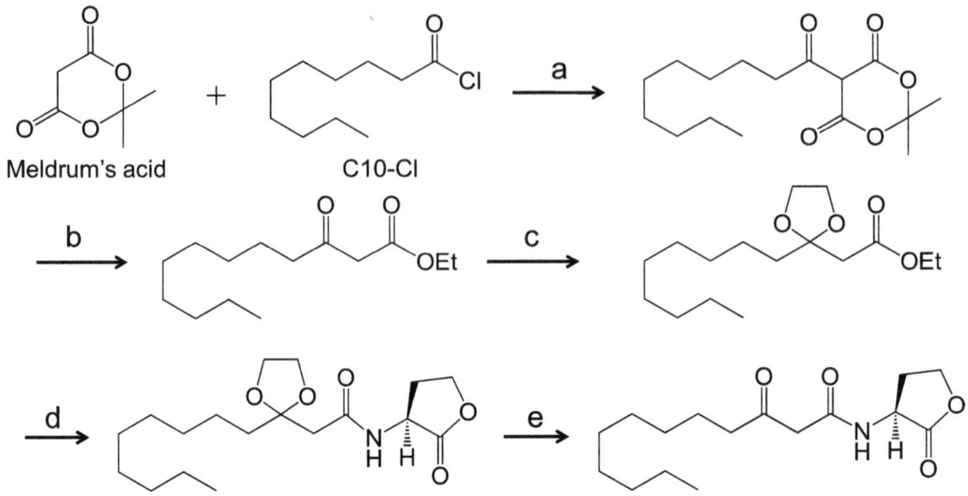

Fig. 5. Synthesis of 3-oxo-C_{12}-HSL. (**a**) triethylamine, dichloromethane, 0°C; (**b**) ethanol, reflux; (**c**) ethylene glycol, p-TsOH, benzene, reflux; (**d**) $LiOH·H_2O$, tetrahydrofuran and water, 75°C followed by HSL,EDC, DMAP, DIPEA, DMF, room temperature; (**e**) TFA, DMF, 0°C.

9. Reflux a mixture of ethyl 3-oxo-dodecanoate (1.0 eq), ethylene glycol (10 eq), and a catalytic amount of p-TsOH in benzene overnight.
10. Remove the solvent under reduced pressure and dilute the residue with ethyl acetate.
11. Wash the solution with saturated aqueous $NaHCO_3$ and brine.
12. Dry the organic phase over $MgSO_4$ and concentrate to give ethyl 3,3-ethylenedioxo-dodecanoate (Fig. 5c). The crude product can be used for the next reaction without further purification.
13. Gently stir a mixture of ethyl 3,3-ethylenedioxo-dodecanoate (1.0 eq) and $LiOH·H_2O$ (100 eq) in a 1:1 ratio of tetrahydrofuran and water and heat at 75°C for 20 h.
14. Remove the solvent under reduced pressure.
15. Purify the residue by flash chromatography (ethyl acetate/methanol) to give 3,3-ethylenedioxo-dodecanoic acid in 60% yield.
16. Stir a mixture of 3,3-ethylenedioxo-dodecanoic acid (1.0 eq), L-homoserine lactone hydrochloride (1.0 eq), EDC (1.1 eq), DMAP (1.0 eq), DIPEA (1.2 eq), and dimethylformamide (DMF) overnight at room temperature.
17. Remove the solvent under reduced pressure and dilute the residue with ethyl acetate.

18. Wash the solution with saturated aqueous $NaHCO_3$ and brine.
19. Dry the organic phase over $MgSO_4$ and concentrate.
20. Purify the residue by flash chromatography (ethyl acetate/methanol) to give 3,3-ethylenedioxo-C_{12}-HSL in 80% yield (Fig. 5d).
21. Dissolve 3,3-ethylenedioxo-C_{12}-HSL in DMF and cool to 0°C.
22. Add approximately 1 ml of TFA and water (9:1) dropwise to the solution with continuous stirring for 1 h.
23. Remove the solvent under reduced pressure and dissolve the residue in ethyl acetate.
24. Wash the solution with saturated aqueous $NaHCO_3$ and brine.
25. Dry the organic layer over $MgSO_4$ and concentrate using a vacuum evaporator.
26. Purify the resulting residue by flash chromatography (ethyl acetate/methanol) to give desired 3-oxo-C_{12}-HSL in 90% yield.
27. The NMR readings are as follows:

    ^{1}H NMR (500 MHz, $CDCl_3$) δ 0.88 (t, *J*=7.2 Hz, 3H), 1.20–1.38 (broad, 12H), 1.50–1.65 (m, 2H), 2.20–2.30 (m, 1H), 2.54 (t, *J*=7.2 Hz, 2H), 2.72–2.84 (m, 1H), 3.47 (s, 2H), 4.24–4.32 (m, 1H), 4.48 (t, *J*=9.2 Hz, 1H), 4.58–4.64 (m, 1H), 7.67 (broad, 1H).

3.3. Synthesis of AI-1 Agonists, N-2-Oxocyclopentyl-butanamide and N-(trans-2-Hydroxycyclohexyl)-3-Oxododecanamide (Fig. 3)

1. Couple butyryl chloride (1.0 eq) with 2-aminocyclopentanol hydrochloride (1.0 eq) and triethylamine (2.2 eq) in dichloromethane to give the desired C4-2-aminocyclopentanonol in 80% yield, as described above in Subheading 3.1.
2. Treat the coupled C4-amino alcohol derivatives with Dess–Martin periodinane to generate the desired *N*-2-oxo-cyclopentylbutanamide in 70–75% yield.
3. The NMR readings are as follows:

    ^{1}H NMR (500 MHz, $CDCl_3$) δ 0.954 (t, J=7.5 Hz, 3H), 1.56–1.70 (m, 3H), 1.82–1.94 (m, 1H), 2.02–2.10 (m, 1H), 2.18–2.26 (m, 3H), 2.38–2.44 (m, 1H), 2.60–2.68 (m, 1H), 4.10–4.20 (m, 1H), 6.00 (s, 1H).
4. Couple *trans*-2-aminocyclohexanol hydrochloride (instead of L-homoserine lactone hydrochloride) with 3,3-ethylenedioxo-dodecanoic acid as described above in Subheading 3.2.3, to give the *N*-(*trans*-2-hydroxycyclohexyl)-3-oxododecanamide in75% yield.

5. The NMR readings are as follows:

   ^1H NMR (500 MHz, $CDCl_3$) δ 0.88 (t, J=7.5 Hz, 3H), 1.18–1.40 (m, 16H), 1.58 (m, 2H), 1.72 (m, 2H), 1.90–2.10 (m, 2H), 2.53 (t, J=7.5, 2H), 3.35 (m, 1H), 3.43 (s, 2H), 3.67 (m, 1H), 7.19 (d, J=5 Hz, 1H).

3.4. Synthesis of AI-1 Antagonists, N-2-Hydroxyphenyl-butanamide, N-(2-Hydroxyphenyl)-3-Oxododecanamide (Fig. 3)

1. Couple butyryl chloride (1.0 eq) with 2-aminophenol (1.0 eq) and triethylamine (2.2 eq) in dichloromethane, as described above in Subheading 3.1, to give the desired *N*-2-hydroxyphenylbutanamide in 85% yield.
2. The NMR readings are as follows:

   ^1H NMR (500 MHz, $CDCl_3$) δ 0.80 (t, J=7.2 Hz, 3H), 1.49 (q, J=7.2 Hz, 2H), 2.15 (t, J=7.2 Hz, 2H), 6.91 (m, 1H), 6.98–7.10 (m, 2H), 7.18 (m, 1H).
3. Couple 2-aminophenol (instead of L-homoserine lactone hydrochloride) with 3,3-ethylenedioxo-dodecanoic acid, as described above in Subheading 3.2, to give *N*-(2-hydroxyphenyl)-3-oxododecanamide in 78% yield.
4. The NMR readings are as follows:

   ^1H NMR (500 MHz, $CDCl_3$) δ 0.91 (t, J=7.2 Hz, 3H), 1.28–1.34 (broad, 12H), 1.65 (m, 2H), 2.50 (t, 2H), 3.53 (s, 2H), 6.91 (m, 1H), 6.98–7.10 (m, 2H), 7.18 (m, 1H), 9.89 (broad, 1H).

4. Notes

1. All chemicals can be purchased from Sigma-Aldrich.
2. All solvents should be of anhydrous grade or distilled before use.
3. Anhydrous $MgSO_4$ was used as a drying agent for all products.
4. All reactions were carried out in oven-dried glassware under an argon atmosphere, except the ones containing water.

Acknowledgments

This work was supported by Japan Science and Technology Innovative Technology Development Fund awarded to H.S. and Otsuka Chemical Corporation, Ltd.

References

1. Chen X, Schauder S, Potier N, Van Dorsselaer A, Pelczer I, Bassler BL et al (2002). Structural identification of a bacterial quorum-sensing signal containing boron. Nature **415**, 545–549.
2. Passador L, Cook JM, Gambello MJ, Rust L, Iglewski BH (1993). Expression of Pseudomonas aeruginosa virulence genes requires cell-to-cell communication. Science, **260**, 1127–1130.
3. Van Delden C, Iglewski B (1998). Cell-to-cell signaling and Pseudomonas aeruginosa infections. Emerg Infect Dis, **4**, 1–13.
4. Fuqua C, Greenberg EP (2002). Listening in on bacteria: acyl-homoserine lactone signaling. Nat Rev Mol Cell Biol, **3**, 685–695.
5. Hilgers MT, Ludwig ML (2001). Crystal structure of the quorum-sensing protein LuxS reveals a catalytic metal site. Proc Natl Acad Sci U S A, **98**, 11169–11174.
6. Ruzheinikov SN, Das SK, Sedelnikova SE, Hartley A, Foster SJ, Horsburgh MJ, Cox AG, McCleod CW, Mekhalfia A, Blackburn GM, Rice DW, Baker PJ (2001). The 1.2 A structure of a novel quorum-sensing protein, Bacillus subtilis LuxS. J Mol Biol, **313**, 111–122.
7. Bottomley MJ, Muraglia E, Bazzo R, Carfi A (2007). Molecular insights into quorum sensing in the human pathogen Pseudomonas aeruginosa from the structure of the virulence regulator LasR bound to its autoinducer. J Biol Chem, **282**, 18, 13592–13600.
8. More MI, Finger DL, Stryker JL, Fuqua C, Eberhard A, Winans SC (1996). Enzymatic synthesis of a quorum-sensing autoinducer through use of defined substrates. Science, **272**, 1655–1658.
9. Val DL, Cronan JE Jr (1998). In vivo evidence that Sadenosylmethionine and fatty acid synthesis intermediates are the substrates for the LuxI family of autoinducer synthases. J Bacteriol, **180**, 2644–2651.
10. Lithgow JK, Wilkinson A, Hardman A, Rodelas B, Wisniewski-Dye F, Williams P, Downie JA (2000). The regulatory locus cinRI in Rhizobium leguminosarum controls a network of quorumsensing loci. Mol Microbiol, **37**, 81–97.
11. Eberhard A, Burlingame AL, Eberhard C, Kenyon GL, Nealson KH, Oppenheimer NJ (1981). Structural identification of autoinducer of Photobacterium fischeri luciferase. Biochemistry, **20**, 2444–2449.
12. Pearson JP, Gray KM, Passador L, Tucker KD, Eberhard A, Iglewski BH, Greenberg EP (1994). Structure of the autoinducer required for expression of Pseudomonas aeruginosa virulence genes. Proc Natl Acad Sci U S A, **91**, 197–201.
13. de Kievit TR, Iglewski BH (2000). Bacterial quorum sensing in pathogenic relationships. Infect Immun, **68**, 4839–4849
14. Huber B, Riedel K, Köthe M, Givskov M, Molin S, Eberl L (2002). Genetic analysis of functions involved in the late stages of biofilm development in Burkholderia cepacia H111. Mol Microbiol, **2**, 411–426.
15. Smith KM, Bu Y, Suga H (2003). Induction and inhibition of Pseudomonas aeruginosa quorum sensing by synthetic autoinducer analogs. Chem Biol, **10**, 81–89
16. Smith KM, Bu Y, Suga H (2003). Library screening for synthetic agonists and antagonists of a Pseudomonas aeruginosa autoinducer. Chem Biol, **10**, 563–571.
17. Jog GJ, Igarashi J, Suga H (2006). Stereoisomers of brief communication P. aeruginosa autoinducer analog to probe the regulator binding site. Chem Biol, **13**, 123–128.
18. Rasmussen TB, Bjarnsholt T, Skindersoe ME, Hentzer M, Kristoffersen P, Köte M, Nielsen J, Eberl L, Givskov M (2005). Screening for quorum-sensing inhibitors (QSI) by use of a novel genetic system, the QSI selector. J Bacteriol, **187**, 1799–1814.
19. Müh U, Schuster M, Heim R, Singh A, Olson ER, Greenberg EP (2006). Novel Pseudomonas aeruginosa quorum-sensing inhibitors identified in an ultra-high-throughput screen. Antimicrob Agents Chemother, **50**, 3674–3679.
20. Glansdorp FG, Thomas GL, Lee JJK, Dutton JM, Salmond GPC, Welch M, Spring DR.(2004). Synthesis and stability of small molecule probes for Pseudomonas aeruginosa quorum sensing modulation. Org Biomol Chem, **2**, 3329–3336.
21. Castang S, Chantegrel B, Deshayes C, Dolmazon R, Gouet P, Haser R, Reverchon S, Nasser W, Hugouvieux-Cotte-Pattat N, Doutheau A (2004). N-Sulfonyl homoserine lactones as antagonists of bacterial quorum sensing. Bioorg Med Chem Lett, **14**, 5145–5149.
22. Geske GD, Wezeman RJ, Siegel AP, Blackwell HE (2005). Small molecule inhibitors of

bacterial quorum sensing and biofilm formation. J Am Chem Soc, **127**, 12762–12763.

23. Geske GD, O'Neill JC, Miller DM, Mattmann ME, Blackwell HE (2007). Modulation of bacterial quorum sensing with synthetic ligands: systematic evaluation of N-acylated homoserine lactones in multiple species and new insights into their mechanisms of action. J Am Chem Soc, **129**, 13613–13625.

Chapter 20

Heterologous Overexpression, Purification, and In Vitro Characterization of AHL Lactonases

Pei W. Thomas and Walter Fast

Abstract

Quorum-quenching enzymes are useful as biochemical tools and possible therapeutic proteins. One of the best-characterized families of these catalysts is the *N*-acyl-L-homoserine lactone (AHL) lactonases, which rely on a dinuclear metal ion active site to hydrolytically cleave the autoinducer's lactone bond and inactivate signaling. A detailed understanding of how this enzyme works can help in the design of more selective and efficient reagents. To facilitate these studies, we describe a methodology to heterologously express, purify, and conduct in vitro characterization of several metalloforms of the AHL lactonase from *Bacillus thuringiensis* (AiiA). These procedures should be applicable to similar enzymes and will facilitate the production of more useful quorum-quenching reagents for biochemical studies and possible therapeutic applications.

Key words: Quorum quenching, Lactonase, *N*-acyl-L-homoserine lactone, Protein purification, Metalloprotein, Enzyme kinetics

1. Introduction

Enzymes capable of disrupting *N*-acyl-L-homoserine (AHL)-based signaling have proven to be useful biochemical tools for investigating quorum-sensing pathways. Due to the ability of these catalysts to block the virulence of some plant and human pathogens, they also hold promise as potential therapeutic proteins (1). Some advantages of using enzymes as quorum-quenching agents include their substrate specificity, their ability to be genetically encoded, and their catalytic nature (in contrast to quenching by autoinducer-binding antibodies (2, 3)). A detailed understanding of how these enzymes work can help in the development of more selective and effective quorum-quenching agents, a process

Kendra P. Rumbaugh (ed.), *Quorum Sensing: Methods and Protocols*, Methods in Molecular Biology, vol. 692,
DOI 10.1007/978-1-60761-971-0_20, © Springer Science+Business Media, LLC 2011

facilitated by effective production, purification, and characterization of these proteins.

Although enzymes capable of degrading several different types of quorum-sensing signals have been reported (4, 5), the best characterized are those that target AHLs by hydrolysis of either the amide (acylases) or the lactone (lactonases). Many of these enzymes have not been characterized kinetically, but the specific activities of some have been reported, and k_{cat} values can be calculated (for various substrates): 0.09 s^{-1} for porcine kidney acylase I (C_4-HSL), 0.1 s^{-1} for the *Streptomyces* sp. strain M664 AhlM acylase (C_8-HSL), 8.7 s^{-1} for the human PON2 lactonase (3-oxo-C_{12}-HSL), and 91 and 510 s^{-1} for the dizinc and dicobalt metalloforms of AHL lactonase from *Bacillus thuringiensis* (C_6-HSL) (6–9). It should be noted that some of these measurements may not have been conducted under saturating conditions and so should be considered a lower limit. However, since most cells will first encounter AHLs at subsaturating concentrations, the more relevant catalytic constants are perhaps k_{cat}/K_M values, which are only available for a few quorum-quenching catalysts: 1 $M^{-1}s^{-1}$ for porcine kidney acylase I (C_4-HSL), 3.5 $M^{-1}s^{-1}$ for the catalytic antibody lactonase XYD-11G2 (3-oxo-C_{12}-HSL), 5.3×10^4 $M^{-1}s^{-1}$ for an engineered N226Y *Mycobacterium avium* MCP lactonase (C_{12}-HSL), and 1.6×10^4 and 1.4×10^6 $M^{-1}s^{-1}$ for dizinc and dicobalt AHL lactonase (C_6-HSL) from *B. thuringiensis* (6, 9–11). Notably, in the case of AHL lactonase, the metal content of the enzyme can significantly enhance catalytic rate constants. These catalysts display a range of effectiveness, and the AHL lactonases are suggested for further development.

The AHL lactonases are classified as part of the metallo-β-lactamase superfamily of proteins based on homology of sequence, structure, and mechanism (12–16). Historically, the metal content of these superfamily members has been particularly vexing. True to form, some of the pioneering studies of AHL lactonase predicted a metal-free enzyme, ostensibly stemming from a copurified apo form of the enzyme (17, 18). However, structural and functional studies have since demonstrated a dinuclear zinc center in AHL lactonase that is important for both protein stability and catalysis of the ring opening reaction (9, 13). The ability to manipulate the metal ion content can increase the k_{cat}/K_M value up to 125-fold over dizinc metalloforms (9).

Herein, we describe a methodology for the heterologous overexpression, purification, and characterization of several metalloforms of the AHL lactonase from *B. thuringiensis*. These methods are generally applicable to related enzymes and should contribute to the successful production, evaluation, and development of more effective quorum-quenching reagents for biochemical and therapeutic applications.

2. Materials

2.1. Cloning the Coding Sequence of AiiA into an Expression Vector

1. Expression vector pMAL-c2X (New England Biolabs, Ipswich, MA).
2. A PCR template containing the coding sequence for AiiA (autoinducer inactivator A), the AHL lactonase from *B. thuringiensis*, such as pMAL-c2x-*aiiA* (18).
3. Forward and reverse primers designed to incorporate a tobacco etch virus (TEV) protease cleavage site between the maltose binding protein and AHL lactonase; Forward: 5′-aacctcgggGAAAACCTGTATTTTCAGGGAaggatttcaatgaca-3′ containing an *Ava*I restriction site (underlined) followed by 21 nucleotides encoding the TEV protease recognition site (upper case) and the last 15 bases that match the pMAL-c2X multiple cloning site and the first two codons in *aiiA*; Reverse: 5′-gtcgaattcctcaacaagatactcctaatg-3′ containing an *Eco*RI restriction site (underlined).
4. QIAquick PCR purification kit (Qiagen, Valencia, CA).
5. *Ava*I and *Eco*RI Restriction enzymes (New England Biolabs, Ipswich, MA) and T4 DNA ligase (Fisher, Pittsburgh, PA).
6. High-fidelity triplemaster Polymerase Mix (Eppendorf, Westbury, NY).
7. MJ Research PTC 200 thermocylcer (Waltham, MA).
8. Electrocompetent Dh5αE and BL21 (DE3) *E. coli* cells prepared according to Sambrook and Russel (19) and a BioRad MicroPulser.

2.2. Expression and Purification of AHL Lactonase (AiiA)

1. Deionized water treated with a Barnstead NanoPure Diamond analytical ultrapure water system with a resistivity of at least 18.2 MΩ cm.
2. Chelex 100 cation exchange resin (BioRad Laboratories, Hercules, CA).
3. Low-pressure BioLogic LP protein purification systems (BioRad).
4. Autoclave-sterilized stock solutions of:
   (a) 1 M $MgSO_4$.
   (b) 1 M $CaCl_2$.
   (c) 5× M9 salts: 240 mM Na_2HPO_4, 110 mM KH_2PO_4, 43 mM NaCl, and 94 mM NH_4Cl.
5. Various stock solutions, all filter-sterilized using a presterilized 0.22-μM filter and stored at 4°C:

(a) Glucose (25% w/v).

(b) 0.2 M $ZnSO_4$.

(c) 0.2 M $CoCl_2$.

6. M9 media (1 L): 200 mL of 5× M9 salts, 2 mL of 1 M $MgSO_4$, 16 mL of glucose (25%, w/v), and 0.1 mL of 1 M $CaCl_2$ combined with sterile deionized H_2O to a total volume of 1 L.
7. Ampicillin (50 mg/mL) stock solution, filter-sterilized and stored in aliquots at −20°C.
8. 0.5 M IPTG stock solution, filter-sterilized and stored in aliquots at −20°C.
9. Sonic dismembrator Model 500 (Fisher Scientific).
10. Buffer A: 20 mM Tris–HCl and 100 mM NaCl at pH 7.4.
11. Buffer B: 20 mM Tris–HCl and 1 M NaCl at pH 7.4.
12. Elution Buffer: 20 mM Tris–HCl, 100 mM NaCl and 10 mM maltose at pH 7.4.
13. Amylose–agarose resin (New England BioLabs).
14. DEAE-Sepharose FF column (GE Healthcare, Piscataway, NJ).

2.3. Proteolytic Cleavage to Remove the Affinity Tag

1. TEV protease, either purchased commercially (Invitrogen, Carlsbad, CA) or purified in-house (20, 21).
2. 1 M Tris–HCl at pH 8.5.

2.4. Pretreatment of Concentrators to Remove Extraneous Metal Ions

1. Amicon ultracentrifugal filter devices (Millipore) with a 10,000 Da molecular weight cutoff (MWCO, Millipore, Billerica, MA).
2. Membrane washing buffer: 20 mM Hepes, 5 mM NaCl, and 3 mM 1,-10-phenanthroline at pH 7.
3. Swinging bucket centrifuge with 50-mL tube capacity.

2.5. Steady-State Kinetic Assay of AHL Lactonase (AiiA) Activity

1. A 2.5 mM phenol red stock solution in methanol.
2. 10 mM Hepes buffer at pH 7.5 (see Note 1), stored at 4°C.
3. 0.5 M Na_2SO_4 stock solution, stored at 4°C.
4. Freshly prepared 2× assay solution: 2 mM Hepes buffer, 200 mM Na_2SO_4 at pH 7.5, and 80 μM phenol red (see Note 2).
5. Stock solutions of AHL substrates (typically 0.1–1 M) prepared in methanol (see Note 3).
6. Kaleidagraph (Synergy Software; Reading, PA).

2.6. Native and Denaturing Polyacrylamide Gel Electrophoresis

1. Denaturing SDS-PAGE gels with 4% stacking and 12% separating layers prepared using Tris–HCl buffer.
2. Native PAGE gels purchased commercially (BioRad) or prepared with SDS and DTT omitted.

3. GelCode Blue stain reagent (Pierce, Rockford, IL).
4. BioRad protein assay kit (BioRad).

3. Methods

3.1. Cloning the Coding Sequence of aiiA into an Expression Vector

1. The gene encoding AiiA, the AHL lactonase from *B. thuringiensis*, is cloned into the expression vector pMAL-c2X between the *Xmn*I and *Eco*RI sites. We found that using Factor X to cleave the MBP tag from AHL lactonase is not efficient at low temperatures, and the fusion protein is prone to denaturation at higher temperatures. Therefore, an alternative cleavage site is installed, as described below.
2. A TEV protease cleavage site (ENLYFQ↓G) is introduced into the site between maltose-binding protein and AHL lactonase as follows, where ↓ designates the predicted site of proteolytic cleavage. Forward and reverse primers (see Subheading 2) and the pMAL-aiiA template are combined, and a PCR is run for 5 min at 95°C followed by 30 cycles of 30 s at 95°C, 30 s at 55°C, and 72 s at 72°C using an MJ Research PTC 200 thermocylcer and high-fidelity triplemaster Polymerase Mix (see Note 4).
3. The PCR product is purified by using a Qiaquick kit. This and the expression vector pMAL-c2X are digested with *Ava*I and *Eco*RI endonucleases for 2 h at 37°C.
4. Both reaction products are purified using a Qiaquick kit after insuring complete digestion by agarose gel electrophoresis.
5. A ligation is carried out in a total volume of 10–20 μL using T4 DNA ligase at room temperature for 1–2 h with an insert-to-vector ratio of 5:1 to yield pMAL-t-aiiA.
6. 2–4 μL of ligation mixture is subsequently used to transform electrocompetent *E. coli* DH5αE cells using a BioRad MicroPulser (see Note 5).
7. Several of the resulting colonies grown on Amp (50 μg/mL) LB plates are then regrown, the plasmid DNA purified, and the insert fully sequenced to insure that there are no undesired mutations.

3.2. Overexpression of AHL Lactonase in E. coli Growing on Minimal Media Supplemented with Various Metal Salts

1. The expression plasmid pMAL-t-aiiA is transformed into *E. coli* BL21(DE3) cells, which are used for expression, and the resulting transformants selected for resistance on LB Amp (50 μg/mL) plates.
2. Single colonies are then grown to saturation overnight in a 25-mL culture containing LB AMP (50 μg/mL) medium and

then used to inoculate 2 L of M9 minimal media supplemented with ampicillin (100 μg/ml) in 4 L culture flasks. M9 minimal media is used for expression cultures of BL21(DE3) (pMAL-t-*aiiA*) *E. coli*. The 5× M9 salts are diluted into water before the other stocks are added (see Note 6). The cultures are incubated at 37°C with shaking until cells reach an OD_{600} of 0.5–0.6 (about 5 h). Cells are pelleted by centrifugation (8,275 × *g*) at 4 C and then resuspended in fresh room temperature M9 medium containing either 0.5 mM $ZnSO_4$ or 0.1 mM $CoCl_2$ (see Note 7), ampicillin (100 μg/ml), and 0.3 mM IPTG to start induction, which is continued with shaking for an additional 16 h at 25°C.

3. Cells are harvested by centrifugation at 8,275 × *g* for 7 min at 4°C. The resulting cell pellet can be used immediately for purification, or stored at −20°C for later use.

3.3. Purification of AHL Lactonase

1. All purification procedures are carried out at 4°C unless otherwise noted. For some metal-ion-sensitive experiments, water is additionally treated using Chelex 100 cation exchange resin to scavenge divalent metal ions, according to the manufacturer's instructions. All chromatographic procedures are performed using low-pressure BioLogic LP protein purification systems.
2. Cell pellets resulting from 2 L of culture are resuspended in 200 mL of ice-cold Buffer A (see Note 8) and transferred into a Pyrex beaker.
3. Cells are lysed by sonication using a sonic dismembrator with six pulses of 20 s each (for 200 mL) and 120 s cooling intervals (see Note 9). Sonication is either carried out at 4°C, or the beaker containing the cell suspension is suspended in an ice bath during the procedure.
4. Cell debris is removed by centrifugation at 34,500 × *g* for 30 min at 4°C.
5. The resulting supernatant is loaded onto a DEAE-Sepharose FF column (2.5 × 20 cm) equilibrated with Buffer A at a flow rate of 3.0 mL/min using a peristaltic pump (see Note 10).
6. 10 mL of Buffer A is used to wash the column walls and left to drain by gravity flow. An additional 5 mL of Buffer A is gently added to the column, and the column is then connected to the BioLogic system and washed with an additional 200 mL of Buffer A at a flow rate of 2.0 mL/min until the Abs_{280} stabilizes to a constant value, indicating that no additional proteins are eluting. Fractions are collected at 8 mL/tube while increasing the NaCl concentration from 100 to 300 mM using a linear gradient (0–20% Buffer B). The active fraction of metal-containing MBP–AHL lactonase elutes at a

conductivity of approximately 19 mS/cm (approximately 180–190 mM NaCl), while apo MBP–AHL lactonase elutes at a higher salt concentration with a conductivity of approximately 26 mS/cm. When MBP–AHL lactonase is expressed with $CoCl_2$ supplemented growth media, most of the enzyme elutes from the DEAE column as the metal-ion-bound, active enzyme. However, when expressed in unsupplemented TB media or in the presence of $ZnSO_4$, the resulting AHL lactonase is a mixture of aggregates (which elute in the flow-through), active metal-bound enzyme, and inactive apoenzyme (Fig. 1) (18).

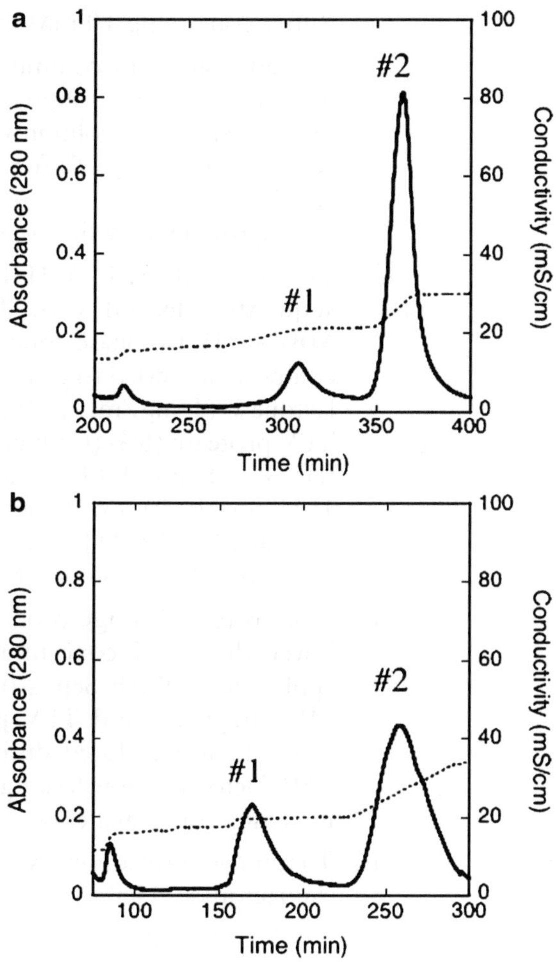

Fig. 1. Ion-exchange chromatography separates dinculear metal ion bound AHL lactonase from metal-free apo lactonase. Solid lines show Abs_{280}, and dashed lines show conductivity. (**a**) Initial DEAE column chromatography after growth in TB media. (**b**) Initial DEAE column chromatography after growth in TB media supplemented with 1 mM $ZnSO_4$. In both cases, *peak 1* contains active dinuclear zinc MBP–AHL lactonase, and *peak 2* contains apoprotein. Retention times are somewhat different because salt concentrations were adjusted manually. Reprinted with permission from *Biochemistry*. Copyright 2005 American Chemical Society (18).

7. The enzymic activity due to AHL lactonase in each fraction is tested (method described below).
8. The fractions containing AHL lactonase activity are combined, and an equal volume of Buffer A is added. Depending on how much fusion protein is collected, one to two columns (2.5 × 10 cm) of amylose–agarose resin are used for the purification (see Note 11). The flow rate during sample loading is kept at less than 1 mL/min.
9. After the column is washed with two to three column volumes of Buffer A at a flow rate of 0.9 mL/min, the fusion protein is eluted at a flow rate of 1.5 mL/min using Elution Buffer containing 10 mM maltose.
10. The amylose resin columns can be regenerated by washing with three column volumes of water, three column volumes of 0.1% SDS, one column volume of water, and five column volumes of Loading Buffer. If the column regeneration process is carried out as soon as possible, each column can be used approximately five to ten times.
11. The purified MBP–AHL lactonase fusion protein is subsequently cleaved with TEV protease as follows. The MBP–AHL lactonase concentration is adjusted and reaction components added to give final concentrations of MBP–AHL lactonase (1 mg/mL), 50 mM Tris–HCl, 1 mM DTT, and TEV protease (8%) ((milligram protease/milligram total protein) × 100) at pH 8.0. Incubation of the resulting mixture at 10°C for 16–20 h with gentle shaking results in specific proteolysis of MBP–AHL lactonase and separates the MBP tag from AHL lactonase (see Note 12).
12. The protein cleavage products are diluted with Buffer A to lower the NaCl concentration to 100–140 mM and are applied to a DEAE-Sepharose FF column (see Note 13). The MBP fragment and TEV protease both elute in the flow-through or are eluted during a wash with 140 mM NaCl. AHL lactonase, now lacking the MBP affinity tag, is eluted by increasing the NaCl concentration to 140–200 mM.
13. To remove trace amounts of uncleaved MBP–AHL lactonase, the lactonase fractions eluted above are loaded onto a small amylose–agarose affinity column (1.5 × 10 cm), and AHL lactonase lacking an MBP affinity tag is collected in the flow-through and wash fractions.
14. Protein yields of purified AHL lactonase are dependent on the type of metal ion bound. Typically, starting with 2 L cultures, about 52 mg of dicobalt (Table 1) or 7 mg of dizinc AHL lactonase can be purified (see Note 14). The protein samples collected after each purification step are analyzed for purity by SDS-PAGE gel (Fig. 2).

Table 1
Purification of the dicobalt metalloform of AHL lactonase from 2 L of growth media

Purification step	Total protein[a] (mg)	Total MBP–AiiA[b] or AiiA[b] (mg)	Purity (%)	Purification yield (%)
Crude	800	214	27	100
DEAE ion-exchange column	240	160	67	75
Amylose–agarose column	150	153	99	72
DEAE ion-exchange column	58	55	95	67
Amylose–agarose column and concentration	52	52	99	63

[a]Total protein amount was determined using BioRad protein assay
[b]Protein concentrations of MBP–AiiA or AiiA were estimated according to the band densities of SDS-PAGE gels

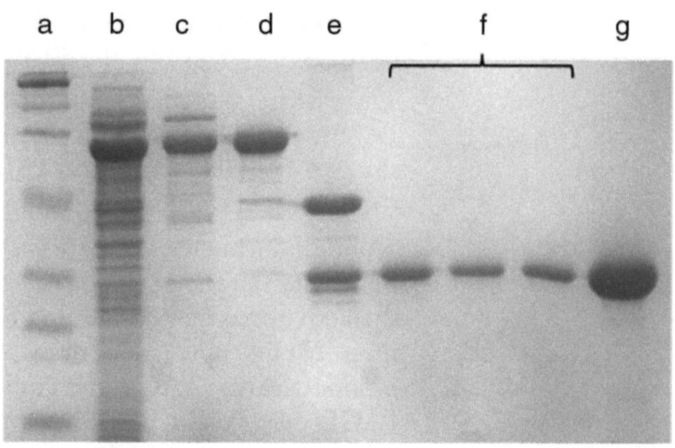

Fig. 2. Purification of the dicobalt metalloform of AHL lactonase. (**a**) 12% SDS-PAGE denaturing protein gel is used to resolve the proteins in (**a**) MW markers (203, 120, 90, 52, 34, 28 and 20 kDa), (**b**) crude soluble extract, (**c**) eluate from DEAE anion-exchange column, (**d**) eluate from agarose affinity column, (**e**) after cleavage by TEV protease, (**f**) after DEAE chromatography, and (**g**) after amylose affinity chromatography.

3.4. Storage and Characterization of the AHL Lactonase

1. After removal of the MBP affinity tag, AHL lactonase is concentrated using a prewashed Amicon Ultra-15 centrifugal filter device filter with 10,000 Da molecular weight cutoff.
2. Prior to use, the concentrator is extensively washed to remove extraneous metal ions. Membrane washing buffer and the

metal chelator 1,-10-phenanthroline (3 mM) is passed through the filter by centrifugation at 2,500 × *g* for 10 min. The same washing buffer is added, and the filter is incubated at room temperature for 1 h, and again centrifuged. For the third wash, the buffer is added and incubated overnight at room temperature and then centrifuged for 20 min. The excess chelator is then rinsed by four separate additions of metal-ion-depleted water, each followed by centrifugation at 2,500 × g for 15 min. To detect removal of the chelator, the concentration of phenanthroline can be determined by measuring the UV–Vis spectrum of the flow-through (ε_{265} = 31,500 $M^{-1} cm^{-1}$) (22).

3. Using the washed filtration device, approximately 14-mL aliquots of the purified protein fractions can be concentrated by centrifugation for 20–30 min at 2,500 × g, and this is typically repeated three times to concentrate the protein to a final volume of 500 μL with a typical concentration of 10 mg/mL. To insure that purified protein is not lost through a leaky membrane, the flow-through is monitored by Abs_{280} (see Note 15).
4. For short-term storage, the concentrated protein is kept at 4°C. For long-term storage, the protein is made to 10% glycerol before flash freezing in aliquots (typically 0.5–1 mL) using $N_{2(l)}$ and stored at −80°C.
5. The purity of protein is judged by Coomassie-stained SDS-PAGE, and the latter typically shows a single band. Gels consist of a 4% stacking gel and 12% separating gel and are run at room temperature for 35–40 min at 200 V.
6. Nondenaturing PAGE gels are used as a quick test to determine whether the purified enzyme contains metal ions because the active enzyme (containing two metal ions) migrates faster than inactive apo (metal-free) AHL lactonase (Fig. 3) (18). Nondenaturing PAGE gels contain the same reagents as SDS-PAGE gels except that SDS and reducing agent (DTT) are omitted, and the samples are subjected to electrophoresis for 100 min at 14 mA at 4°C.
7. Both SDS-PAGE and native gels are stained using GelCode Blue stain reagent.
8. The concentration of the protein is determined using the BioRad protein assay kit.
9. To more accurately determine the final metal-ion content of the purified lactonase, protein samples and the flow-through of the final concentration step are diluted with Buffer A to give a final protein concentration of 0.1 mg/mL, and both samples are analyzed by inductively coupled plasma mass spectrometry (ICP-MS). The difference of the concentration

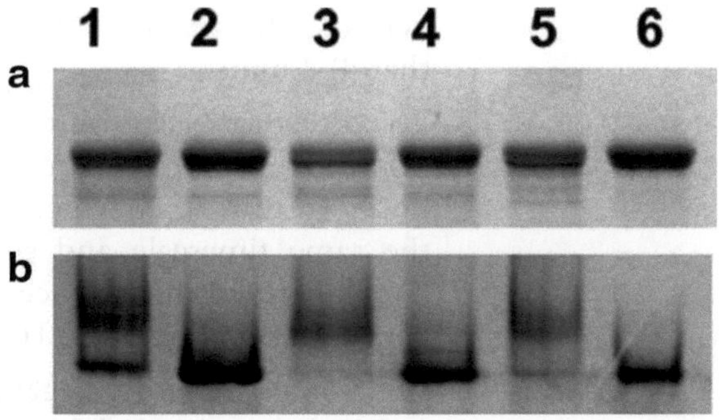

Fig. 3. PAGE characterization of AHL lactonase preparations. (**a**) 12% denaturing SDS-PAGE gel. (**b**) 12% native PAGE gel. Lanes are the same in both. (1) Protein mixture containing both apo- and dinuclear zinc MBP–AHL (2) The dinuclear zinc metalloform of MBP–AHL lactonase found in peak 1 of the DEAE column (3) The metal-free apoprotein form of MBP–AHL lactonase found in peak 2 from the DEAE column (4) The purified dinuclear zinc metalloform of MBP–AHL lactonase (5) Purified MBP–AHL lactonase after removal of metal ions (6) MBP–AHL lactonase after reconstitution of the apoprotein with zinc. Reprinted with permission from *Biochemistry.* Copyright 2005 American Chemical Society (18).

of each metal ion in the flow-through from that in the protein sample is divided by the protein's concentration to yield the equivalents of each metal ion bound to the protein. Purified proteins typically bind 1.8–2.0 total equivalents of metal ions.

3.5. Steady-State Kinetic Assay of AHL Lactonase Activity

1. Hydrolysis of AHLs yields one equivalent of *N*-acyl-L-homoserine and, at pH values significantly above 4 (the approximate pK_a value of the product's carboxylic acid), one equivalent of a proton. This net production of a proton can be detected spectrophotometrically by incorporating a colorimetric pH indicator into the assay buffer. This assay has been previously described for carbonic anhydrase and haloalkane dehalogenase (23–25), among other enzymes.
2. A typical kinetic assay contains 702.5 μL deionized water, 750 μL of 2× dye solution, 10 μL of an AHL lactonase stock solution, and 37.5 μL of an AHL substrate stock dissolved in methanol. It is possible to assay at different pH values by changing the indicator/buffer pair. Here, the 40 μM pheonol red/1 mM Hepes pair is used to monitor reactions starting at pH 7.4–7.6. Concentrations of the indicators can be adjusted experimentally to yield final assay solutions with a starting absorbance of 1 at the absorbance maximum (577 nm for phenol red). Here, initial rates for hydrolysis of AHLs were monitored as a decrease in absorbance at 557 nm during approximately 0.2 min at 25°C. Over our typical assay time

periods, the pH of the reaction mixture changes by less than 0.2 units.

3. To correct for background nonenyzmic hydrolysis rates, reaction mixtures are prepared by substituting 10 μL of storage buffer instead of enzyme, and initial rates are recorded over the same timescale and subsequently subtracted from the rates observed in presence of the enzyme at each substrate concentration (see Note 16).

4. Initial rates are plotted against substrate AHL concentrations (mM), here using Kaleidagraph (Fig. 4), and fit directly to a form of the Michaelis–Menten equation to derive V_{max} and K_M values. The k_{cat} value is calculated using the total concentration of AHL lactonase in the assay, the V_{max} value (Δ Abs units/s), and a θ value to convert the observed change in absorbance to proton concentration, using the following equation: k_{cat} (s^{-1}) = V_{max} (Δ Abs units/s) ÷ (θ (Δ Abs units/μM) × $[AiiA]_T$ (μM)).

5. The term θ (Δ Abs units/μM) is the slope of a standard curve measured using the assay mixture and additions of an HCl standard solution. Standard solutions can be commercially purchased (Sigma–Aldrich) or validated by titration using a gravimetric standard. In our hands, a typical θ value is 0.0013 Abs units/μM.

6. A typical example of metal-ion content and steady-state kinetic constants for a set of AHL lactonase metalloforms is summarized in Table 2.

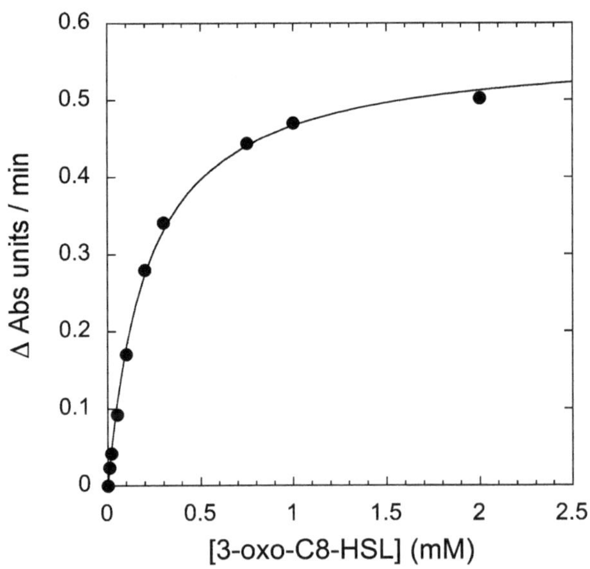

Fig. 4. Steady-state kinetics of 3-oxo-C_8-HSL hydrolysis by the dicobalt metalloform of AHL lactonase. The *solid line* is a fit to data using the Michaelis–Menten equation (k_{cat} = 1,100 ± 20 s^{-1}; K_M = 220 ± 10 μM).

Table 2
Steady-state kinetic constants for hydrolysis of C_6-HSL by various metalloforms of AHL lactonase (9)

Metal supplement	Metal ion content[a] (equiv)	k_{cat} (s^{-1})[b]	K_M (mM)	k_{cat}/K_M ($M^{-1} s^{-1}$)
$MnSO_4$	Mn^{2+} (1.8), Zn^{2+} (0.1)	330 ± 20	0.18 ± 0.04	1.8×10^6
$CoCl_2$	Co^{2+} (1.9), Zn^{2+} (0.2)	510 ± 10	0.36 ± 0.04	1.4×10^6
$ZnSO_4$	Zn^{2+} (2.0)	91 ± 3	5.6 ± 0.6	1.6×10^4
$CdCl_2$	Cd^{2+} (2.0)	480 ± 20	0.24 ± 0.03	2.0×10^6

[a]All proteins were analyzed for Mn^{2+}, Co^{2+}, Zn^{2+}, and Cd^{2+} using ICP-MS, and the metal-ion content is shown for all ions ≥ 0.1 equiv

[b]Reactions were carried out at pH 7.4 and at 28°C

4. Notes

1. The pH of 10 mM Hepes buffer can decrease during storage, so the pH should be verified before use.
2. For longer-term storage, the 2× working solution can be kept at 4°C and warmed before using. If stored at room temperature, the phenol red indicator gradually turns orange, indicating a more acidic pH, presumably due to dissolved CO_2 forming bicarbonate. For qualitative activity determinations, this can be reversed by simply adding a few drops of 1 M NaOH to increase the pH of solution until Abs_{557} of the 1× solution is 1 Abs unit. For quantitative analysis, 2× buffer should be made fresh. A new HCl standard curve should be generated for each batch of 2× working solution, although we observe minimal differences between batches.
3. Typically, stock solutions of AHLs are prepared in methanol at the following concentrations: 1 M C_4-HSL, 0.5 M C_6-HSL, 0.5 M 3-oxo-C_6-HSL, 0.2 M C_8-HSL, 0.5 M 3-oxo-C_8-HSL, 0.1 M C_{10}-HSL, and 0.2 M *N*-carbobenzyloxy-HSL (CBZ-HSL). In the aqueous assay solutions, the maximum AHL concentrations are limited by their solubility, with precipitation noted above the following concentrations: 2 mM C_8-HSL, 0.4 mM C_{10}-HSL, and 3 mM CBZ-HSL.
4. The DNA can be eluted from the purification column using either water or TE buffer. Although TE buffer gave higher yields of DNA recovery, this buffer also inhibits the subsequent restriction enzyme digest. Therefore, water is used instead of buffer to elute DNA from the column.

5. For electrotransformation using the BioRad MicroPulser, the total salt content of the reaction should be kept low to optimize the transformation efficiency.
6. Upon addition of 1 M $CaCl_2$, a white precipitate appears which dissolves after further mixing. Addition of $ZnSO_4$ can also result in some precipitation, but this does not interfere with cell growth.
7. The dimanganese or dicadmium metalloforms of AHL lactonase (AiiA) can be prepared by supplementing the M9 minimal expression medium with 0.5 mM $MnSO_4$ or 0.5 mM $CdCl_2$ salts during induction. However, expression levels of these two metalloforms are typically lower than dicobalt, and the alternative metal ions are easily replaced by low concentrations of contaminating zinc during purification. Therefore, if alternative metalloforms are desired, the MBP–AHL lactonase fusion protein rather than the cleaved version is used to minimize purification steps. For example, purified MBP–AHL lactonase typically retains 90% of the alternative metal ions, but purified AHL lactonase only retains about half of the alternative metal ion with the other half substituted by zinc ions.
8. If the cell pellets are stored, not fresh, then addition of a protease inhibitor cocktail is recommended.
9. Longer time is needed if the volume of extraction solution is >200 mL. The efficiency of the sonication step is lower if the volume is >300 mL. To optimize the procedure, the extraction solution can be divided in half for batchwise processing.
10. A substitution of 20 mM Tris–HCl at pH 7.4 with 5 mM NaCl can also be used and results in increased protein binding. However, an extended washing step is required.
11. For the dizinc metalloform of AHL lactonase, one amylose–agarose resin column (2.5 × 10 cm) is sufficient, but two to three columns may be needed for dicobalt AHL lactonase. Batchwise binding methods can be substituted if desired. The MBP–AHL lactonase fusion protein that is eluted from the DEAE column (approximately 50 mg) can be combined with 50 mL of amylose–agarose resin and gently rocked for 2 h at 4°C. The resin mixture is then poured into a column, washed, and eluted as described above.
12. If the fusion protein is in 20 mM Tris–HCl, pH 7.6, and approximately 180 mM NaCl, then 30 mM of additional Tris–HCl (at pH 8.5) is required to obtain a final concentration of 50 mM at a pH between 7.6 and 8. If needed, TEV protease can be added up to a 16% ratio during the cleavage step.
13. The optimum size of the DEAE column depends on the amount of protein loaded. Typically, for dicobalt AHL

lactonase, a 2.5 × 10 cm column is used. For the lower-yielding dizinc AHL lactonase, a 1.5 × 10 cm column is used.

14. As an alternative method, the dicobalt AHL lactonase can be converted into the dizinc metalloform by dialyzing against buffer containing $ZnSO_4$.
15. Because of the washing steps, the Ultra-14 concentrator is used multiple times. The manufacturer suggests centrifugation at 3,000 × *g*, but we found that after multiple spins, the use of this force often results in a leaky membrane. However, centrifugation at 2,500 × *g*, for 20–30 min, allows the concentrator to be used for up to approximately 20 spins. To insure the membrane is not torn during concentration, the Abs_{280} of the flow-through solution is monitored to insure that protein is not present. Centrifugation over an extended period of time can cause protein precipitation, so the protein is removed from the concentrator after three spins. If a higher concentration of protein is needed, the process can be repeated.
16. Background hydrolysis rates appear to be somewhat substrate-dependent and can be ignored if minimal and not concentration-dependent. If background hydrolysis rates are slow, the quantity of AHL substrates used can be minimized by first measuring the background hydrolysis rate in the absence of enzyme for the first 0.2 min and then adding enzyme to the same cuvette to measure the enzyme-catalyzed hydrolysis rates.

Acknowledgments

This work was supported in part by the Texas Advanced Research Program (Grant 003658-0018-2006) and the Robert A. Welch Foundation (Grant F-1572).

References

1. Dong, Y. H., Wang, L. H., and Zhang, L. H. (2007) Quorum-quenching microbial infections: mechanisms and implications, *Philos. Trans. R. Soc. Lond. B Biol. Sci.* **362**, 1201–1211.
2. Kaufmann, G. F., Park, J., and Janda, K. D. (2008) Bacterial quorum sensing: a new target for anti-infective immunotherapy, *Expert Opin. Biol. Ther.* **8**, 719–724.
3. Kapadnis, P. B., Hall, E., Ramstedt, M., Galloway, W. R., Welch, M., and Spring, D. R. (2009) Towards quorum-quenching catalytic antibodies, *Chem. Commun. (Camb.)* **5**, 538–540.
4. Pustelny, C., Albers, A., Buldt-Karentzopoulos, K., Parschat, K., Chhabra, S. R., Camara, M., Williams, P., and Fetzner, S. (2009) Dioxygenase-mediated quenching of quinolone-dependent quorum sensing in *Pseudomonas aeruginosa*, *Chem. Biol.* **16**, 1259–1267.
5. Roy, V., Fernandes, R., Tsao, C. Y., and Bentley, W. E. (2010) Cross species quorum quenching using a native AI-2 processing enzyme, *ACS Chem. Biol.* **5**(2), 223–232.

6. Xu, F., Byun, T., Deussen, H. J., and Duke, K. R. (2003) Degradation of *N*-acylhomoserine lactones, the bacterial quorum-sensing molecules, by acylase, *J. Biotechnol.* **101**, 89–96.
7. Park, S. Y., Kang, H. O., Jang, H. S., Lee, J. K., Koo, B. T., and Yum, D. Y. (2005) Identification of extracellular *N*-acylhomoserine lactone acylase from a *Streptomyces* sp. and its application to quorum quenching, *Appl. Environ. Microbiol.* **71**, 2632–2641.
8. Teiber, J. F., Horke, S., Haines, D. C., Chowdhary, P. K., Xiao, J., Kramer, G. L., Haley, R. W., and Draganov, D. I. (2008) Dominant role of paraoxonases in inactivation of the *Pseudomonas aeruginosa* quorum-sensing signal *N*-(3-oxododecanoyl)-L-homoserine lactone, *Infect. Immun.* **76**, 2512–2519.
9. Momb, J., Thomas, P. W., Breece, R. M., Tierney, D. L., and Fast, W. (2006) The quorum-quenching metallo-gamma-lactonase from *Bacillus thuringiensis* exhibits a leaving group thio effect, *Biochemistry* **45**, 13385–13393.
10. De Lamo Marin, S., Xu, Y., Meijler, M. M., and Janda, K. D. (2007) Antibody catalyzed hydrolysis of a quorum sensing signal found in Gram-negative bacteria, *Bioorg. Med. Chem. Lett.* **17**, 1549–1552.
11. Chow, J. Y., Wu, L., and Yew, W. S. (2009) Directed evolution of a quorum-quenching lactonase from *Mycobacterium avium* subsp. *paratuberculosis* K-10 in the amidohydrolase superfamily, *Biochemistry* **48**, 4344–4353.
12. Dong, Y. H., Xu, J. L., Li, X. Z., and Zhang, L. H. (2000) AiiA, an enzyme that inactivates the acylhomoserine lactone quorum-sensing signal and attenuates the virulence of *Erwinia carotovora*, *Proc. Natl. Acad. Sci. U. S. A.* **97**, 3526–3531.
13. Liu, D., Lepore, B. W., Petsko, G. A., Thomas, P. W., Stone, E. M., Fast, W., and Ringe, D. (2005) Three-dimensional structure of the quorum-quenching *N*-acyl homoserine lactone hydrolase from *Bacillus thuringiensis*, *Proc. Natl. Acad. Sci. U. S. A.* **102**, 11882–11887.
14. Liu, D., Momb, J., Thomas, P. W., Moulin, A., Petsko, G. A., Fast, W., and Ringe, D. (2008) Mechanism of the quorum-quenching lactonase (AiiA) from *Bacillus thuringiensis*. 1. Product-bound structures, *Biochemistry* **47**, 7706–7714.
15. Momb, J., Wang, C., Liu, D., Thomas, P. W., Petsko, G. A., Guo, H., Ringe, D., and Fast, W. (2008) On the mechanism of the quorum-quenching lactonase (AiiA) from *Bacillus thuringiensis*: 2. Substrate modeling and active site mutations, *Biochemistry* **47**, 7715–7725.
16. Kim, M. H., Choi, W. C., Kang, H. O., Lee, J. S., Kang, B. S., Kim, K. J., Derewenda, Z. S., Oh, T. K., Lee, C. H., and Lee, J. K. (2005) The molecular structure and catalytic mechanism of a quorum-quenching *N*-acyl-L-homoserine lactone hydrolase, *Proc. Natl. Acad. Sci. U. S. A.* **102**, 17606–17611.
17. Wang, L. H., Weng, L. X., Dong, Y. H., and Zhang, L. H. (2004) Specificity and enzyme kinetics of the quorum-quenching *N*-Acyl homoserine lactone lactonase (AHL-lactonase), *J. Biol. Chem.* **279**, 13645–13651.
18. Thomas, P. W., Stone, E. M., Costello, A. L., Tierney, D. L., and Fast, W. (2005) The quorum-quenching lactonase from *Bacillus thuringiensis* is a metalloprotein, *Biochemistry* **44**, 7559–7569.
19. Sambrook, J., and Russell, D. W. (2001) *Molecular cloning : a laboratory manual*, 3rd ed., Cold Spring Harbor Laboratory Press, Cold Spring Harbor, NY.
20. Kristelly, R., Earnest, B. T., Krishnamoorthy, L., and Tesmer, J. J. (2003) Preliminary structure analysis of the DH/PH domains of leukemia-associated RhoGEF, *Acta. Crystallogr. D Biol. Crystallogr.* **59**, 1859–1862.
21. Kapust, R. B., Tozser, J., Fox, J. D., Anderson, D. E., Cherry, S., Copeland, T. D., and Waugh, D. S. (2001) Tobacco etch virus protease: mechanism of autolysis and rational design of stable mutants with wild-type catalytic proficiency, *Protein Eng.* **14**, 993–1000.
22. Vallee, B. L., Rupley, J. A., Coombs, T. L., and Neurath, H. (1960) The role of zinc in carboxypeptidase, *J. Biol. Chem.* **235**, 64–69.
23. Khalifah, R. G. (1971) The carbon dioxide hydration activity of carbonic anhydrase. I. Stop–flow kinetic studies on the native human isoenzymes B and C, *J. Biol. Chem.* **246**, 2561–2573.
24. Hurt, J. D., Tu, C., Laipis, P. J., and Silverman, D. N. (1997) Catalytic properties of murine carbonic anhydrase IV, *J. Biol. Chem.* **272**, 13512–13518.
25. Schindler, J. F., Naranjo, P. A., Honaberger, D. A., Chang, C. H., Brainard, J. R., Vanderberg, L. A., and Unkefer, C. J. (1999) Haloalkane dehalogenases: steady-state kinetics and halide inhibition, *Biochemistry* **38**, 5772–5778.

Chapter 21

High-Performance Liquid Chromatography Analysis of N-Acyl Homoserine Lactone Hydrolysis by Paraoxonases

John F. Teiber and Dragomir I. Draganov

Abstract

Mammalian paraoxonases (PONs) are a unique, highly conserved family of calcium-dependent esterases consisting of PON1, PON2, and PON3. The PONs can hydrolyze the lactone ring of a range of N-acyl-L-homoserine lactone (AHL) quorum sensing signaling molecules, rendering them inactive. This chapter describes a method that utilizes high-performance liquid chromatography analysis with UV detection for determining the rate of AHL hydrolysis in cell lysates, tissue homogenates, serum, and with purified proteins. Also described are the techniques used to prepare cell culture lysates and tissue homogenates for analysis and the use of class-specific enzyme inhibitors to determine the contribution of PONs to AHL hydrolysis in the samples.

Key words: Acyl Homoserine Lactone, High-performance liquid chromatography, Lactonase, Paraoxonases, Quorum sensing

1. Introduction

The paraoxonase (PON) enzymes hydrolyze a broad range of lactones and have overlapping, but distinct, substrate specificities (1). PON1 also hydrolyzes aromatic esters and phosphotriesters, whereas PON2 and PON3 have limited arylesterase and no phosphotriesterase activities (1). Although the physiological function(s) and natural substrates for the PONs are uncertain, accumulating evidence indicates that the lactones may be their natural substrates (2). A range of N-acyl-L-homoserine lactone (AHL) quorum sensing signaling molecules produced by gram-negative bacteria are substrates for the PONs (1, 3). PONs hydrolyze the lactone ring of the AHLs into their open hydroxyl acid products, rendering them biologically inactive. Only the natural L-isomers of the AHLs appear to be hydrolyzed by PONs.

Kendra P. Rumbaugh (ed.), *Quorum Sensing: Methods and Protocols,* Methods in Molecular Biology, vol. 692, DOI 10.1007/978-1-60761-971-0_21, © Springer Science+Business Media, LLC 2011

Recent studies have shown that PONs, and particularly PON2, are the primary enzymes inactivating the *Pseudomonas aeruginosa* quorum sensing signal *N*-(3-oxododecanoyl)-L-homoserine lactone (3-oxo-C_{12}-HSL) in lysates from a number of human and mouse cell lines, as well as in mouse lung and liver homogenates (3, 4). This has led investigators to suggest that PONs may be an important component of the host defense against some gram-negative bacterial infections. This chapter describes an HPLC method with UV detection for determination of AHL hydrolysis by PONs and the methods employed to prepare tissue samples and cultured cells for analysis.

2. Materials

2.1. Preparation of Cell Culture Lysates and Tissue Homogenates

1. Phosphate-buffered saline (PBS).
2. Tris–HCL (25 mM), pH 7.4 with and without 1 mM $CaCl_2$.
3. Solution of trypsin (0.25%, Mediatech, Inc., Herndon, VA).
4. Cell lysis buffer: 25 mM Tris–HCL (pH 7.4), 1 mM $CaCl_2$, and 0.05% *n*-dodecyl-b-maltoside (Dojindo Molecular Technologies, Gaithersburg, MD), and the following protease inhibitors from Calbiochem, San Diego, CA; 2 mg/ml aprotinin, 5 mg/ml leupeptin, 1 mg/ml pepstatin A.
5. Tissue lysis buffer: cell lysis buffer without *n*-dodecyl-b-maltoside.
6. 100 mM 4-(2-Aminoethyl)benzenesulfonylfluoride (AEBSF; Calbiochem, San Diego, CA) in methanol.
7. Stock EDTA solution: 500 mM EDTA, pH 8.0.
8. Sonicator.
9. Polytron Homogenizer.
10. Potter-Elvehjem Homogenizer.

2.2. PON AHL Hydrolysis Assay and HPLC Analysis

1. Milli-Q or deionized-distilled water.
2. 3-oxo-C_{12}-HSL (Cayman Chemical, Ann Arbor, MI).
3. 2.5 mM Tris–HCL, pH 7.4.
4. Reaction buffer: 2.5 mM Tris–HCL, pH 7.4, 1 mM $CaCl_2$.
5. Stock solution of 5 mM AHL in methanol, stored at –20°C.
6. Acetonitrile (HPLC grade).
7. HPLC with a UV detector set at 205 nm.
8. HPLC mobile phase: 75% acetonitrile, 0.2% acetic acid.
9. Restek Pinnacle II C-18 column (250×4.6 mm, 5 μm particles) equipped with a pre-column Restek cap frit (4 mm, 0.5 μm).
10. 37°C water bath.

3. Methods

Determination of rates of AHL hydrolysis by purified enzymes, serum, cellular lysates, or tissue homogenates have been performed with the described reversed-phase HPLC technique. With detection at 205 nm, both the lactone substrate and the hydroxyl acid product formed by AHL lactonase hydrolysis can be detected simultaneously (Fig. 1). The amount of AHL hydrolyzed is calculated from the ratio of the area of the AHL to the area of the hydroxyl acid after a response factor to account for the differences in peak area per mole of compound between the AHL-acid and AHL has been determined. The following method describes the assay conditions developed for analysis of the *P. aeruginosa* quorum sensing signal molecule 3-oxo-C_{12}-HSL. Although the method is similar over a range of AHLs, minor modifications might be necessary when using different AHL substrates (see Note 1). The optimal HPLC conditions and response factor for each AHL and corresponding acid must be determined experimentally (see Note 2).

In addition to PONs, other esterases can hydrolyze AHLs. To distinguish between hydrolysis catalyzed by PONs and other esterases, inhibition studies with the PON inhibitor EDTA and/ or a serine esterase inhibitor such as AEBSF can be performed. PONs require calcium for their stability and enzymatic activity. Therefore, treatment of the samples with a metal chelator such as EDTA renders them hydrolytically inactive. In contrast, PONs are not inhibited by serine esterase inhibitors. Comparison of samples treated with EDTA or AEBSF and nontreated samples generally allows for the determination of the contribution of serine esterases versus PONs to AHL hydrolysis in the sample. It is possible that non-PON metal-dependent lactonases may be present which contribute to AHL hydrolysis, although such mammalian lactonases with significant AHL lactonase activity have not been identified.

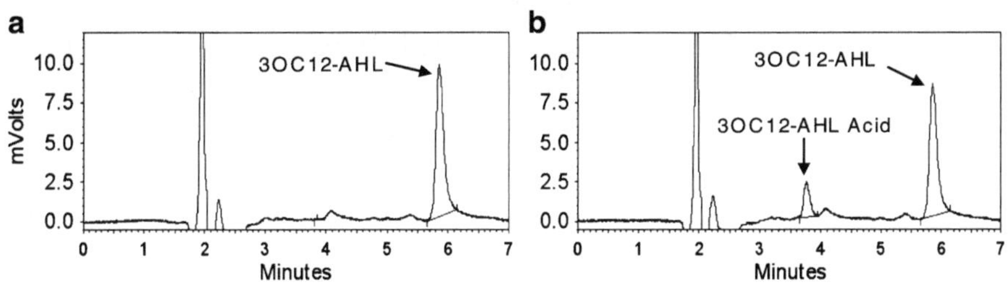

Fig. 1. HPLC chromatograms showing substrate and product peaks of 3-oxo-C_{12}-HSL incubated with mouse liver homogenate. 3-oxo-C_{12}-HSL was incubated at a concentration of 25 μM in the presence of (**a**) 5 μl of lysis buffer (**b**) 0.1 mg/ml of homogenate. Reactions were stopped after 20 min, and supernatants were analyzed by HPLC as described.

3.1. Preparation of Cell Culture Lysates

1. Detach plated cells, roughly 100,000–200,000 with trypsin, transfer to a 1.5-ml eppendorf tube, and pellet by centrifuging at low speed.
2. Carefully aspirate the trypsin and resuspend the cells in about 1 ml of PBS.
3. Pellet the cells by centrifuging at low speed and carefully aspirate the PBS. At this stage, pellets may be frozen and stored at −80°C until analysis.
4. Resuspend pellet in 75 μl of lysis buffer.
5. Pellets are sonicated in ice water by subjecting them to eight, 1-s bursts on a Branson Sonifier 250 equipped with a microtip. The Sonifier is set to an output control of 4.5.
6. Centrifuge the lysates for 2 min at 14,000 × *g* at 4°C. Transfer the supernatant to a clean eppendorf tube being careful not to disturb the pellet.
7. Protein concentration in the supernatant is determined by the BCA method per the manufacturer's protocol (Pierce, Rockford, IL).
8. Supernatants are analyzed for AHL activity or can be stored at −80°C. No loss of PON activity is apparent after several freeze–thaw cycles.

3.2. Preparation of Tissue Homogenates

1. Common procedures for preparation of tissue homogenates can be used; however, buffers used should not contain calcium chelators and have to be supplemented with calcium chloride to maintain PON activity. Also, tissues must be thoroughly perfused to remove any PONs present in blood. This section describes the protocol used in our laboratory (see Note 3).
2. Perfused tissue (freshly obtained or stored at −80°C) is thawed and rinsed with ice-cold Tris–HCL (25 mM), pH 7.4 containing 1 mM $CaCl_2$ and weighed.
3. Keeping the tissue on ice, mince coarsely with scissors or a razor blade.
4. To the tissue, add approximately twice the amount (grams) of ice-cold tissue lysis buffer.
5. Homogenize, on ice, with 10 passes on a polytron homogenizer followed by 15 passes on a Potter-Elvehjem homogenizer.
6. Centrifuge homogenates for 10 min at 600 × *g* at 4°C to remove nuclei and larger particles and transfer the supernatant to a clean eppendorf tube being careful not to disturb the pellet.
7. Protein concentration in the supernatant is determined by the BCA method per the manufacturer's protocol (Pierce, Rockford, IL).

8. Supernatants are analyzed for AHL activity or can be stored at −80°C. No loss of PON activity is apparent after several freeze–thaw cycles.

3.3. AHL Hydrolysis Assay and HPLC Analysis

1. This section describes a typical reaction. Thaw prepared supernatants at room temperature and place on ice. For the negative control, add 94 μl of the reaction buffer followed by 5 μl of appropriate lysis buffer into an eppendorf tube. When analyzing purified enzymes or serum, the negative control should contain the enzyme storage buffer or PBS, respectively. For the PON-containing samples, add 94 μl of the reaction buffer followed by 5 μl of the sample supernatant into an eppendorf tube (see Note 4).
2. Initiate the reaction by adding 1 μl of the stock AHL solution (make sure that the AHL is completely dissolved in the methanol before adding it to the reaction). Gently mix the reaction and place in 37°C water bath.
3. To stop the reaction, add 70 μl of ice-cold acetonitrile, briefly mix, and place on ice for 2 min to precipitate protein (see Note 5).
4. Briefly vortex-mix the sample and centrifuge for 2 min at 14,000 × *g*. Transfer the supernatant to a clean tube, being careful not to disturb the precipitated protein pellet. Samples should be analyzed by HPLC on the same day.
5. Equilibrate the HPLC with mobile phase at a flow rate of 1 ml/min and the UV detector set at 205 nm.
6. Typically, 20 μl of the sample is injected onto the HPLC column, and run time is set to 30 min (see Note 6).
7. Determine AHL and AHL-acid peak areas. Retention times for 3-oxo-C_{12}-HSL and 3-oxo-C_{12}-HSL-acid at the described HPLC conditions were 5.8 and 3.8 min, respectively (see Fig. 1) (see Note 7).

3.4. Preparation of AHL-Acids and Calculation of the Response Factor

1. The AHL-acid is prepared by making a 25 μM solution (from a 5 mM stock solution of AHL in methanol) of the lactone in 0.5 mM NaOH. A blank sample that does not contain the AHL is prepared by adding an equivalent volume of methanol to the 0.5 mM NaOH.
2. The solution is thoroughly mixed and incubated at room temperature. Approximately 30 min later, the NaOH is neutralized by adding an equal volume of acetonitrile containing 1 mM HCL, and the sample is mixed.
3. Typically, 20 μl of the neutralized hydrolysate is injected onto the HPLC column and the area of the acid determined. No peak corresponding to the lactone should be present.

If the lactone is still present, the AHL must be incubated in NaOH for a long enough time to allow for complete hydrolysis of the lactone ring. The negative control sample is run to ensure that no background peaks are present that interfere with the AHL-acid or AHL peak.

4. The lactone is prepared by making a 25 μM solution (from the same 5 mM stock solution of AHL in methanol) of the AHL in water. Immediately, add an equal volume of acetonitrile containing 1 mM HCL and 1 mM NaOH, mix, and inject 20 μl onto the HPLC column.
5. The response factor (RF) is calculated by dividing the average peak area of the AHL-acid by the average peak area for the AHL.

3.5. Calculation of AHL Hydrolysis Rates

1. The fraction of AHL hydrolyzed is determined by the following equation:

   Fraction of AHL hydrolyzed = (AHL-acid peak area)/[AHL-acid peak area + (AHL peak area × RF)]

   where RF is the response factor.
2. The fraction of AHL hydrolyzed in the negative control sample (containing lysis buffer) due to spontaneous hydrolysis is subtracted from the fraction of AHL hydrolyzed in the PON-containing samples.
3. The rate of the reaction is then calculated using the formula:

   (Fraction of AHL hydrolyzed) × (nanomoles of AHL in the reaction)/reaction time/ml of sample added to the reaction

   This gives the rate of the reaction in nmol/min/ml of sample.

3.6. Inhibition Assays

1. To inhibit PONs, prepare lysis buffer, without $CaCl_2$, containing 0.5 mM EDTA using the stock EDTA solution.
2. To inhibit serine esterases, prepare lysis buffer containing 1 mM AEBSF.
3. Tissue or cells are processed as described above (see Subheadings 3.1 and 3.2) using either the lysis buffer containing EDTA, for PON inhibition, or the AEBSF-containing buffer for serine esterase inhibition. Control (uninhibited) samples are prepared using lysis buffer. The supernatants are then incubated overnight at 4°C to allow for complete inhibition.
4. AHL activity is analyzed as described above (see Subheading 3.3) except that assays with EDTA-treated samples are performed in 2.5 mM Tris–HCL, pH 7.4 instead of reaction buffer.

4. Notes

1. Generally, the longer the acyl chain on the AHL, the more lipophilic the compound is. We have found that a 50 μM concentration of 3-oxo-C_{12}-HSL is soluble in the reaction buffer. Less lipophilic AHLs are more soluble and can be analyzed at higher concentrations, whereas more lipophilic compounds may not be soluble at a concentration of 50 μM.
2. The optimal amount of acetonitrile in the HPLC mobile phase for 3-oxo-C_{12}-HSL analysis on our HPLC system is 75%. More lipophilic AHLs will require a higher percent of acetonitrile, whereas more water soluble AHLs will require lower concentrations of acetonitrile in the mobile phase. The amount of acetonitrile can be increased or decreased to provide appropriate retention times and minimum background interference at the elution times of the lactone and the acid.
3. For tissues that have little connective tissue and are easily disrupted, such as brain and spinal cord, roughly 1-mg aliquots may be processed in the same manner as cell pellets (see Subheading 3.1).
4. Many compounds originating from the cell lysates or tissue homogenates may absorb at 205 nm, and some may interfere with AHL analysis. Therefore, it is best to keep the volume of supernatant added as low as possible. The volume of supernatant needed to detect AHL hydrolysis will depend upon the amount of PON or other AHL lactonases in the sample. Initial experiments can be performed to determine the optimal amount of supernatant to be added, and/or the HPLC conditions can be modified to reduce interference.
5. To estimate initial reaction rates, reactions are run under conditions in which less than 10% of the substrate is hydrolyzed. Reaction rates will depend upon the amount of PON(s) or other AHL lactonases in the sample and the amount of sample added to the reaction. Preliminary studies can be performed to optimize the reaction time.
6. Run time must be long enough to allow the detergent, *n*-dodecyl-b-maltoside, to come off the column and the column to equilibrate before the next run. Because of the low wavelength, 205 nm, used for detection, the baseline is very sensitive to minor pressure changes and changes in the constitution of the mobile phase (due to buffer in the injected sample).
7. The sum of the area of the AHL-acid, after adjustment with the response factor, and AHL should be the same for the negative controls (containing lysis buffer) and the samples.

If the sum of the areas are lower in the sample compared to the negative control, this suggests that the AHL is being metabolized/degraded through pathways other than just lactone hydrolysis.

References

1. Draganov DI, Teiber JF, Speelman A, Osawa Y, Sunahara R, La Du BN. (2005) Human paraoxonases (PON1, PON2, and PON3) are lactonases with overlapping and distinct substrate specificities. J Lipid Res; 46(6):1239–1247.
2. Draganov D, Teiber J. (2008) PONs' natural substrates-the key for their physiological roles. In: Mackness M, Aviram M, Parah G, editors. Protein and Cell Regulation, Volume 6: The Paraoxonases: Their Role in Disease Development and Xenobiotic Metabolism. Dordrecht: Springer: 297–306.
3. Stoltz DA, Ozer EA, Ng CJ et al. (2007) Paraoxonase-2 deficiency enhances Pseudomonas aeruginosa quorum sensing in murine tracheal epithelia. Am J Physiol Lung Cell Mol Physiol; 292(4):L852–L860.
4. Teiber JF, Horke S, Haines DC et al. (2008) Dominant role of paraoxonases in inactivation of the Pseudomonas aeruginosa quorum-sensing signal N-(3-oxododecanoyl)-L-homoserine lactone. Infect Immun; 76(6):2512–2519.

Chapter 22

Generation of Quorum Quenching Antibodies

Gunnar F. Kaufmann, Junguk Park, Alexander V. Mayorov, Diane M. Kubitz, and Kim D. Janda

Abstract

The exchange of information within and among bacterial populations using small diffusible molecules has been termed "quorum sensing" (QS). Due to the extracellular distribution of the QS autoinducer molecules and the evolutionary highly conserved nature of signaling components, microbial QS systems represent an excellent target for anti-infective immunotherapy. Recently, we have described the generation of quorum quenching monoclonal antibodies (mAbs) against acyl homoserine lactones (AHL) used by *Pseudomonas aeruginosa* as well as Staphylococcal autoinducing peptides (AIP). These mAbs suppressed QS signaling in bacteria and neutralized AHL-mediated cytotoxic effects in vitro, as well as protected animals in *Staphylococcus aureus* infection models.

Key words: Quorum sensing, Monoclonal antibodies, Autoinducers, Hybridoma, Quorum quenching, *Pseudomonas aeruginosa*, *Staphylococcus aureus*

1. Introduction

Bacterial quorum sensing (QS) has become an attractive research area for the discovery of new antivirulence therapeutics. Notably, chemical research has focused on the synthesis and characterization of autoinducer analogs, mostly focusing on acyl homoserine lactone (AHL)-based QS systems (1–7). However, we and others have reported on the biochemical effects exerted by AHLs on the mammalian host, including subversion of the innate immune system as well as direct cytotoxicity against mammalian cells (8–11). Furthermore, the intrinsic cytotoxic activity of AHLs provides a suitable foundation to regard these autoinducers as bacterial small molecule toxins.

Based on these observations, the need for alternative therapeutic approaches to interfere with QS signaling as well as to neutralize the

Kendra P. Rumbaugh (ed.), *Quorum Sensing: Methods and Protocols*, Methods in Molecular Biology, vol. 692,
DOI 10.1007/978-1-60761-971-0_22, © Springer Science+Business Media, LLC 2011

cytotoxicity of the autoinducing molecules has emerged. Recently, such an immunotherapeutic approach for QS quenching was pioneered by generation of the anti-AHL monoclonal antibody (mAb) RS2-1G9 elicited against a synthetic 3-oxo-C_{12}-HSL analog (12). The inherent hydrolytic instability of AHLs required the design and synthesis of AHL analog haptens that, while retaining the overall AHL structure, including stereochemistry, possessed improved hydrolytic stability. We accomplished this goal by substituting the homoserine lactone moiety with a lactam group. Notably, the new hapten also did not undergo a previously observed Claisen-like rearrangement to a tetramic acid (13). RS2-1G9 efficiently suppressed QS signaling in *Pseudomonas aeruginosa* as well as conferred protection upon mammalian cells via neutralization of 3-oxo-C_{12}-HSL in vitro (12, 14). The crystal structure of RS2-1G9 impressively illustrates how the mAb specifically and with high affinity binds to the AHL, even though AHLs lack desirable features of haptens, such as aromaticity and charge (15).

With regards to staphylococcal autoinducing peptides (AIPs), the generation of anti-AIP mAbs was accomplished by using a hapten in which the hydrolytically labile thiolactone was replaced with a more stable lactone moiety (16). The obtained mAbs were evaluated in vitro as well as in vivo and displayed potent quorum quenching abilities, including the protection of mice from an otherwise lethal *S. aureus* infection.

2. Materials

2.1. Immunoconjugate Generation

1. Haptens: the desired quorum sensing molecule analogs with appropriate reactive group for conjugation to the carrier protein, e.g., carboxyl, or thiol, have to be synthesized according to published protocols (12, 16). AIP analogs can be obtained as custom-synthesized peptides from commercial peptide sources.
2. Carrier proteins: keyhole limpet hemocyanin (KLH) as well as bovine serum albumin (BSA) are available from commercial sources (Pierce, Sigma).
3. Linkers and coupling reagents: The linker, namely, sulphosuccinimidyl 4-(*N*-maleimidomethyl)-cyclohexane-1-carboxylate (sulpho-SMCC), required for the conjugation of AIP haptens to the appropriate carrier protein is commercially available (Pierce). The coupling reagents, namely, 1-ethyl-3-(3-dimethylaminopropyl) carbodiimide (EDC) and sulpho-*N*-hydroxysuccinimide (sulpho-NHS), for the activation of carrier proteins prior to coupling of the AHL haptens is commercially available (Pierce).

4. Coupling buffer. Phosphate-buffered saline (PBS): prepare 10× stock from powder concentrate (Fisher). Dilute 1:10 into water before use and adjust the pH to 7.4. The working solution will contain: 137 mM NaCl, 2.7 mM KCl, 10 mM Na_2HPO_4, and 2 mM KH_2PO_4.
5. Equipment. LC/MS apparatus: Agilent 1100 MSD LC/MS system.

2.2. Immunization and Fusion

1. Mice: BALB/c mice (8–12 weeks old) are routinely used for the generation of monoclonal antibodies.
2. Adjuvants: Sigma Adjuvant system (SAS) and Alum commercially available (Sigma).
3. ELISA reagents: 96-well 1/2-area EIA plate (Corning 3696), blocking solution 5% w/v powdered skim milk in PBS; goat anti-mouse–horseradish peroxidase (HRP) conjugate (Southern Biotech), HRP substrate (Pierce), Myeloma cell line. P3.X63-Ag8.653 (ATCC CRL-1580).
4. Cell culture media: *Wash medium* – RPMI 1640 media (supplemented with 2 mM l-glutamine, 1 mM sodium pyruvate, 10 mM HEPES, and 50 μg/mL gentamycin); *Selection medium* – HAT media (RPMI-1640 supplemented with 2 mM l-glutamine, 1 mM sodium pyruvate, 10 mM HEPES, 50 μg/mL gentamycin, 10% FCS, 0.1 mM hypoxanthine, 0.4 μM aminopterin, and 16 μM thymidine); *Growth medium* – HT media (RPMI-1640 supplemented with 2 mM l-glutamine, 1 mM sodium pyruvate, 10 mM HEPES, 50 μg/mL gentamycin, 10% FCS, 0.1 mM hypoxanthine, and 16 μM thymidine); *freezing media* – 90% fetal calf serum and 10% dimethyl sulfoxide (DMSO).
5. Fusion reagent: PEG 1500 – prepare 50% solution in dH_2O.
6. Antibody purification: Gammabind™ Plus Sepharose™ (GE Healthcare); Hi Trap Protein G HP 5-mL prepacked columns (GE Healthcare); elution buffer – 0.1 M acetic acid (pH 3.0); Amicon ultra-15 centrifugal filter units (Millipore).
7. Miscellaneous disposable supplies: 1-mL syringes, 18g and 23g needles, 96-well plates (Corning), cryovials (Nalgene).
8. Equipment: ELISA reader SpectraMax M2 (Molecular Devices); ÄKTA-FPLC (GE Healthcare).

2.3. Relative Quantification of α-Hemolysin Production by Western Blotting

1. *Staphylococcus aureus* (*agr* positive/*hla* positive strain, e.g., RN6390B).
2. CYGP medium: Casamino acids (Difco) 10 g/L, yeast extract (Difco) 10 g/L, glucose 5 g/L, NaCl 5.9 g/L, and 1.5 M β-glycerophosphate (add after autoclaving) 40 mL/L.
3. Blood BHI agar plate (Hardy Diagnostics).

4. Cell-culturing tubes. 5 and 15 mL (BD).
5. Millex-GV filter unit (0.22 μm, Millipore).
6. SDS-PAGE: NuPAGE 10% Bis-Tris Gel (Invitrogen), SDS sample loading buffer (Bio-Rad), Prestained protein marker (Bio-Rad), MOPS SDS running buffer (Invitrogen).
7. Western blot analysis: PVDF membrane (0.2 μm, Invitrogen), blocking solution – 5% w/v powdered skim milk in TBS buffer (pH 7.2), Tween 20 (10%, Bio-Rad), TBST buffer: 1 L of Tris-buffered saline (TBS) buffer containing 10 mL of 10% Tween 20 solution, HRP-conjugated sheep polyclonal α-hemolysin antibody (Abcam), and SuperSignal West Dura Extended Duration Substrate (Thermo Scientific).
8. Equipment: Microcentrifuge; 5415R (Eppendorf), FluoroChem 8900 Imager (Alpha Innotech).

2.4. Relative Quantification of Pyocyanin Production by *P. aeruginosa*

1. *P. aeruginosa* (e.g., PAO I).
2. Growth medium and plates: LB medium: Bacto-tryptone (Difco) 10 g/L, yeast extracts (Difco) 5 g/L, NaCl 10 g/L; LB agar plates – LB broth containing 1.5% agar.
3. Tubes: 15-mL cell-culturing tubes (BD), 15 mL conical tubes (Corning).
4. Extraction reagents: Chloroform, 0.2 N HCl (aq) (Sigma).
5. Equipment: Centrifuge; GS-6R (Beckman), UV/Vis spectrophotometer; SpectraMax M2 (Molecular Devices).

3. Methods

3.1. Generation of Immunoconjugates

1. *Generation of AIP hapten–carrier protein conjugates.* Preparation of the AIP immunoconjugates relied on the maleimide–thiol conjugation chemistry, for which the immunogenic protein carrier (KLH) as well as BSA are preactivated with sulpho-SMCC. The AIP hapten is constructed such to bear a maleimide-reactive cysteine residue (Fig. 1) (see Note 1).
2. 5 mg of the carrier protein is resuspended in 1 mL of PBS (pH 7.4).
3. To this solution, 1 mg of the linker sulfo-SMCC is added.
4. The solution is stirred for 6–8 h.
5. The protein–linker conjugate is purified by dialysis in PBS at 4°C.
6. To the protein–linker conjugate in PBS, 100 μL DMF containing 2 mg of the hapten is added. The solution is shaken overnight. Due to a low solubility of the AIP4 hapten

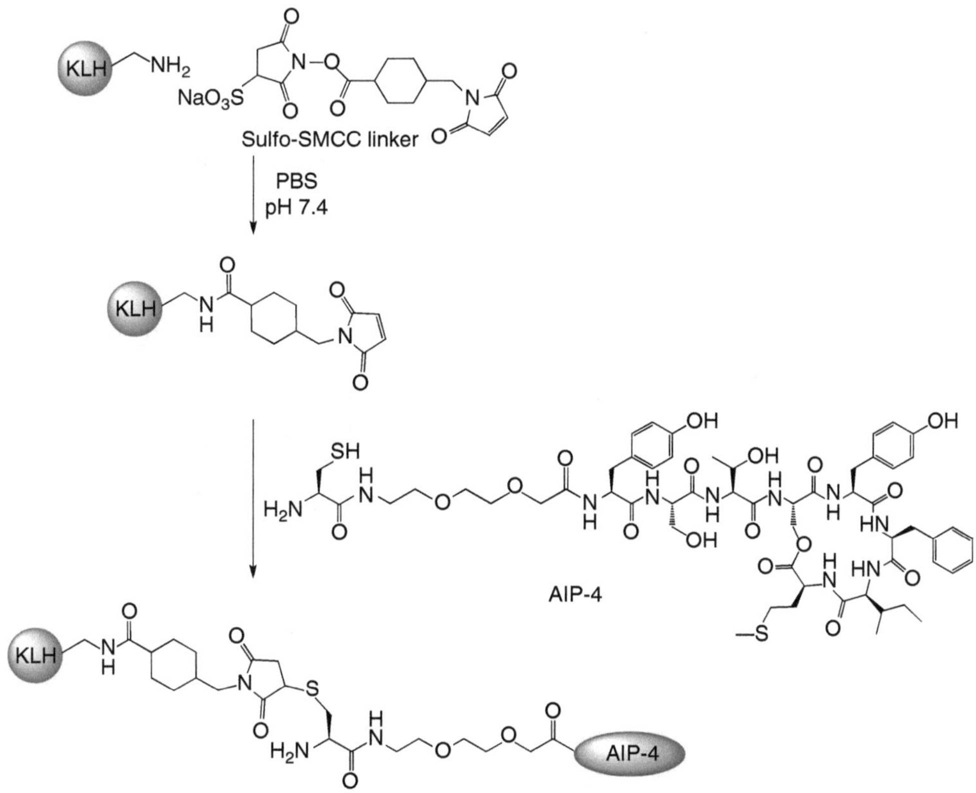

Fig. 1. Maleimide–thiol chemistry of the AIP hapten–protein carrier conjugation.

in aqueous media, the conjugation required extended reaction times (10 h+, RT).

7. The protein–hapten conjugate is purified by dialysis.
8. The efficacy of the conjugation is monitored by MALDI-TOF MS analysis of the corresponding BSA conjugate (see Note 2).
9. *Generation of AHL hapten–carrier protein conjugates*: The protein conjugation of the homoserine lactone hapten is accomplished via a conventional carbodiimide/*N*-hydroxysuccinimide coupling chemistry (Fig. 2).
10. The hapten is preactivated with EDC/Sulfo-NHS (0.4 mg EDC and 1.1 mg of sulfo-NHS are added to 1 mL of hapten solution in PBS (1 mg/mL) and incubated for 5–15 min at room temperature).
11. The crude water-soluble sulfo-NHS hapten ester is coupled directly to the carrier proteins in pH 7.4 PBS buffer.
12. The protein–hapten conjugate is purified by dialysis in PBS at 4°C.
13. The extent of the conjugation is also monitored by MALDI-TOF MS analysis of the RS2–BSA conjugate.

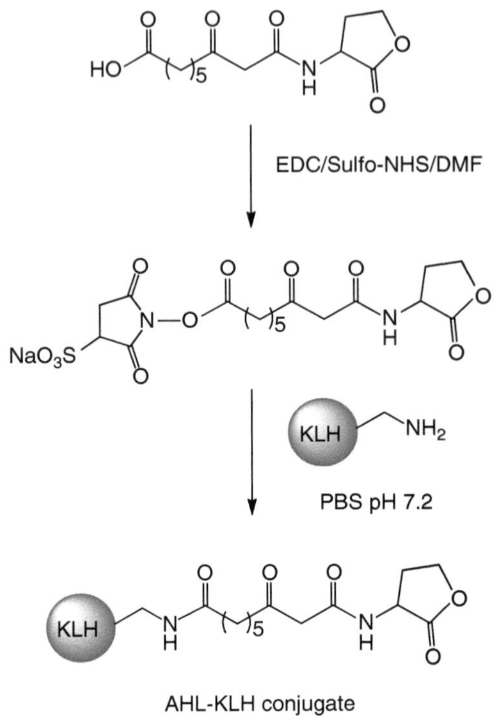

Fig. 2. Synthesis of an AHL hapten–KLH conjugate.

3.2. Immunization

1. Three mice per hapten are immunized with 100 μg of hapten–KLH-conjugate in 100 μL of PBS mixed with an equal volume of SAS reconstituted in PBS per mouse. The mice are injected intraperitoneally (IP) using a 23g needle.
2. After 2 weeks, they are given a second injection (100 μg per mouse in SAS, i.p.).
3. 7–10 days later, the serum titers are determined by ELISA (see below). If the titer, i.e., the dilution giving 50% of the maximum absorbance is less than 6,400, the mice are given additional boosts of 50–100 μg after 2–3 weeks until the titer is 12,800–25,600 or more. One month after a sufficient titer has been reached, the mice receive a final injection of 50 μg of KLH-conjugate in 100 μL PBS intravenously (i.v.) in the lateral tail vein. The spleen is removed 3 days later for fusion.

3.3. Hybridoma Production and Screening

1. At least 1 week prior to the fusion, a fresh vial of myeloma cells P3.X63-Ag8.653 is thawed (see Note 3). It is critical to ensure that the myeloma cells are healthy and in log phase on the day of the fusion for it to be successful. Cells that have been growing too long and have come out of log phase or cells that have not been growing long enough to achieve log

growth will not fuse optimally. The cell density should be kept between 3 and 6×10^5/mL.

2. The spleen cells from the hyperimmunized mouse and myeloma cells are washed three times with 30 mL RPMI 1640 media supplemented with 2 mM l-glutamine, 1 mM sodium pyruvate, 10 mM HEPES, 50 μg/mL gentamycin and mixed together at a 5:1 (spleen and myeloma) ratio in a 50-mL conical.
3. After the final spin, the media is carefully aspirated.
4. The pellet is spread out by gently tapping the tube. It is important that the cells be spread along the bottom edge of the conical so that all the cells have equal access to the PEG.
5. 1 mL of 50% PEG 1500, prewarmed to 37°C, is added dropwise over 1 min using a 1-mL syringe and 18g needle while gently rotating and tapping the tube to resuspend the cells.
6. The PEG is then slowly diluted out with 1 mL of RPM1-1640 media over 1 min and then 8 mL over 2 min. The cell membranes are very fragile at this point.
7. The cells are then placed in a 37°C water bath for 10 min and then centrifuged.
8. The supernatant is decanted and the cells gently resuspended in 5 mL of RPM1-1640 with 10% FCS.
9. The cells are again placed in a 37°C water bath for 10 min and then added to 225 mL of HAT media.
10. They are plated onto 15 96-well plates (150 μL/well).
11. The fusion is fed 50 μL/well with HT media on day 4, 8, and 12. By this time, macroscopic colonies should be visible. Generally, growth is expected in 40–60% of the wells. Supernatant directly from the 96-well plates is used for antigen binding.

3.4. ELISA Screening

1. The ELISA protocol can be modified or adapted if necessary depending on the nature of the hapten being used. Since a KLH-conjugate has been used for immunization, it is important to utilize another carrier protein for screening purposes. BSA is easy to conjugate, does not aggregate, and precipitates easily, to achieve even coating of the ELISA plate. The hapten–BSA conjugate is diluted to 50 μg/mL in PBS and plated at 25 μL/well in to a 96-well 1/2-area EIA plate and allowed to dry overnight at 37°C.
2. The antigen is fixed to the plate with 50 μL/well methanol for 5 min.
3. The methanol is then removed by shaking and the plates allowed to air-dry for 5–10 min.

4. Nonspecific binding sites are blocked with 50 μL/well of blocking solution for 30 min at room temperature. The excess blocking solution is removed. Importantly, the plates should not be allowed to dry out from this point on to avoid nonspecific binding.
5. The cell supernatant or purified quorum quenching antibody is immediately added to the well. Add 25 μL/well of primary antibody diluted in blocking solution, (1:1 for tissue culture supernatant, 1:100 for mouse sera, and 1:10,000 for purified antibody).
6. The plates are incubated for 1–2 h at 37°C and then washed ten times with deionized water (see Note 4).
7. The excess water is removed and 25 μL/well of the secondary goat anti mouse–HRP antibody conjugate are added – diluted 1:2,000 in blocking solution.
8. The plates are incubated for 1 h at 37°C and washed ten times with deionized water.
9. To develop the ELISA, 50 μL/well of the HRP substrate is added.
10. The plates are then read at 414 nm after 30 min.

3.5. Subcloning of Antibody-Producing Hybridomas

1. A series of dilutions is made that result in plating out approximately 0.5–5 cells per well in 200 μL in a 96-well plate, 48 wells at each density.
2. It will take about 10–14 days for macroscopic colonies to grow.
3. Pick 8–16 different clones per hybridoma to move up into a 48-well plate – depending on the initial titer of the hybridoma (see Notes 5–7).
4. Pick the six best to transfer to a 24-well plate. At this point, all the subclones are titered.
5. The best producers are picked and frozen down.
6. The whole process is repeated a second and third time until all clones from a 96-well plate are positive for binding. At this point, it is highly likely that a stable monoclonal cell line has been generated. The cells should be frozen with six identical aliquots per hybridoma.
7. To ensure viability of the frozen stock, it is important to freeze the cells while in log phase. The cells are pelleted, resuspended in 3 mL of freezing media, and aliquoted at 0.5 mL per cryovial. The vials are placed immediately into a cryovial at –80°C, which allows for a controlled, slow rate of freezing (approximately 1°C per minute).
8. After 4–6 h, the vials are placed into a liquid-nitrogen storage tank.

9. After 2–3 days, one of the vials is thawed and revived to ensure viability and sterility. The vials are thawed quickly in a 37°C water bath, and the cells placed in 10 mL of complete RPMI-10% FCS. The cells are then pelleted to remove traces of the DMSO and then resuspended in 5 mL of media.
10. The cells are cultured for at least 48 h, and then the cell supernatant is assayed for antibody titer.

3.6. Antibody Production

1. There are a number of different techniques that can be used to produce the monoclonal antibodies depending upon the amount of antibody required, the facilities available, the time frame, and the available budget. If 30 mg or less of purified antibody is needed, the production in T-flasks is sufficient. Hybridoma cells will typically produce 2–50 μg/mL of antibody/mL of culture depending upon the isotype of the clone and the cell density. Antibody levels can be maximized by growing 500–1,000 mL of cells and allowing them to grow until they start dying naturally (approximately 10–14 days).
2. The supernatant is then collected, centrifuged, and filtered through a 0.2-μm filter.
3. The produced antibodies are affinity-purified using Protein G [Gammabind™ Plus Sepharose™; (GE Healthcare)] and Hi Trap Protein G HP 5-mL prepacked columns (GE Healthcare) depending upon the amount of antibody being purified. The antibody (in PBS, pH 7.4) is bound to the Protein G column and washed with 3 column volumes PBS to remove unbound proteins.
4. The bound antibodies are then eluted with elution buffer 0.1 M acetic acid pH 3.0 and immediately neutralized with 1 M TRIS pH 9.0.
5. The antibody is then dialyzed into PBS pH 7.4, concentrated (Amicon ultra-15 centrifugal filter units), sterile-filtered, and stored at 4°C. For long-term storage, we aliquot the antibody and freeze it at −20°C.

3.7. Evaluation of Anti-AIP Quorum Quenching Antibodies: Relative Quantification of α-Hemolysin Production by Western Blotting

1. Cell culture: After overnight growth on a blood BHI agar plate at 37°C, a single colony of *S. aureus* is inoculated into 3 mL of CYGP medium and grown overnight (18 h) (17). The overnight cultured cells are diluted to $OD_{600}=0.03$ in fresh CYGP medium and distributed into 5-mL polystyrene cell-culturing tubes, where each tube contains 0.5 mL of the diluted cells and appropriate anti-AIP antibodies or control antibody.
2. Preparation of the *S. aureus* secretome: After growth for 20–24 h at 37°C without agitation, the samples are transferred to

1.5-mL microcentrifuge tubes and centrifuged at 15,700 rcf (relative centrifugal force) for 5 min. The supernatants that contain the *S. aureus* secretome are sterilized by filtration through a Millex-GV filter unit (0.22 μm).

3. SDS-PAGE and Western Blotting: The sterilized secretome (20 μL) is separated by SDS-PAGE (NuPAGE 10% Bis-Tris Gel; Invitrogen) with a prestained protein marker and transferred to a PVDF membrane for Western Blotting. After transfer, the membrane is washed three times for 5 min with 10 mL of TBST buffer and soaked in the blocking buffer for 45 min at room temperature. The remaining blocking buffer is washed away with 10 mL of TBST buffer three times for 5 min, and the membrane is cut at approximately 50 kDa (indicated by the prestained marker) and only the lower half of the membrane (≤50 kDa) is used for α-hemolysin analysis.
4. α-Hemolysin analysis: The lower part of the membrane is incubated with 10 mL of the blocking buffer containing 10 μL of HRP-conjugated sheep polyclonal anti-α-hemolysin antibody (Abcam) for 3–16 h at room temperature. Then, the membrane is washed five times with 10 mL of TBST buffer for 5 min and developed using SuperSignal West Dura Extended Duration Substrate. The resulting chemiluminescence is recorded with a FluoroChem 8900 Imager (see Note 8).

3.8. Evaluation of Anti-AHL Quorum Quenching Antibodies: Relative Quantification of Pyocyanin Production by *P. aeruginosa*

1. Cell culture: A single colony (PAO1) is picked after overnight growth on a LB agar plate and grown for 18 h in 5 mL of LB medium. Fresh LB medium (10 mL) is inoculated with the overnight cultured *P. aeruginosa* ($OD_{600} = 0.05$) and incubated for 3–4 h to mid-log cell growing phase ($OD_{600} = 0.4–0.6$). New fresh LB medium (100 mL) is inoculated with the mid-log phase cells to OD = 0.05 and distributed into 15-mL cell-culturing tubes, where each tube contained 5 mL of the diluted cells and the appropriate antibodies.
2. Pyocyanin extraction: After growth for 16–18 h, the samples are centrifuged at 3,000 rpm for 15 min and the supernatants are transferred to 15-mL conical tubes. To extract pyocyanin, add 3 mL of chloroform and vigorously vortex the tube for 30 s. The samples are centrifuged at 2,000 rcf for 5 min, and the upper layer – aqueous layer – is carefully discarded. In order to reextract pyocyanin from the organic phase into the aqueous layer, add 1 mL of HCl (0.2 N, aq) to the samples and vigorously vortex for 15 s, followed by centrifugation at 3,000 rpm for 5 min. At this time, pyocyanin will be in upper

aqueous layer (pink), so carefully recover the upper phase and measure the absorbance at 520 nm using a UV/Vis spectrophotometer (see Note 9).

4. Notes

1. Alternatively, to activate KLH and BSA with sulpho-SMCC, pre-activated proteins are commercially available (Pierce).
2. The MS analysis of the BSA conjugate is required as KLH is a high molecular weight protein (or protein complex) that makes the required resolution of conjugation efficiency impossible. To determine the conjugation efficiency, the mass of BSA and BSA conjugate are determined and the mass difference is then divided by the molecular weight of the hapten. The resulting number equates the number of hapten copies per carrier protein.
3. It is important to choose a myeloma line that has been selected to be a nonproducer of IgG. We use the X63-Ag8.653 line because it is a nonproducer and possesses high fusion efficiency.
4. Alternate the wash direction, so no wells are missed.
5. The hybridomas should be subcloned at least two generations to ensure monoclonality. The subcloning relies on limiting dilution. If monoclonality is not achieved after two generations or the hybridoma does not appear stable, one or two more generations are recommended. Notably, newly established hybridomas do not grow well when plated out at limiting dilutions. However, it is important to start subcloning the hybridomas as soon as possible to avoid the cells being taken over by nonproducers.
6. Repeated titering of the individual wells while still being expanded is used to monitor the specific activity. Our immunizations usually result in a large number of hybridomas to subclone; thus, we subclone by limiting dilution without prior counting of the cells.
7. Try to take wells that appear to have only one clone. At this point, many of the clones will not be positive by ELISA.
8. For the strains that produce low amount of α-hemolysin, concentrating the secretome using the Microcon Centrifugal Filter Device (3,000 MWCO, Millipore) is recommended.
9. The method is based on the previous report by Smith et al., and no mathematical conversion is required for relative quantification (18).

Acknowledgments

This work was supported by the National Institutes of Health (AI080715 to G.F.K). This is manuscript number 20549 from The Scripps Research Institute.

References

1. Hjelmgaard, T., Persson, T., Rasmussen, T. B., Givskov, M., and Nielsen, J. (2003) Synthesis of furanone-based natural product analogues with quorum sensing antagonist activity, *Bioorganic & Medicinal Chemistry* 11, 3261–3271.
2. Hentzer, M., Riedel, K., Rasmussen, T. B., Heydorn, A., Andersen, J. B., Parsek, M. R., Rice, S. A., Eberl, L., Molin, S., Hoiby, N., Kjelleberg, S., and Givskov, M. (2002) Inhibition of quorum sensing in Pseudomonas aeruginosa biofilm bacteria by a halogenated furanone compound, *Microbiology* 148, 87–102.
3. Smith, K. M., Bu, Y. G., and Suga, H. (2003) Induction and inhibition of Pseudomonas aeruginosa quorum sensing by synthetic autoinducer analogs, *Chemistry & Biology* 10, 81–89.
4. Geske, G. D., O'Neill, J. C., Miller, D. M., Mattmann, M. E., and Blackwell, H. E. (2007) Modulation of bacterial quorum sensing with synthetic ligands: Systematic evaluation of N-acylated homoserine lactones in multiple species and new insights into their mechanisms of action, *Journal of the American Chemical Society* 129, 13613–13625.
5. Muh, U., Schuster, M., Heim, R., Singh, A., Olson, E. R., and Greenberg, E. P. (2006) Novel Pseudomonas aeruginosa quorum-sensing inhibitors identified in an ultra-high-throughput screen, *Antimicrobial agents and chemotherapy* 50, 3674–3679.
6. George, E. A., Novick, R. P., and Muir, T. W. (2008) Cyclic peptide inhibitors of staphylococcal virulence prepared by Fmoc-based thiolactone peptide synthesis, *Journal of the American Chemical Society* 130, 4914–4924.
7. Lyon, G. J., Mayville, P., Muir, T. W., and Novick, R. P. (2000) Rational design of a global inhibitor of the virulence response in Staphylococcus aureus, based in part on localization of the site of inhibition to the receptor-histidine kinase, AgrC, *Proceedings of the National Academy of Sciences of the United States of America* 97, 13330–13335.
8. Kravchenko, V. V., Kaufmann, G. F., Mathison, J. C., Scott, D. A., Katz, A. Z., Wood, M. R., Brogan, A. P., Lehmann, M., Mee, J. M., Iwata, K., Pan, Q., Fearns, C., Knaus, U. G., Meijler, M. M., Janda, K. D., and Ulevitch, R. J. (2006) N-(3-oxo-acyl) homoserine lactones signal cell activation through a mechanism distinct from the canonical pathogen-associated molecular pattern recognition receptor pathways, *The Journal of biological chemistry* 281, 28822–28830.
9. Tateda, K., Ishii, Y., Horikawa, M., Matsumoto, T., Miyairi, S., Pechere, J. C., Standiford, T. J., Ishiguro, M., and Yamaguchi, K. (2003) The Pseudomonas aeruginosa autoinducer N-3-oxododecanoyl homoserine lactone accelerates apoptosis in macrophages and neutrophils, *Infection and immunity* 71, 5785–5793.
10. Shiner, E. K., Terentyev, D., Bryan, A., Sennoune, S., Martinez-Zaguilan, R., Li, G., Gyorke, S., Williams, S. C., and Rumbaugh, K. P. (2006) Pseudomonas aeruginosa autoinducer modulates host cell responses through calcium signalling, *Cellular microbiology* 8, 1601–1610.
11. Kravchenko, V. V., Kaufmann, G. F., Mathison, J. C., Scott, D. A., Katz, A. Z., Grauer, D. C., Lehmann, M., Meijler, M. M., Janda, K. D., and Ulevitch, R. J. (2008) Modulation of gene expression via disruption of NF-kappaB signaling by a bacterial small molecule, *Science (New York, N.Y.)* 321, 259–263.
12. Kaufmann, G. F., Sartorio, R., Lee, S. H., Mee, J. M., Altobell, L. J., 3rd, Kujawa, D. P., Jeffries, E., Clapham, B., Meijler, M. M., and Janda, K. D. (2006) Antibody interference with N-acyl homoserine lactone-mediated bacterial quorum sensing, *Journal of the American Chemical Society* 128, 2802–2803.

13. Kaufmann, G. F., Sartorio, R., Lee, S. H., Rogers, C. J., Meijler, M. M., Moss, J. A., Clapham, B., Brogan, A. P., Dickerson, T. J., and Janda, K. D. (2005) Revisiting quorum sensing: Discovery of additional chemical and biological functions for 3-oxo-N-acylhomoserine lactones, *Proceedings of the National Academy of Sciences of the United States of America* 102, 309–314.
14. Kaufmann, G. F., Park, J., Mee, J. M., Ulevitch, R. J., and Janda, K. D. (2008) The quorum quenching antibody RS2-1G9 protects macrophages from the cytotoxic effects of the Pseudomonas aeruginosa quorum sensing signalling molecule N-3-oxo-dodecanoyl-homoserine lactone, *Molecular immunology* 45, 2710–2714.
15. Debler, E. W., Kaufmann, G. F., Kirchdoerfer, R. N., Mee, J. M., Janda, K. D., and Wilson, I. A. (2007) Crystal structures of a quorum-quenching antibody, *Journal of molecular biology* 368, 1392–1402.
16. Park, J., Jagasia, R., Kaufmann, G. F., Mathison, J. C., Ruiz, D. I., Moss, J. A., Meijler, M. M., Ulevitch, R. J., and Janda, K. D. (2007) Infection control by antibody disruption of bacterial quorum sensing signaling, *Chemistry & biology* 14, 1119–1127.
17. Novick, R. P. (1991) Genetic systems in staphylococci, *Methods in enzymology* 204, 587–636.
18. Smith, K. M., Bu, Y., and Suga, H. (2003) Library screening for synthetic agonists and antagonists of a Pseudomonas aeruginosa autoinducer, *Chemistry & biology* 10, 563–571.

INDEX

A

B

Kendra P. Rumbaugh (ed.), *Quorum Sensing: Methods and Protocols*, Methods in Molecular Biology, vol. 692, DOI 10.1007/978-1-60761-971-0, © Springer Science+Business Media, LLC 2011

P

Q

R

S

T

V

MIX
Papier aus verantwortungsvollen Quellen
Paper from responsible sources
FSC® C105338

If you have any concerns about our products,
you can contact us on
ProductSafety@springernature.com

In case Publisher is established outside the EU,
the EU authorized representative is:
Springer Nature Customer Service Center GmbH
Europaplatz 3, 69115 Heidelberg, Germany

Printed by Libri Plureos GmbH
in Hamburg, Germany